Recent Advances in Engineering Mathematics and Physics

Mohamed Hesham Farouk • Maha Amin Hassanein

Editors

Recent Advances in Engineering Mathematics and Physics

Proceedings of the International Conference
RAEMP 2019

 Springer

Editors
Mohamed Hesham Farouk
Faculty of Engineering
Cairo University
Giza, Egypt

Maha Amin Hassanein
Faculty of Engineering
Cairo University
Giza, Egypt

ISBN 978-3-030-39849-1 ISBN 978-3-030-39847-7 (eBook)
https://doi.org/10.1007/978-3-030-39847-7

This Springer imprint is published by the registered company Springer Nature Switzerland AG
The registered company address is: Gewerbestrasse 11, 6330 Cham, Switzerland

Preface

Engineering mathematics and physics are branches of applied mathematics and physics concerned with results, methods, and techniques used in engineering and industrial applications. Much of the research work produced by engineers may be taken to fall under the rubric of engineering mathematics and physics, provided that it passes some reasonable level of theoretical sophistication. Foundational research in the mathematical and physical underpinnings of engineering problems continues to be fascinating, deep, and indispensable.

This volume contains the papers presented at the International Conference on Recent Advances in Engineering Mathematics and Physics (RAEMP2019) organized by the Department of Engineering Mathematics and Physics at Cairo University and held in Cairo, Egypt, in December 2019. The papers discuss some of the latest advances in the applications of mathematics and physics in different engineering disciplines, including information technology, control, mechanical engineering, electromagnetic waves, and solid-state electronics. Contributions come from both academia and the industry.

All submissions were reviewed by two reviewers on average, with all reviewers affiliated to diverse reputable international institutes. Out of 65 papers which were submitted to the conference, 42 were selected by the program committee, while only 27 made it to the final conference proceedings.

We would like to express our sincere thanks to the plenary speakers, session chairs, and all reviewers for helping us produce a rich technical program. We would also like to extend our sincere appreciation for the outstanding work contributed over many months by the organizing committee.

Giza, Egypt　　　　　　　　　　　　　　　Mohamed Hesham Farouk
Giza, Egypt　　　　　　　　　　　　　　　　Maha Amin Hassanein

Organizing Committee

Prof. M. Hesham, Cairo University, Egypt.
Prof. Maha Hassanein, Cairo University, Egypt.
Prof. Doaa Shawky, Cairo University, Egypt.
Prof. Mohamed Abdel-Aziz El-Beltagy, Cairo University, Egypt.
Dr. Haythem Ismail, Cairo University, Egypt.

Technical Reviewers

Prof. AboElMagd Nour-El-Din, Queen's Univ., Canada
Prof. Abdelrazik Sebak, Concordia University, Canada
Prof. Ahmad Osman, Univ. of Appl. Sciences, Germany
Prof. Ahmed Tewfik, The University of Texas at Austin
Dr. Ahmed El-Shiekh, Cairo University, Egypt
Dr. Ahmed Abdelsamea, Nile University
Prof. Alaa AbelMaged, Cairo University, Egypt.
Dr. Amr Guily, Nile University, Egypt
Dr. Amr Essawi, Stanford University, USA.
Prof. Amr Sharawi, American University in Cairo (AUC)
Dr. Hazem Gheith, ARCADIS, USA.
Dr. Islam Kamel, Toronto University, Canada
Prof. Mohamed Moustafa Hassan, Cairo University
Prof. Nabila Philip, Cairo University, Egypt.
Dr. Nihal Yassin,, Cairo University, Egypt.
Prof. Sahar El-Nagar, Cairo University, Egypt.
Prof. Sayed Kasseb, Cairo University, Egypt.
Dr. Soumaya Mahmoud, NARSS, Egypt.
Dr. Tamer Ashour, Zewail University, Egypt.
Dr. Tamer Kassem, Nile University, Egypt.
Dr. Wassim Alexan, German University in Cairo (GUC)
Dr. Yasser Fathi, ElSewedy Electrometer Inc., Egypt

Abstracts of Invited Talks

Towards Man–Machine Symbiosis

A. H. Tewfik

Department of Electrical and Computer Engineering, University of Texas, Austin, TX, USA

Numerous articles in the general press warn against a dark future in which evermore powerful machines will displace humans. Yet, empirical evidence establishes that properly designed human–machine systems outperform man and machine and have the potential of increasing human creativity and cognitive abilities. In this talk, I will provide an overview of cognitive biases in human decision-making, give examples of man–machine symbiosis, and review our recent work in the area. In particular, I will focus on machine-assisted human decision-making and the use of brain–machine interfaces to improve speech recognition, recognize the audio source a person is listening to, and whether the person is listening to her mother tongue. Time permitting, I will describe some of the work that we have been performing on reducing the amount of data needed to train support vector machines and deep neural networks.

Ahmed H. Tewfik received his B.Sc. degree from Cairo University, Cairo, Egypt, in 1982 and his M.Sc., E.E., and Sc.D. degrees from MIT, in 1984, 1985, and 1987, respectively. He is the Cockrell Family Regents Chair in Engineering and the Chairman of the Department of Electrical and Computer Engineering at the University of Texas, Austin. He was the E. F. Johnson professor of Electronic Communications with the department of Electrical Engineering at the University of Minnesota until September 2010. Dr. Tewfik worked at Alphatech, Inc. and served as a consultant to several companies.

From August 1997 to August 2001, he was the President and CEO of Cognicity, Inc., an entertainment marketing software tools publisher that he cofounded, on partial leave of absence from the University of Minnesota. His current research interests are in cognitive augmentation through man–machine symbiosis and mobile computing, low energy broadband communications, applied machine learning, and brain computing interfaces. Prof. Tewfik is a Fellow of the IEEE. He was a Distinguished Lecturer of the IEEE Signal Processing Society in 1997–1999. He received the IEEE Third Millennium Award in 2000 and the IEEE Signal Processing Society Technical Achievement Award in 2017. He was elected to the positions of President-Elect of the IEEE Signal Processing Society in 2017 and VP Technical Directions of that Society in 2009. He served as VP from 2010 to 2012 and on the board of governors of that Society from 2006 to 2008. He has given several plenary and keynote lectures at IEEE conferences.

http://www.ece.utexas.edu/people/faculty/ahmed-tewfik

Frequency Selective Surfaces for mm Wave Antennas Gain Enhancement and Radar Scattering Control

Abdel Razik Sebak

Electrical and Computer Engineering, Concordia University, Montreal, QC, Canada

Abstract: Recently, there has been increasing interest and rapid growth in millimeter (mm)-wave antennas and devices for use in diverse applications, services, and technologies such as short-range communication, future mm-wave mobile communication for the fifth-generation (5G) cellular networks, and sensor and imaging systems. Due to the corresponding smaller wavelength, mm-wave frequencies offer the advantage of physically smaller antennas and circuits as well as the availability of much wider bandwidth compared to microwave frequencies. In addition, they provide additional spectrum for wireless communications. The planned 5G cellular networks base stations and mobile devices will essentially make use of mm-wave frequency bands to meet consumers' ever growing demand for high data rate and capacity from wireless service providers.

Millimeter-wave antenna design is considered as the first step for realizing mm-wave wireless communication and imaging systems. Design requirements for such antennas include highly directional patterns—for long transmission range and high detection sensitivity—and size reduction with a suitable impedance matching bandwidth. Frequency selective surface (FSS) technology is recently employed to enhance the performance of radiation and scattering properties of antennas used in different sectors such as aerospace, medical, and microwave industry. Therefore, it is appropriate and attractive to propose the use of FSS technology to design practical and efficient high gain antennas. This talk will address the market demand for compact high efficient antennas for next-generation wireless communications, sensing and imaging systems. The main part of the talk will focus on investigation and development of mm-wave high gain broadband antenna elements and arrays that cover multiple mm-wave frequency bands to serve several applications. Antennas with high gain produce very directive narrow beam for high resolution sensing as well as reduce the demand for power requirements and consumptions by wireless systems.

The talk will also discuss the development of frequency selective surface (FSS), and their diverse applications in mm-wave electromagnetic spectrum including: (a) an approach to enhance circularly polarized (CP) antenna gain (b) a linear to circular polarization converter which is based on multilayer FSS slab, and (c) a wideband FSS metasurface for radar cross section (RCS) reduction based on a polarization conversion is proposed.

Dr. Abdel Sebak is a Tier I Concordia University Research Chair in mm-wave antennas and systems. Before joining Concordia University, he was a professor at the University of Manitoba. He was also with Cairo University and worked with the Canadian Marconi Company on the design of microstrip phased array antennas. Dr. Sebak's recent research activities cover two streams: Antenna Engineering, and Analytical and Computational Electromagnetics. Applied and sponsored projects include high gain mm-wave antennas, advanced composite materials for aerospace shielding and antenna applications, microwave sensing and imaging, ultra-wideband antennas, and microwave beamforming. Dr. Sebak's original research contributions and technical leadership have been extensive and resulted in over 500 publications in prestigious refereed journals and international conference proceedings.

Dr. Sebak was inducted as a Fellow of the Institute of Electrical and Electronics Engineers in 2009. He is also a Fellow of the Engineering Institute of Canada. Dr. Sebak is a member of Concordia University Provost's Circle of Distinction for his career achievements. For his joint efforts in establishing one of the most advanced electromagnetic computational and antennas labs at the University of Manitoba, Dr. Sebak received the Rh Award for Outstanding Contributions to Scholarship and Research. Dr. Sebak received the 1992 and 2000 University of Manitoba Merit Award for outstanding Teaching and Research. In 1996, Dr. Sebak received the Faculty of Engineering Superior Academic Performance. Dr. Sebak has also received the IEEE Antennas and Propagation Society Best Chapter Award.

Dr. Sebak is serving as the Co-Chair of Publicity Committee of the 2020 IEEE Antennas and Propagation Symposium and has served as the General Chair of IEEE-ANTEM 2016 Symposium and Co-Chair of the IEEE ICUWB 2015. He has also served as Chair for the IEEE Canada Awards and Recognition Committee (2002–2004), IEEE Canada Conference Committee (2000–2002), and as the Technical

Program Chair for the 2002 IEEE CCECE Conference and the 2006 URSI-ANTEM Symposium. He has also served as a member (2002–2004) of the IEEE RAB Awards and Recognition Committee. Dr. Sebak has served as Associate Editor, *Journal of Applied Computational Electromagnetic Society*, Associate Editor, *International Journal of Antennas and Propagation*, Associate Editor, *J. Engineering Research*. He is a member of the Canadian National Committee of International Union of Radio Science (URSI) Commission B.

Contents

About the Editors

Mohamed Hesham Farouk received his Ph.D. in 1994 on digital processing of speech and is now a full-time professor at the Engineering Mathematics and Physics Department, Cairo University. He has a long experience in the field of engineering physics since 1990 when he participated as a research assistant in a project funded by research agency at Egypt for upgrading the control system of Coke-calciner at EGYPALUM. Prof. Hesham also participated in another funded project on developing intelligent Automatic Testing Equipment (ATE) for PCBs during the mid-1990s. He has many publications in the field of acoustic scattering and wavelet-based machine learning. He ~~also,~~ recently supervised theses on a related topic. He was the principal-investigator (P.I) of a funded research in the field of intelligent processing of instructional video in 2007. In 2017, he finalized, as a P.I., another funded project on electronic-circuits inspection using thermography. Prof. Hesham is the author and coauthor of more than 30 papers and one book published by Springer.

Faculty of Engineering, Cairo University, Giza, Egypt

Maha Amin Hassanein is a professor at the Department of Engineering Mathematics and Physics—Cairo University, Egypt (CUFE), where she has been a faculty member since 1989. Maha completed her Ph.D. and M.Sc. degrees in Engineering Mathematics in 1999 and 1994, respectively, and her B.Sc. degree in Communication and Electronic Engineering in 1989, all at CUFE. Her research interests lie in the area of numerical linear algebra, parallel computing on Graphics Processing Units (GPU), interval analysis, and wavelets theory. She is the author and coauthor of 16 research papers related to her research interests. Maha is a college educator with years of experience teaching a variety of mathematics courses for engineering students. Maha has interests in the quality of education where she attended a variety of workshops for the quality assurance and accreditation (QAA) and has an ABET certificate in program assessment in 2019. She held a position at the QAA unit at CUFE from 2014 until 2017.

Faculty of Engineering, Cairo University, Giza, Egypt

Part I
Engineering Mathematics

State-Dependent Parameter PID+ Control Applied to a Nonlinear Manipulator Arm

H. Sayed, E. M. Shaban, and A. Abdelhamid

Abstract The successful implementation of the PID+ control persuades the authors to investigate the same controller using the state-dependent parameter transfer function (SDP-TF) model so as to improve the performance of PID+ control when used on nonlinear systems. For this reason, this chapter introduces the SDP-PID+ control, for which the PID+ control is used on the four-degrees-of-freedom manipulator arm when modeled using SDP model structure, where the parameters of the TF change as a function of the state variables. Here, the additional input and proportional compensators that exist in the PID+ approach fight the effect of the discrete-time SDP-TF associated with samples time delay greater than unity and order greater than two. These additional compensators enable the use of the full-state feedback to develop the time-variant state variable feedback (SDP-SVF) control action for the SDP-PID+ controller. In this work, two tuning techniques are introduced for the SDP-PID+ controller; they are the LQ cost function through using the SDP-NMSS of the SDP-TF model and the pole placement. Both approaches provide an appropriate performance in addition to reject output and input disturbance with retrieving the zero steady-state error in a suitable time.

Keywords Discrete PID+ control · Discrete-time nonlinear system · State-dependent parameter (SDP) model · SDP pole placement · Non-minimal state space (NMSS) form · Linear quadratic (LQ) optimization

1 Introduction

Proportional–integral–derivative (PID) control is one of the earlier control strategies. It has been popular in the industry; about 94% of all control loops are in the category of PID [1], because it provides reliable control performance for almost all industrial

H. Sayed (✉) · E. M. Shaban · A. Abdelhamid
Faculty of Engineering (Mattaria), Mechanical Design Department, Helwan University, Cairo, Egypt

© Springer Nature Switzerland AG 2020
M. H. Farouk, M. A. Hassanein (eds.), *Recent Advances in Engineering Mathematics and Physics*, https://doi.org/10.1007/978-3-030-39847-7_1

applications, when PID gains are tuned correctly [2, 3]. The reliability and robustness of PID controller have been tested in several applications, e.g., [1, 3–6]. However, tuning the PID controller has even now been a significant problem because several industrial models in many cases have characteristics, like time delay, higher order, and nonlinearities [7], which can result in instability and poor performance of a system [8], cyclic/slow response, bad robustness, and collapse of the whole system [9].

In this regard, Shaban et al. [10] introduced a PID+ controller for linear discrete-time TF, where additional proportional and input compensators are utilized to counteract the effects of orders more than two as well as the effects of samples time delay greater than unity.

The successful implementation of PID+ control [10] persuades the authors to further improve the effectiveness of the PID+ through exploiting the nonlinear state-dependent parameter (SDP) model to create the SDP-PID+ controller. Here, the nonlinear system is modeled utilizing a quasi-linear design in which the parameters of the TF change as functions associated with the state variables, i.e., the SDP-TF model describes the nonlinear behavior of the system dynamics, where the parameters are generally functionally dependent upon other variables within the system [11].

The linear-like, affine design with the SDP-TF model ensures that the SDP-TF model can be considered to be a *frozen* linear system at every sampling instant [12]; therefore, the SDP formulation is similar to the linear formulation. This formulation helps to design the SDP-PID+ control law utilizing linear system style and design techniques, like pole assignment or linear quadratic (LQ). Subsequently, the SDP-PID+ control law consists of time-variant state feedback compensators which are state dependent in its nature.

The development of nonlinear SDP-PID+ control is the contribution of this chapter with practical implementation for a laboratory manipulator arm of four DOF, described in [13]. Primarily, the work shades the light to the linear PID+ control, then demonstrating the effect involving SDP variables. The two tuning techniques, PID+/LQ through using the NMSS-SDP model and also PID+/pole assignment, are developed and implemented to the laboratory manipulator arm to verify their applicability. The results show an appropriate closed-loop performance with satisfactory output and input disturbance rejection along with retrieving the zero steady-state error in an acceptable time when used on the manipulator arm [13].

2 Linear PID+ Control Design

To be able to develop the linear discrete-time PID+ control algorithm, the linear representation from the system dynamics is required, which may take the subsequent form

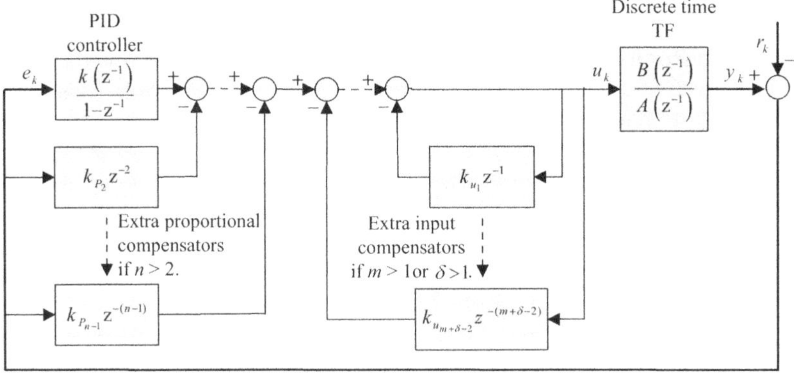

Fig. 1 The regulator structure of discrete PID+ controller [10]

$$y_k = -\sum_{i=1}^{n} a_i y_{k-i} + \sum_{j=1}^{m} b_{j+\delta-1} u_{k-(j+\delta-1)} \tag{1}$$

The incremental description of the system dynamics Eq. (1) may be displayed throughout discrete-time TF form, by making use of backward shift operator z^{-1}, as

$$y_k = \frac{\sum_{j=1}^{m} b_{j+\delta-1} z^{-(j+\delta-1)}}{1 + \sum_{i=1}^{n} a_i z^{-i}} u_k = \frac{B(z^{-1})}{A(z^{-1})} u_k \tag{2}$$

Here y_k, u_k, n, m, and δ are the output of the system, the input of the system, the TF order, the number of input parameters $\{b_\delta, \ldots, b_{m+\delta-1}\}$, and the samples time delay, respectively. The polynomials $A(z^{-1})$ and $B(z^{-1})$ are generally defined within the backward shift operator z^{-1}, for which $z^{-i} y_k = y_{k-i}$. As indicated throughout model (Eq. 2), any pure time delay samples, i.e., $\delta > 1$, may be accounted through setting the $\delta - 1$ leading parameters in $B(z^{-1})$ polynomial to zero, i.e., $b_1, \ldots, b_{\delta-1} = 0$. It is worth remembering that data-based modeling may be used to find the model structure of Eq. (2), i.e., the best values for the traid $\{n, m, \delta\}$, along with the best estimation the parameters of discrete-time TF, i.e., $\{a_1, \ldots, a_n\}$ and $\{b_\delta, \ldots, b_{m+\delta-1}\}$ [13].

PID+ control is an extension to the conventional PID control [10] for which the controller performs its purpose by means of additional $n - 2$ proportional compensators in order to fight the effect from the higher order of the TF ($n > 2$), in addition to extra $m + \delta - 2$ input compensators to help fight the effect of over one numerator of the TF ($m > 1$) as well as samples times delay greater than unity ($\delta > 1$), as stated within Fig. 1. The figure represents the regulator form of PID+ that the set point resets to zero, i.e., $e_k = -y_k$, and the reference signal, r_k, is considered as a disturbance.

Here, the polynomial associated with the conventional discrete PID controller is $k(z^{-1}) = k_1 - k_2 z^{-1} + k_3 z^{-2}$, which exhibits the interactions among the PID compensators $\{k_{P_1}, k_I, k_D\}$, for which $k_1 = k_{P_1} + k_I + k_D$, $k_2 = k_{P_1} + 2k_D$, and $k_3 = k_D$, giving that $\{k_{P_1}, k_I, k_D\}$ are the proportional, integral, and derivative compensators, respectively, of the conventional discrete PID controller.

The regulator form, depicted in Fig. 1, combined with the additional compensators enables the exploiting of the full-state feedback to develop the state variable feedback (SVF) control action to the PID+ controller, as

$$u_k = -\mathbf{k}^+ \mathbf{x}_k \tag{3}$$

for which $\mathbf{x}_k$ is the $n + m + \delta - 1$ full-state feedback vector, $\forall n \geq 2$, and defined as

$$\mathbf{x}_k = \begin{bmatrix} z_k & e_k & \Delta e_k & \vdots & e_{k-2} & e_{k-3} & \cdots & e_{k-(n-1)} & \vdots & u_{k-1} & u_{k-2} & \cdots & u_{k-(m+\delta-2)} \end{bmatrix}^T \tag{4}$$

and $\mathbf{k}^+$ is the PID+ control vector, which defined as

$$\mathbf{k}^+ = -\begin{bmatrix} k_I & k_{P_1} & k_D & \vdots & k_{P_2} & k_{P3} & \cdots & k_{P_{n-1}} & \vdots & -k_{u_1} & -k_{u_2} & \cdots & -k_{u_{m+\delta-2}} \end{bmatrix} \tag{5}$$

The PID+ gains have been provided using two tuning approaches; these are the LQ cost function by means of applying the non-minimal state space (NMSS) representation, PID+/LQ, and also the pole placement, PID+/pole placement.

Direct algebra shows that the model represented in Eq. (1) could be defined using the following NMSS representation:

$$\begin{aligned} \mathbf{x}_k &= \mathbf{F}\mathbf{x}_{k-1} + \mathbf{g}u_{k-1} \\ y_k &= \mathbf{h}\mathbf{x}_k \end{aligned} \tag{6}$$

Here $\mathbf{x}_k$ is the full-state feedback vector, defined in Eq. (4). The typical form intended for the state transient square matrix $\mathbf{F}$, input column vector $\mathbf{g}$, and output row vector $\mathbf{h}$ may be defined as follows:

$$\mathbf{F} = \begin{bmatrix} 1 & -(a_1+a_2) & a_2 & -a_3 & -a_4 & \cdots & -a_n & -b_2 & -b_3 & \cdots & -b_{m+\delta-1} \\ 0 & -(a_1+a_2) & a_2 & -a_3 & -a_4 & \cdots & -a_n & -b_2 & -b_3 & \cdots & -b_{m+\delta-1} \\ 0 & -(a_1+a_2+1) & a_2 & -a_3 & -a_4 & \cdots & -a_n & -b_2 & -b_3 & \cdots & -b_{m+\delta-1} \\ 0 & 1 & -1 & 0 & 0 & \cdots & 0 & 0 & 0 & \cdots & 0 \\ 0 & 0 & 0 & 1 & 0 & \cdots & 0 & 0 & 0 & \cdots & 0 \\ \vdots & \vdots & \vdots & \cdots & \ddots & \cdots & \vdots & \vdots & \vdots & \ddots & \vdots \\ 0 & 0 & 0 & 0 & 0 & \cdots & 0 & 0 & 0 & \cdots & 0 \\ 0 & 0 & 0 & 0 & 0 & \cdots & 0 & 0 & 0 & \cdots & 0 \\ 0 & 0 & 0 & 0 & 0 & \cdots & 0 & 1 & 0 & \cdots & 0 \\ \vdots & \vdots & \vdots & \vdots & \vdots & \ddots & \vdots & \cdots & \ddots & \cdots & \vdots \\ 0 & 0 & 0 & 0 & 0 & \cdots & 0 & 0 & 0 & \cdots & 0 \end{bmatrix}, \mathbf{g} = \begin{bmatrix} -b_1 \\ -b_1 \\ -b_1 \\ 0 \\ 0 \\ \vdots \\ 0 \\ 1 \\ 0 \\ \vdots \\ 0 \end{bmatrix},$$

$$\mathbf{h} = \begin{bmatrix} 0 & -1 & 0 & \vdots & 0 & 0 & \cdots & 0 & \vdots & 0 & 0 & \cdots & 0 \end{bmatrix} \tag{7}$$

The optimal SVF-PID+ control vector might be obtained through minimizing the following LQ type of performance criterion, i.e.,

$$J = \sum_{k=0}^{\infty} \left[\mathbf{x}_k^T \mathbf{Q} \mathbf{x}_k + R u_k^2 \right] \tag{8}$$

This shows the typical formula of the infinite time optimal LQ servomechanism cost function for SISO system. Right here, the matrix $\mathbf{Q} = \mathrm{diag}\left\{ q_z \quad q_{e_1} \quad q_{\Delta e} \,\vdots\, q_{e_2} \quad \cdots \quad q_{e_{n-1}} \,\vdots\, q_{u_1} \quad \cdots \quad q_{u_{m+\delta-2}} \right\}$ is the weighting square symmetric positive matrix of the states defined in Eq. (4), and R is the positive scalar.

Consider the NMSS-PID+ system described in Eq. (7), as $\{\mathbf{F}, \mathbf{g}\}$, the SVF-PID+ gain vector, $\mathbf{k}^+$, may be obtained recursively as being the steady-state solution from the algebraic Riccati equation (ARE) [14]; see Eq. (15).

In case of PID+/pole assignment technique, it is always easy to determine the PID + compensators $\left\{ k_{P_1}, k_I, k_D, k_{P_2}, \ldots, k_{P_{n-1}}, k_{u_1}, \ldots, k_{u_{m+\delta-2}} \right\}$ by developing the reduced closed-loop TF for the PID+ control as follows:

$$y_k = \frac{K^+(z^{-1})B(z^{-1})}{\Delta k_u(z^{-1})A(z^{-1}) + K^+(z^{-1})B(z^{-1})} \, r_k \tag{9}$$

Giving that $\Delta = 1 - z^{-1}$ is the difference operator, the polynomials of the input *plus* and proportional *plus* compensators, $k_u(z^{-1})$ and $K^+(z^{-1})$, respectively, are defined in Eq. (20).

3 Nonlinear SDP-PID+ Control Design

The structural form of the SDP-TF model is similar to the linear model of Eq. (2) and may take the form

$$y_k = \frac{B_k(\boldsymbol{\chi}_k, z^{-1})}{A_k(\boldsymbol{\chi}_k, z^{-1})} u_k \tag{10}$$

where the state-dependent polynomials $A_k(\boldsymbol{\chi}_k, z^{-1})$ and $B_k(\boldsymbol{\chi}_k, z^{-1})$ are time-varying equivalents of the linear polynomials $A(z^{-1})$ and $B(z^{-1})$ which existed in Eq. (2) and may be written in the following incremental form:

$$\begin{aligned} A_k(\boldsymbol{\chi}_k, z^{-1}) &= 1 + a_1(\boldsymbol{\chi}_k)z^{-1} + \cdots + a_n(\boldsymbol{\chi}_k)z^{-n} \\ B_k(\boldsymbol{\chi}_k, z^{-1}) &= b_1(\boldsymbol{\chi}_k)z^{-1} + \cdots + b_m(\boldsymbol{\chi}_k)z^{-m} \end{aligned} \tag{11}$$

The parameters $a_i(\boldsymbol{\chi}_k)$ $\forall 1 \leq i \leq n$ and $b_j(\boldsymbol{\chi}_k)$ $\forall 1 \leq j \leq m$ are themselves functions of the lagged system variables, i.e., the parameters are nonlinear functions of the vector $\boldsymbol{\chi}_k$ where, in general, $\boldsymbol{\chi}_k$ is defined in terms of any measured variables.

The nonlinear SDP-TF model (Eq. 10) within the PID+ control methodology leads normally to gains that are themselves state dependent; this justifies the name SDP-PID+ control. One possible benefit of the nonlinear SDP-PID+ control design is that it exploits the same theories and procedures applied to the linear PID+ control design. Therefore, the regulator form of SDP-PID+ takes the same structure of the linear PID+, shown in Fig. 1, replacing the constant compensators with state-dependent compensators, recalling that the external command input resets to zero and the main disturbance is considered as the change within the reference signal, r_k. Consequently, the SDP-PID+ control vector can be defined as

$$\mathbf{k}_k^+ = -\begin{bmatrix} k_{I,k} & k_{P_1,k} & k_{D,k} & \vdots & k_{P_2,k} & k_{P_3,k} & \cdots & k_{P_{n-1},k} & \vdots & -k_{u_1,k} & -k_{u_2,k} & \cdots & -k_{u_{m+\delta-2},k} \end{bmatrix}$$

$$(12)$$

The regulator form with the additional compensators enables the exploiting of the full-state feedback to construct the SDP-SVF control law for the SDP-PID+ controller, as

$$u_k = -\mathbf{k}_k^+ \mathbf{x}_k \tag{13}$$

for which $\mathbf{x}_k$ is the full-state feedback vector defined in Eq. (4).

Similar to linear control design, the nonlinear SDP-PID+ control design, for both approaches SDP-PID+/pole placement and SDP-PID+/LQ, can be utilized to find the SDP-PID+ gains of Eq. (12), given that the nonlinear discrete-time SDP-TF model of the process is controllable [15].

3.1 SDP-PID+/LQ Tuning Approach

The definition of the nonlinear SDP-TF model enables for the exploitation of many aspects of linear systems theory, such as the representation of NMSS-SDP, as follows:

$$\begin{aligned} \mathbf{x}_k &= \mathbf{F}_k x_{k-1} + \mathbf{g}_k u_{k-1} \\ y_k &= \mathbf{h} \mathbf{x}_k \end{aligned} \tag{14}$$

recalling that the NMSS-SDP representation of the SDP-TF model provides a complete description of the nonlinear dynamical behavior of the system, which is established from the nonlinear SDP-TF model.

The state transition matrix $\mathbf{F}_k$ and the input vector $\mathbf{g}_k$ may be defined as depicted in Eq. (7), by appropriately placing the SDP parameters instead of those parameters in the linear TF parameters, correspondingly.

Given the NMSS system description of the nonlinear SDP-PID+ controller $\{\mathbf{F}_k, \mathbf{g}_k\}$, weighting matrix $\mathbf{Q}$, and the scalar control input weighting R, the SVF of the nonlinear SDP-PID+ gain vector, $\mathbf{k}_k^+$, can then be obtained recursively from the steady-state solution of the ARE [14], derived from the standard LQ cost function Eq. (8), at each sample instant k as follows:

$$
\mathbf{k}_k^+ = \left[\mathbf{g}_k^T \mathbf{P}^{(i+1)} \mathbf{g}_k + R \right]^{-1} \mathbf{g}_k^T \mathbf{P}^{(i+1)} \mathbf{F}_k
$$
$$
\mathbf{P}^{(i)} = \mathbf{F}_k^T \mathbf{P}^{(i+1)} \left[\mathbf{F}_k - \mathbf{g}_k \mathbf{k}_k \right] + \mathbf{Q}
$$
(15)

for which $\mathbf{P}$ is a symmetrical positive definite matrix with the initial value $\mathbf{P}^{(i+1)}$ equals to the weighting matrix $\mathbf{Q}$, and $\mathbf{k}_k^+$ is the SVF of the SDP-PID+ control gain vector defined in Eq. (12). The SVF of nonlinear SDP-PID+ control action can then be implemented via Eq. (13).

It is worth noting here that for simplicity, the frozen parameter system for $\{\mathbf{F}_k', \mathbf{g}_k'\}$ can be used for which the frozen system is defined as a single sample member of the family of $\{\mathbf{F}_k, \mathbf{g}_k\}$ [16]. Consequently, the matrix $\mathbf{P}$ becomes time-invariant symmetrical positive-definite matrix solution of the ARE, by merging the two equations in (Eq. 15) as follows:

$$
\mathbf{P}^{(i)} = \mathbf{F}_k'^T \mathbf{P}^{(i+1)} \left[\mathbf{F}_k' - \mathbf{g}_k' \left(\mathbf{g}_k'^T \mathbf{P}^{(i+1)} \mathbf{g}_k' + R \right)^{-1} \mathbf{g}_k'^T \mathbf{P}^{(i+1)} \mathbf{F}_k' \right] + \mathbf{Q}
$$
(16)

Given the values of the positive-definite matrix $\mathbf{P}$, using Eq. (16), the value of SDP gain vector $\mathbf{k}_k^+$ can be obtained from Eq. (15) as

$$
\mathbf{k}_k^+ = \left[\mathbf{g}_k^T \mathbf{P} \mathbf{g}_k + R \right]^{-1} \mathbf{g}_k^T \mathbf{P} \mathbf{F}_k
$$
(17)

while keeping the system matrices $\{\mathbf{F}_k, \mathbf{g}_k\}$ unfrozen.

3.2 SDP-PID+/Pole Placement Tuning Approach

Similar to linear approach, the evaluation of the nonlinear SDP-PID+ compensators $\left\{ k_{P_1,k}, k_{I,k}, k_{D,k}, k_{P_2,k}, k_{P_3,k}, \ldots, k_{P_{n-1},k}, k_{u_1,k}, k_{u_2,k}, \ldots, k_{u_{m+\delta-2},k} \right\}$ requires the closed-loop control system in TF form. This may be obtained directly by reducing the block diagram shown in Fig. 1 with placing the SDP compensators instead of those compensators in the linear structure. The reduced unity feedback closed-loop SDP-PID+ control system in regulator form is depicted in Fig. 2, which consequently provides the following closed-loop SDP-PID+ control system:

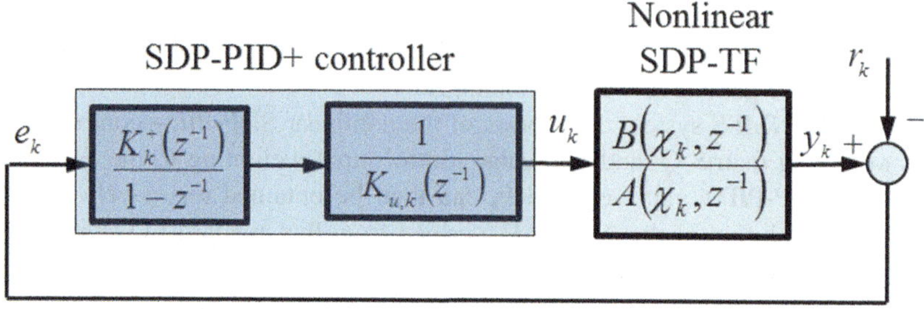

$$K_k^+\!\left(z^{-1}\right)=k_{1,k}-k_{2,k}z^{-1}+k_{3,k}z^{-2}+\sum_{i=4}^{n}k_{i,k}z^{-i+1}\qquad K_{u,k}\!\left(z^{-1}\right)=1+\sum_{j=1}^{m+\delta-2}k_{u_j,k}z^{-j}$$

Fig. 2 The closed-loop regulator structure of the SDP-PID+ control system with unity feedback, showing the two polynomials of the SDP+ compensators polynomials, $K_k^+(z^{-1})$ and $k_{u,\,k}(z^{-1})$

$$y_k=\frac{K_k^+(z^{-1})B_k(\chi_k,z^{-1})}{\Delta k_{u,k}(z^{-1})A_k(\chi_k,z^{-1})+K_k^+(z^{-1})B_k(\chi_k,z^{-1})}\,r_k \tag{18}$$

Equation (18) provides the following characteristic equation for SDP-PID+ control:

$$\Delta k_{u,k}\!\left(z^{-1}\right)A_k\!\left(\chi_k,z^{-1}\right)+K^+\!\left(z^{-1}\right)B_k\!\left(\chi_k,z^{-1}\right) \tag{19}$$

which utilized to obtain the SDP-PID+ gains. Recalling the order of characteristic Eq. (19) is $n+m+\delta-1$, $\forall n\geq 2$, along with minimum value of 3 in the case of discrete-time TF with traid$\{1,1,1\}$. Also, the polynomials of the proportional *plus* and input *plus* compensators, $K_k^+(z^{-1})$ and $k_{u,\,k}(z^{-1})$, respectively, can be defined as

$$K_k^+\!\left(z^{-1}\right)=k_{1,k}-k_{2,k}z^{-1}+k_{3,k}z^{-2}+\sum_{i=4}^{n}k_{i,k}z^{-i+1}$$

$$K_{u,k}\!\left(z^{-1}\right)=1+\sum_{j=1}^{m+\delta-2}k_{u_j,k}z^{-j} \tag{20}$$

for which

$$\begin{aligned}
k_{1,k}&=k_{P_1,k}+k_{I,k}+k_{D,k}\\
k_{2,k}&=k_{P_1,k}+2k_{D,k}\\
k_{3,k}&=k_{D,k}+k_{P_{i-1},k} \qquad\qquad i=3 \text{ if } n\geq 3\\
&\;\;\vdots\\
k_{i,k}&=-k_{P_{i-2},k}+k_{P_{i-1},k} \qquad 4\leq i\leq n+1
\end{aligned} \tag{21}$$

Applying a simple polynomial algebra for the characteristic Eq. (21) as

$$\Delta k_{u,k}\left(z^{-1}\right)A_k\left(\chi_k,z^{-1}\right)+K_k^{+}\left(z^{-1}\right)B_k\left(\chi_k,z^{-1}\right)$$
$$=\left(1-p_1 z^{-1}\right)\left(1-p_2 z^{-1}\right)\left(1-p_3 z^{-1}\right)\ldots\left(1-p_i z^{-1}\right)\ i=4,\ldots,n+m+\delta-1\ \forall n\geq 2 \tag{22}$$

gives SDP-PID+ gains $\{k_{1,k}, k_{2,k}, k_{3,k}, \ldots, k_{i,k}\ (i=4, \ldots, n+1), k_{u_j,k}\ (j=1, \ldots, m+\delta-2)\}$ at predetermined positions of the poles $\{p_1, p_2, p_3, \ldots, p_i(i=4, \ldots, n+m+\delta-1\ \forall\ n\geq 2)\}$ in the complex z-plane. The control action will then be resulting from Fig. 2 as follows:

$$u_k = u_{k-1} + k_{1,k}e_k - k_{2,k}e_{k-1} + k_{3,k}e_{k-2} + \sum_{i=4}^{n+1} k_{i,k}e_{k-(i-1)}$$
$$-\sum_{j=1}^{m+\delta-2} k_{u_j,k}\left(u_{k-j} - u_{k-(j+1)}\right) \tag{23}$$

Alternatively, the SVF-PID+ control action Eq. (13) can easily be used for which the parameters of SDP-PID+/SVF control vector defined in (Eq. 12) can be obtained utilizing Eq. (21).

4 Application to the Robot Arm

The shoulder joint of the manipulator arm with four DOF [13] is chosen for the sake of practical implementation since it provides a valuable insight to the novel SDP-PID + control, where the SDP-TF of the shoulder joint is first order with four samples time delay ($\delta = 4$) as follows:

$$y_k = \frac{b_{4,k}\,z^{-4}}{1 + a_{1,k}z^{-1}}\,u_k \tag{24}$$

for which $a_{1,\,k} = \alpha_1 y_{k-1} + \alpha_2$, $b_{4,k} = 0.0386$, where $\alpha_1 = -3.6530 \times 10^{-4}$ and $\alpha_2 = -0.969$. The full modeling process and parameter estimation of this joint and the other joints of the arm are discussed in detail in [13]. Here, the sampling rate of 25 samples per second is found suitable regarding both hardware restrictions and accurate modeling [13]. Given the discrete-time SDP-TF Eq. (24), the NMSS of nonlinear SDP-PID+ form of Eq. (14) can be constructed using the definition of Eq. (7) as follows:

$$
\begin{bmatrix} z_k \\ e_k \\ \Delta e_k \\ u_{k-1} \\ u_{k-2} \\ u_{k-3} \end{bmatrix} = \begin{bmatrix} 1 & -a_{1.k} & 0 & 0 & 0 & -b_{4,k} \\ 0 & -a_{1.k} & 0 & 0 & 0 & -b_{4,k} \\ 0 & -(a_{1,k}+1) & 0 & 0 & 0 & -b_{4,k} \\ 0 & 0 & 0 & 0 & 0 & 0 \\ 0 & 0 & 0 & 1 & 0 & 0 \\ 0 & 0 & 0 & 0 & 1 & 0 \end{bmatrix} \begin{bmatrix} z_{k-1} \\ e_{k-1} \\ \Delta e_{k-1} \\ u_{k-2} \\ u_{k-3} \\ u_{k-4} \end{bmatrix} + \begin{bmatrix} 0 \\ 0 \\ 0 \\ 1 \\ 0 \\ 0 \end{bmatrix} u_{k-1} \quad (25)
$$

By freezing $y_{k-1} = 0$ and using ARE (Eq. 16), using weights $Q = $ diag ([1 1 1 1 1 1]) and $R = 1$, it is found that

$$
P = \begin{bmatrix} 13.3047 & 73.8927 & 0 & -2.1778 & -2.5463 & -2.9419 \\ 73.8927 & 653.2557 & 0 & -21.6414 & -23.8188 & -25.9671 \\ 0 & 0 & 1 & 0 & 0 & 0 \\ -2.1778 & -21.6414 & 0 & 3.7431 & 0.8025 & 0.8616 \\ -2.5463 & -23.8188 & 0 & 0.8025 & 2.8789 & 0.9483 \\ -2.9419 & -25.9671 & 0 & 0.8616 & 0.9483 & 2.0354 \end{bmatrix}
$$

$$(26)$$

Then the vector associated with nonlinear SDP-PID+ feedback gains may be obtained using Eq. (17) and gives

$$
\mathbf{K_k^+} = [-0.4592 \ 5.02 a_{1,k} \ 0 \ 0.1692 \ 0.1817 \ 0.1937]^T \quad (27)
$$

It is now possible to apply the control law Eq. (13) to control the shoulder joint of the manipulator arm using SDP-PID+/LQ approach. The practical implementation of control law of Eq. (13) is depicted in Fig. 3 together with the response of the SDP-TF Eq. (24) and the corresponding control action. The figure shows acceptable closed-loop response with satisfactory tracking performance.

The robustness of the novel SDP-PID+ controller may be demonstrated by applying the input and output disturbance rejection tests, as shown in Figs. 4 and 5, respectively. As depicted in these figures, the developed SDP-PID+/LQ denies input/output disturbance and retrieves the zero steady-state error in an acceptable time with satisfactory performance.

For developing the SDP-PID+/pole assignment technique, Eq. (22) may be used as

$$
\begin{aligned}
K_k^+(z^{-1}) &= k_{1,k} - k_{2,k}z^{-1} + k_{3,k}z^{-2} \\
K_{u.k}(z^{-1}) &= 1 + k_{u_1,k}z^{-1} + k_{u_2,k}z^{-2} + k_{u_3,k}z^{-3} \\
A(\chi_k, z^{-1}) &= 1 + a_{1,k}z^{-1} \\
B(\chi_k, z^{-1}) &= b_{4,k}z^{-4} \\
\Delta &= 1 - z^{-1}
\end{aligned} \quad (28)
$$

Subsequently, a straightforward polynomial algebra for the characteristic Eq. (22)

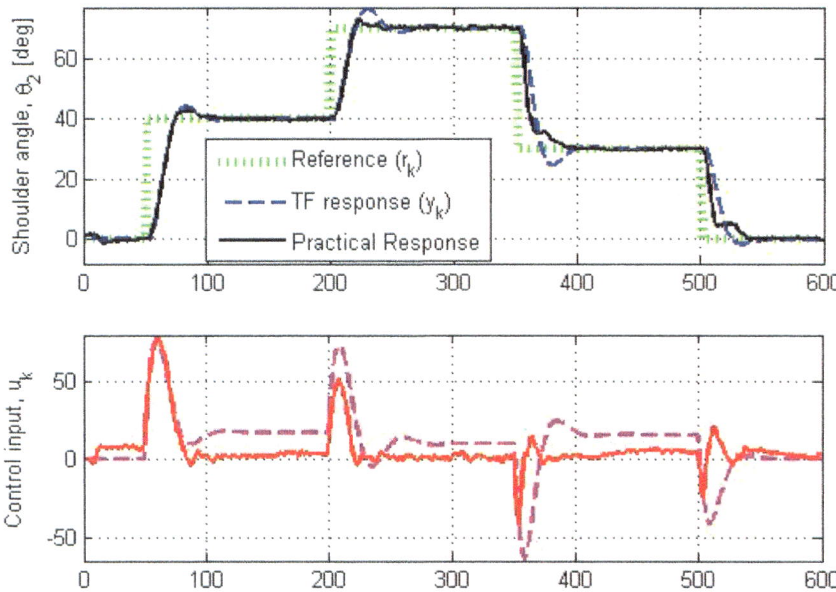

Fig. 3 The practical implementation of the SDP-PID+/LQ control when applied to the shoulder joint of the manipulator arm [13]. Top: practical response (solid), SDP-TF model response (dashed), and command input (dotted). Bottom: the normalized input in percentage, in which the practical control input is displayed as solid and simulation control input as dashed

$$\Delta K_{u.k}(z^{-1})A(\chi_k, z^{-1}) + K_k^+(z^{-1})B(\chi_k, z^{-1})$$
$$= (1 - p_1 z^{-1})(1 - p_2 z^{-1})(1 - p_3 z^{-1})(1 - p_4 z^{-1})(1 - p_5 z^{-1})(1 - p_6 z^{-1})$$

$$(29)$$

may be used to get the SDP-PID+ gains $\{k_{1,k}, k_{2,k}, k_{3,k}, k_{u_1,k}, k_{u_2,k}, k_{u_3,k},\}$ at predetermined positions of the poles $\{p_1, p_2, p_3, p_4, p_5, p_6,\}$ in the complex z-plane. By placing the poles at $p_1 = p_2 = 0.8$, $p_3 = p_4 = 0.5$, and $p_5 = p_6 = -0.25$, the corresponding gains may be obtained as

$$
\begin{aligned}
k_{u_1,k} &= -1.1 - a_{1,k} \\
k_{u_2,k} &= a_{1,k}^2 + 1.1a_{1,k} + 0.1525 \\
k_{u_3,k} &= 0.0425 + a_{1,k}k_{u_1,k} + (1 - a_{1,k})k_{u_2,k} \\
k_{1,k} &= 25.9222[(1 - a_{1,k})k_{u_3,k} + a_{1,k}k_{u_2,k}] - 5.2978 \\
k_{2,k} &= -25.9222a_{1,k}k_{u_3,k} - 0.3888 \\
k_{3,k} &= 0.2592
\end{aligned}
$$

$$(30)$$

The control law Eq. (13) may be applied again to control the shoulder joint of the manipulator arm using the novel SDP-PID+/pole placement approach. A graph of

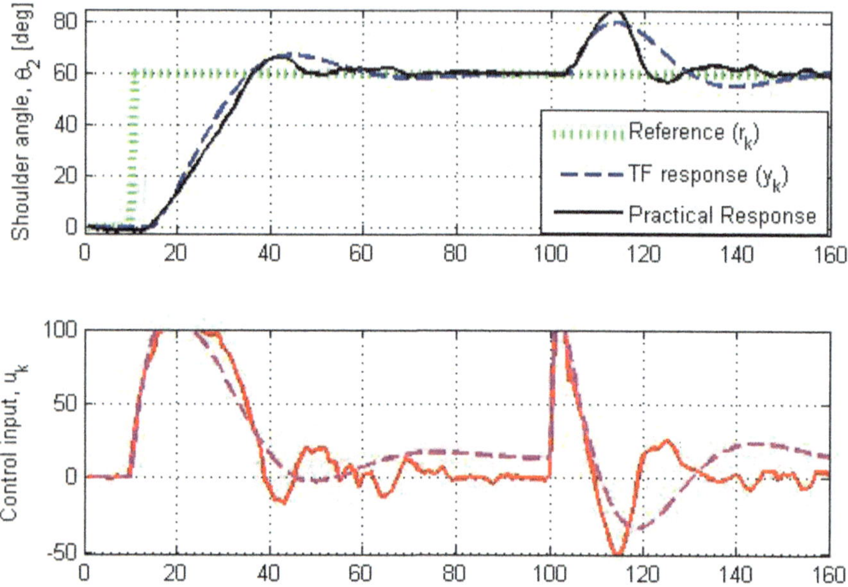

Fig. 4 The practical implementation of input disturbance of NMSS nonlinear SDP-PID+ control when operated for the shoulder joint of the manipulator arm [13]

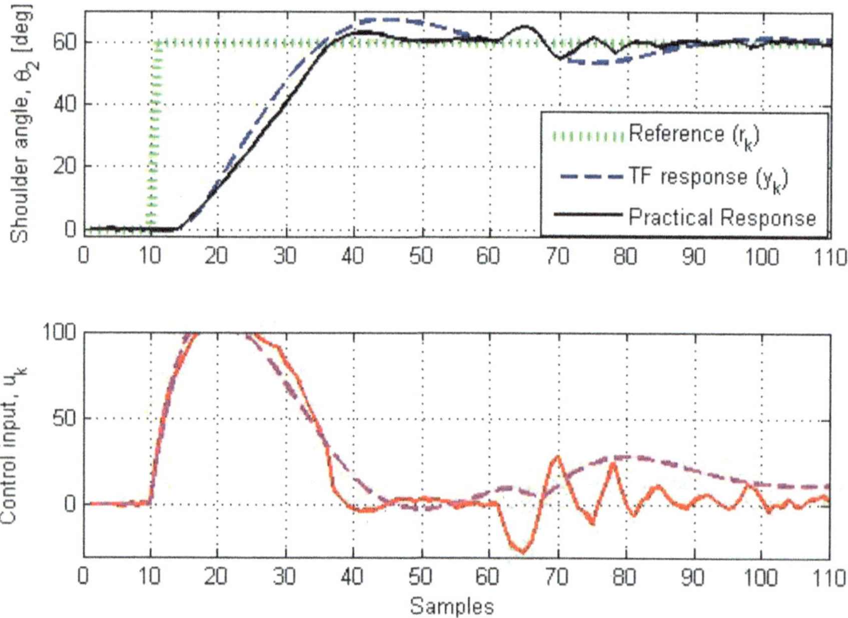

Fig. 5 The practical implementation of output disturbance of NMSS nonlinear SDP-PID+ control when operated for the shoulder joint of the manipulator arm [13]

practical implementation and simulation response of the TF of Eq. (24) is shown in Fig. 6 together with the control action. It is clear from the figure that the novel discrete-time SDP-PID+/pole placement approach provides good tracking response with satisfactory closed-loop performance.

For consistency of this work, the robustness of the SDP-PID+/pole assignment controller is performed using the input/output disturbance rejection tests. The tests are shown in Figs. 7 and 8, respectively, for which the designed PID+/pole assignment controller effectively denies the disturbances and also retrieves the zero steady-state error in an acceptable time with acceptable performance.

A numerical comparison for the performance of the control system is summarized in Table 1. Here, the two tuning approaches, SDP-PID+/LQ and the SDP-PID+/pole placement, are compared regarding the maximum/minimum settling time and maximum/minimum overshoot of both the experimental and simulation data collected for the tracking tests in Figs. 3 and 6. It is worth noting here that the steady-state error for both approaches is always zero.

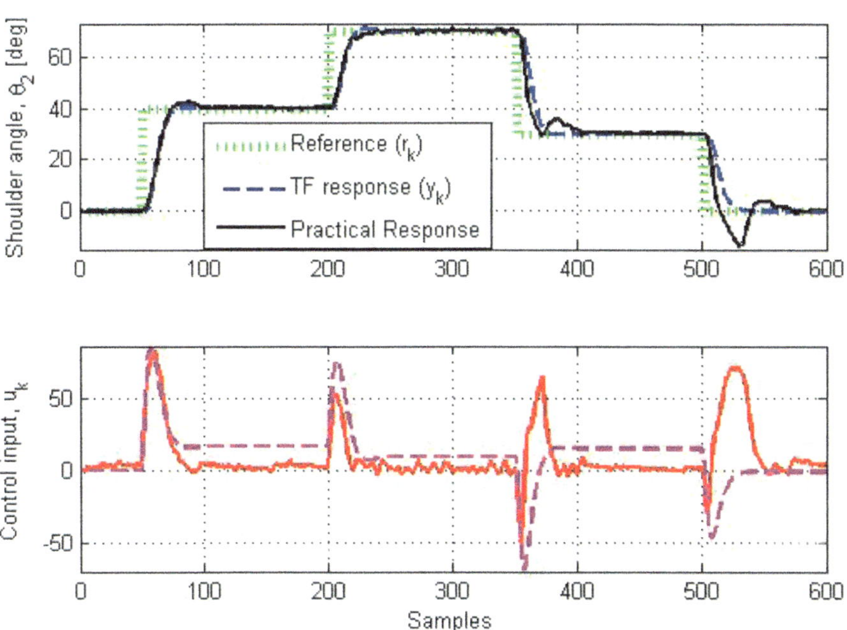

Fig. 6 The practical implementation of the novel discrete-time SDP-PID+/pole assignment control when operated for the shoulder joint of the manipulator arm [13]

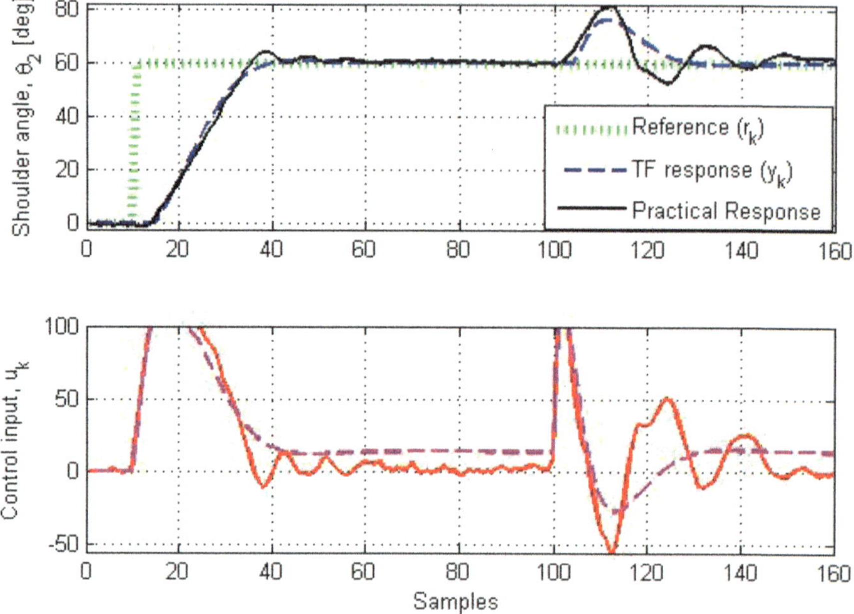

Fig. 7 The practical implementation of the novel discrete-time SDP-PID+/pole assignment control when operated for the shoulder joint of manipulator arm [13] in case of input disturbance

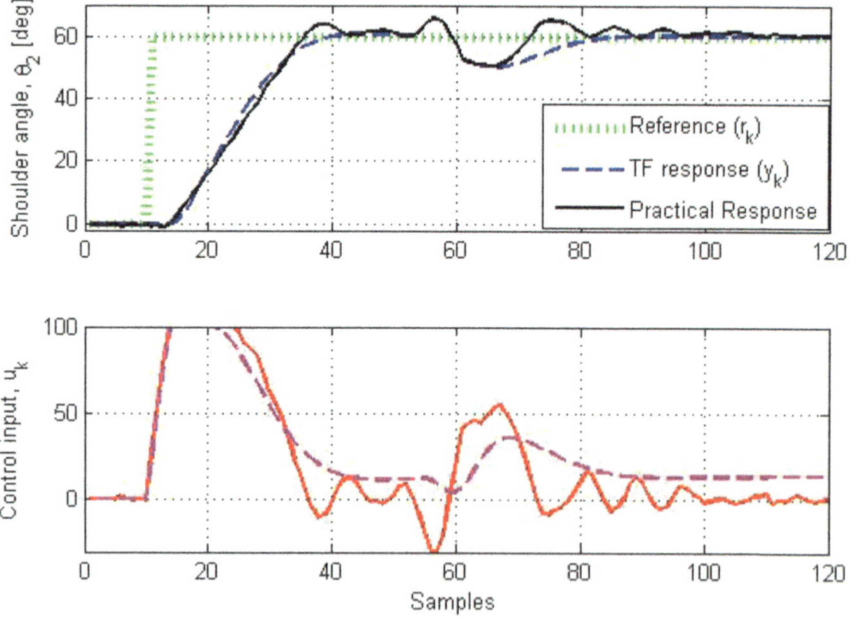

Fig. 8 The practical implementation of the novel discrete-time SDP-PID+/pole assignment control when operated for the shoulder joint of manipulator arm [13] in case of output disturbance

Table 1 A numerical comparison for the performance of the system

Tuning approach	SDP-PID+/LQ				SDP-PID+/pole placement			
	Simulation		Experimental		Simulation		Experimental	
	Max	Min	Max	Min	Max	Min	Max	Min
Settling time (samples)	73	50	66	40	45	28	86	38
% Overshoot	20.5	5.5	9.6	0.9	3.4	0	46	0.6

5 Conclusion

This chapter developed a state-dependent parameter PID+ (SDP-PID+) control for the laboratory four-degrees-of-freedom manipulator arm [13]. Two distinct tuning approaches for the SDP-PID+ controller were presented and successfully implemented; they are the linear quadratic (LQ) cost function by using the novel nonlinear NMSS form and nonlinear pole placement approaches to develop the SDP-PID+ control. The work shows successful implementation for SDP-PID+ control with satisfactory tracking response and input/output rejection, when applied to the four-degrees-of-freedom manipulator arm.

References

1. Sheel, S., & Gupta, O. (2012). New techniques of PID controller tuning of a DC motor—Development of a toolbox. *MIT IJEIE, 2*(2), 65–69.
2. Vaishnav, S. R., & Khan, Z. J. (2010). Performance of tuned PID controller and a new hybrid fuzzy PD+I controller. *World Journal of Modelling and Simulation, 6*(2), 141–149.
3. Kumar, R., Singla, S. K., & Vikram, A. (2013). A comparative analysis of different methods for the tuning of PID controller. *International Journal of Electronics and Communications Electrical Engineering, 3*(2), 1–7.
4. Zhang, R., Wu, S., Lu, R., & Gao, F. (2014). Predictive control optimization based PID control for temperature in an industrial surfactant reactor. *Chemometrics and Intelligent Laboratory Systems, 15*(135), 48–62.
5. Korsane, D. T., Yadav, V., & Raut, K. H. (2014). Pid tuning rules for first order plus time delay system. *International Journal of Innovative Research in Electrical, Instrumentation and Control Engineering, 2*(1), 582–586.
6. Guerrero, J., Torres, J., Creuze, V., Chemori, A., & Campos, E. (2019). Saturation based nonlinear PID control for underwater vehicles: Design, stability analysis and experiments. *Mechatronics, 61*, 96–105.
7. Deepa, S. N., & Sugumaran, G. (2011). Design of PID controller for higher order continuous systems using MPSO based model formulation technique. *International Journal of Electrical and Electronics Engineering, 5*(4), 289–295.
8. Tahmasbi, N., Tehrani, H. A., & Esmaeili, J. (2019). Practical stabilization of time-delay fractional-order systems by parametric controllers. *ISA Transactions, 95*, 211–220.
9. Ali, H., & Wadhwani, S. (2015). Intelligent PID controller tuning for higher order process system. *International Journal of u-and e Service, Science and Technology, 8*(6), 323–330.

10. Shaban, E. M., Sayed, H., & Abdelhamid, A. (2019). A novel discrete PID+ controller applied to higher order/time delayed nonlinear systems with practical implementation. *International Journal of Dynamics and Control, 7*(3), 888–900.
11. Shaban, E., Zied, K., Taylor, C. J., Seward, D. W. (2005). Nonlinear control system design for construction robots: Estimation, partial linearization by feedback and state-dependent-parameter control. *22nd international symposium on automation robotics in construction ISARC 2005,* Ferrara (Italy).
12. Taylor, C. J., Shaban, E. M., Stables, M. A., & Ako, S. (2007). Proportional-integral-plus control applications of state-dependent parameter models. Proceedings of the Institution of Mechanical Engineers, Part I. *Journal of Systems and Control Engineering, 221*(7), 1019–1031.
13. Sayed, H., Abdelhamid, A., & Shaban, E. M. (2018). An experimental validation of finite element method versus data based modelling when applied to the dynamic modelling of spatial manipulator. *Life Science Journal, 15*(2), 73–86. www.lifesciencesite.com.
14. Astrom, K. J., & Wittenmark, B. (1984). *Computer controlled systems: Theory and design* (Prentice-Hall Information and System Sciences Series). Upper Saddle River, NJ: Prentice Hall.
15. Shaban, E. M. (2012). Deadbeat response of nonlinear systems described by discrete-time state dependent parameter using exact linearization by local coordinate transformation. *Journal of American Science, 8*(10), 355–366.
16. Shaban EM, Taylor CJ, Chotai A (2004). State dependent parameter proportional-integral-plus (SDP-PIP) control of a nonlinear robot digger arm. In: *UKACC International Conference on Control*, Glasgow, UK.

Sliding Mode Control with PID Surface for Robot Manipulator Optimized by Evolutionary Algorithms

Fatiha Loucif and Sihem Kechida

Abstract In this study, sliding mode controller (SMC) with PID surface is designed for the trajectory tracking control of robot manipulator using antlion optimization algorithm (ALO) compared with another technique called gray wolf optimizer (GWO). The idea is to determine optimal parameters (K_p, K_i, K_d, and lamda) ensuring best performance of robot manipulator system minimizing the integral time absolute error (ITAE) criterion or the integral time square error (ISTE) criterion; the modeling and the control of the robot manipulator were realized in MATLAB environment. The simulation results prove the superiority of ALO in comparison with GWO algorithm.

Keywords Sliding mode control · PID sliding surface · Nonlinear control · Robot manipulator

1 Introduction

The use of robotic arms in industrial applications has significantly been increased. The robot motion tracking control which required high accuracy, stability, and safety is one of the challenging problems due to highly coupled and nonlinear dynamic. In the presence of model uncertainties such as dynamic parameters (e.g., inertia and payload conditions), dynamic effects (e.g., complex nonlinear frictions), and unmodeled dynamics, conventional controllers have many difficulties in treating these uncertainties. Sliding mode control (SMC) is one of the most robust approaches to overcome this problem. The most distinguished property of the sliding

F. Loucif (✉)
Université 8 Mai 1945, Guelma, Algeria
e-mail: loucif.fatiha@univ-guelma.com

S. Kechida
Laboratoire d'Automatique et Informatique de Guelma LAIG, Guelma, Algeria
e-mail: kechida.sihem@univ-guelma.com

© Springer Nature Switzerland AG 2020
M. H. Farouk, M. A. Hassanein (eds.), *Recent Advances in Engineering Mathematics and Physics*, https://doi.org/10.1007/978-3-030-39847-7_2

mode control lies in its insensitivity to dynamic uncertainties and external disturbances.

However, this approach exhibits high-frequency oscillations called chattering when the system state reaches the sliding surface, which has negative effects on the actuator control and excite the undesirable unmodeled dynamics.

Recently, sliding mode, integral sliding mode controllers (ISMC), and proportional integral sliding mode controllers (PI-SMC) were examined by many researchers as a powerful nonlinear controller in [1–7]. An adaptive sliding mode control is designed in several papers [8–12]. The proportional integral derivative sliding mode controller (PID-SMC) was designed to control the robot manipulator in several works. The investigation of fuzzy logic and neuro-fuzzy logic control to design an adaptive sliding mode control are found in [13–16].

Currently, evolutionary algorithms have appeared as an alternative design method for robotic manipulator. M. Vijay and Debaschisha design a PSO-based backstepping sliding mode controller and observer for robot manipulators [17]. This authors used PSO to tune the sliding surface parameters of SMC coupled with artificial neuro-fuzzy inference system (ANFIS) [18].

The optimization of the PID-SMC parameters using ALO is outperformed in comparison with GA and PSO algorithms by Mokeddem and Draidi to control a nonlinear system [19].

This paper presents the use of novel optimization algorithms to tune SMC with PID surface for the trajectory tracking control of robot manipulator. These algorithms are described in Sect. 3. Section 2 designates the mathematical model of robot manipulator. The principle of SMC and its application on the robot manipulator are titled in Sect. 4. The simulation results are presented in Sect. 5.

2 Dynamic Model of Robot Manipulator

By applying Lagrange's principle, the dynamic model of two-degree-of-freedom (2DOF) robot manipulator is given by

$$\tau = M(q)\ddot{q} + C(q,\dot{q})\dot{q} + G(q) + F(\dot{q}) \tag{1}$$

where q_i, $\dot{q}_i$, and $\ddot{q}_i$ present the link position, velocity, and acceleration vectors, respectively. $M(q)$ is the matrix inertia given by

$$M(q) = \begin{bmatrix} M_{11} & M_{12} \\ M_{21} & M_{22} \end{bmatrix}$$

where

$$M_{11} = m_1 \cdot l_1^2 + m_2 \cdot \left(l_1^2 + 2l_1 \cdot l_2 \cdot \cos\left(q_2\right) + l_2^2 \right)$$

$$M_{12} = M_{12} = m_2 \cdot l_2(l_2 + l_1 \cdot \cos\left(q_2\right))$$

$$M_{22} = m_2 \cdot l_2^2$$

$C(q, \dot{q})$ is the Coriolis centripetal force matrix given by

$$C(q) = \begin{bmatrix} C_{11} & C_{12} \\ C_{21} & C_{22} \end{bmatrix}$$

where

$$C_{11} = -m_2 \cdot l_1 \cdot l_2 \cdot \sin\left(q_2\right) \cdot 2\dot{q}_2$$

$$C_{12} = -m_2 \cdot l_1 \cdot l_2 \cdot \sin\left(q_2\right) \cdot \dot{q}_2$$

$$C_{21} = m_2 \cdot l_1 \cdot l_2 \cdot \sin\left(q_2\right) \cdot \dot{q}_1$$

$$C_{22} = 0$$

The gravity vector $G(q) = [G_{11} \ G_{12}]^T$

is given by

$$G_{11} = (m_1 + m_2) \cdot g \cdot l_1 \cdot \cos\left(q_1\right) + m_2 \cdot g \cdot l_2 \cdot \cos\left(q_1 + q_2\right)$$

$$G_{12} = m_2 \cdot g \cdot l_2 \cdot \cos\left(q_1 + q_2\right)$$

Finally, $F(\dot{q}) = [F_{11}F_{21}]^T$ is the friction force vector given by

$$F_{11} = 2 \cdot \dot{q}_1 + 0.8 \operatorname{sign}\left(\dot{q}_1\right)$$

$$F_{21} = 4 \cdot \dot{q}_2 + 0.1 \operatorname{sign}\left(\dot{q}_2\right)$$

and τ is the vector of the torque control signal, where m_i and l_i are the link mass and length, respectively.

3 Evolutionary Algorithms

3.1 Gray Wolf Optimizer (GWO)

GWO is a recent meta-heuristic optimizer inspired by gray wolves and proposed by [20]. It mimics the leadership hierarchy and the hunting mechanism of gray wolves in nature.

As described in literature, the GWO algorithm includes two mathematical models: encircling prey and hunting prey.

The encircling behavior: During the hunting, the gray wolves encircle prey. The mathematical model is presented in the following equations:

$$\vec{D} = \left| \vec{C} \cdot \vec{X_p}(t) - \vec{X}(t) \right| \tag{2}$$

$$\vec{X}(t+1) = \vec{X_p}(t) - \vec{A}\vec{D} \tag{3}$$

where D is the distance, $\vec{X_p}(t)$ is the position vector of prey, $\vec{X}(t)$ indicates the position of the gray wolf, t indicates the current iteration, and $\vec{A}$ and $\vec{C}$ are coefficient vectors calculated as follows:

$$\vec{A} = 2\vec{a}\vec{r_1} - \vec{a} \tag{4}$$

$$\vec{C} = 2\vec{r_2} \tag{5}$$

where components of a are linearly decreased from 2 to 0 over the course of iterations and r_1 and r_2 are random vectors in [0 1].

The hunting model: Four types of gray wolves participate in chasing prey; alpha, beta, delta, and omega denote the wolf group and are employed as solutions (fittest, best, and candidate) for simulating the leadership hierarchy.

The optimization algorithm is guided by α, β, and δ; with three best solutions obtained so far, the other search agents follow them and update their positions according to the best search agent as follows:

$$\vec{D_\alpha} = \left| \vec{C_1} \cdot \vec{X_\alpha}(t) - \vec{X}(t) \right| \tag{6}$$

$$\vec{D_\beta} = \left| \vec{C_2} \cdot \vec{X_\beta}(t) - \vec{X}(t) \right| \tag{7}$$

$$\vec{D_\delta} = \left| \vec{C_3} \cdot \vec{X_\delta}(t) - \vec{X}(t) \right| \tag{8}$$

$$\vec{X_1} = \vec{X_\alpha}(t) - \vec{A_1} \cdot \vec{D_\alpha} \tag{9}$$

$$\vec{X_2} = \vec{X_\beta}(t) - \vec{A_2} \cdot \vec{D_\beta} \tag{10}$$

$$\vec{X_3} = \vec{X_\delta}(t) - \vec{A_3} \cdot \vec{D_\delta} \tag{11}$$

$$\text{and } \vec{X}(t+1) = \frac{\vec{X_1} + \vec{X_2} + \vec{X_3}}{3} \tag{12}$$

3.2 *The Ant Lion Optimizer*

Another novel nature-inspired algorithm called ant lion optimizer (ALO) mimics the hunting mechanism of antlions in nature [21]. Five main steps of hunting prey such as the random walk of ants, building traps, entrapment of ants in traps, catching preys, and rebuilding traps are implemented in this algorithm.

Since ants move stochastically in nature when searching for food, a random walk is chosen for modeling ants' movement as follows:

$$X(t) = [0, \text{cumsum}(2r(t_1 - 1)), \text{cumsum}(2r(t_2 - 1)), \ldots, \text{cumsum}(2r(t_n - 1))] \tag{13}$$

where cumsum calculates the cumulative sum, n is the maximum number of iteration, t shows the step of random walk (iteration in this study), and $r(t)$ is a stochastic function defined as follows:

$$r(t) = \left\{ \begin{array}{l} 1 \text{ if rand} > 0.5 \\ 0 \text{ if rand} \leq 0.5 \end{array} \right\} \tag{14}$$

and rand is a random number generated with uniform distribution in the interval of [0,1]. The position of ants is saved and utilized during optimization in the following matrix:

$$M_{\text{Ant}} = \begin{bmatrix} A_{1.1} \ A_{1.2} \ldots A_{1.d} \\ A_{2.1} \ A_{2.2} \ldots A_{2.d} \\ \vdots \\ A_{n.1} \ A_{n.2} \ldots A_{n.d} \end{bmatrix} \tag{15}$$

where $A_{i.\,j}$ shows the value of the jth variable (dimension) of ith ant, n is the number of ants, and d is the number of variables. The position of an ant refers to the parameters for a particular solution. A fitness (objective) function is utilized during optimization, and the following matrix stores the fitness value of all ants:

$$M_{\text{OA}} = \begin{bmatrix} f([A_{1.1}, A_{1.2}, \ldots, A_{1.d}]) \\ f([A_{2.1}, A_{2.2}, \ldots, A_{2.d}]) \\ \vdots \\ f([A_{n.1}, A_{n.2}, \ldots, A_{n.d}]) \end{bmatrix} \tag{16}$$

where f is the objective function. In addition to ants, we assume the antlions are also hiding somewhere in the search space. In order to save their positions and fitness values, the following matrices are utilized:

$$M_{\text{Antlion}} = \begin{bmatrix} AL_{1.1} \ AL_{1.2} \ldots AL_{1.d} \\ AL_{2.1} \ AL_{2.2} \ldots AL_{2.d} \\ \vdots \\ AL_{n.1} \ AL_{n.2} \ldots AL_{n.d} \end{bmatrix}$$

$$M_{\text{OAL}} = \begin{bmatrix} f([AL_{1.1}, AL_{1.2}, \ldots, AL_{1.d}]) \\ f([AL_{2.1}, AL_{2.2}, \ldots, AL_{2.d}]) \\ \vdots \\ f([AL_{n.1}, AL_{n.2}, \ldots, AL_{n.d}]) \end{bmatrix} \tag{17}$$

$AL_{i.j}$ shows the jth dimension value of ith antlion, n is the number of antlions, and d is the number of variables (dimension), where M_{OAL} is the matrix for saving the fitness of each antlion.

Random walks of ants: Random walks are all based on the Eq. (1). Ants update their positions with random walk at every step of optimization, which is normalized using the following equation (min–max normalization) in order to keep it inside the search space:

$$X_i^t = \frac{\left(X_i^t - a_i\right) * \left(d_i - c_i^t\right)}{\left(d_i^t - a_i\right)} + c_i \tag{18}$$

where a_i is the minimum of random walk, d_i is the maximum of random walk, c_i^t is the minimum, and d_i^t indicates the maximum of ith variable at tth iteration.

Trapping in antlion's pits: The ants walk in a hypersphere defined by the vectors c and d around a selected antlion are affected by antlions' traps. In order to mathematically model this supposition, the following equations are proposed:

$$c_i^t = \text{Antlion}_j^t + c^t \tag{19}$$

$$d_i^t = \text{Antlion}_j^t + d^t \tag{20}$$

where c^t is the minimum of all variables at tth iteration, d^t indicates the vector including the maximum of all variables at tth iteration, c_i^t is the minimum of all variables for ith ant, d_i^t is the maximum of all variables for ith ant, and Antlion_j^t shows the position of the selected jth antlion at tth iteration.

Building trap: In order to model the antlions's hunting capability, a roulette wheel is employed for selecting antlions based of their fitness during optimization. Ants are assumed to be trapped in only one selected antlion. This mechanism gives high chances to the fitter antlions for catching ants.

Sliding ants towards antlion: Once antlions realize that an ant is in the trap they shoot sands outwards the center of the pit. This behavior slides down the trapped ant that is trying to escape. For mathematically modelling this behavior, the radius of ants's random walks hyper-sphere is decreased adaptively. The following equations are proposed in this regard:

$$c^t = \frac{c^t}{I} \tag{21}$$

$$d^t = \frac{d^t}{I} \tag{22}$$

where I is a ratio, c^t is the minimum of all variables at tth iteration, and d^t indicates the vector including the maximum of all variables at tth iteration. In Eqs. (21) and (22), $I = 10^W \frac{t}{T}$ where T is the maximum number of iterations and w is a constant defined based on the current iteration. Basically, the constant w can adjust the accuracy level of exploitation.

Catching prey and rebuilding the pit: For mimicking the final stage of hunt this process, it is assumed that catching prey occur when ants becomes fitter (goes insides and) than its corresponding antlion, which is required to update its position to the latest position of the hunted ant. The following equation is proposed in this regard:

$$\text{Antlion}_j^t = \text{Ant}_i^t \text{ if } f\left(\text{Ant}_j^t\right) > f\left(\text{Antlion}_j^t\right) \tag{23}$$

where Antlion_j^t shows the position of selected jth antlion at tth iteration and Ant_i^t indicates the position of ith ant at tth iteration.

Elitism: In this algorithm the best antlion obtained so far in each iteration is saved and considered as an elite.

Since the elite is the fittest antlion, it should be able to affect the movements of all the ants during iterations. Therefore, it is assumed that every ant randomly walks around a selected antlion by the roulette wheel and the elite simultaneously as follows:

$$\text{Ant}_i^t = \frac{R_A^t + R_E^t}{2} \tag{24}$$

where R_A^t is the random walk around the antlion selected by the roulette wheel at tth iteration, R_E^t is the random walk around the elite at tth iteration, and Ant_i^t indicates the position of ith ant at tth iteration.

4   Sliding Mode Control

The principle of this type of control consists in bringing, whatever the initial conditions, the representative point of the evolution of the system on a hypersurface of the phase space representing a set of static relationships between the state variables.

The sliding mode control generally includes two terms:

$$U = U_{eq} + U_n \tag{25}$$

U_{eq}: A continuous term, called equivalent command. U_n: a discontinuous term, called switching command.

4.1 Equivalent Command

The method, proposed by Utkin, consists to admit that in sliding mode, everything happens as if the system was driven by a so-called equivalent command. The latter corresponds to the ideal sliding regime, for which not only the operating point remains on the surface but also for which the derivative of the surface function remains zero $\dot{S}(t) = 0$ (that mean, invariant surface over time).

4.2 Switching Control

The switching command requires the operating point to remain at the neighborhood of the surface. The main purpose of this command is to check the attractiveness conditions:

$$U_n = \lambda \mathrm{sign}(S) \tag{26}$$

The gain λ is chosen to guarantee the stability and the rapidity and to overcome the disturbances which can act on the system. The function sign $(S(x, t))$ is defined as

$$\mathrm{Sign}(S(x,t)) = \left\{ \begin{array}{l} 1 \text{ si } S > 0 \\ -1 \text{ si } S < 0 \end{array} \right\} \tag{27}$$

The PID sliding surface for the sliding mode control can be indicated using the following equation:

$$S(t) = k_d \dot{e}(t) + k_p e(t) + k_i \int_0^{tf} e(t)\mathrm{d}t \tag{28}$$

with k_p, k_i, and k_d mentioned as PID parameters. $e(t) = q_d - q, \dot{e}(t) = \dot{q}_d - \dot{q}$. q_d and q are the desired and actual position of the robot articulations. $\dot{q}_d$ and $\dot{q}$ are the desired and actual speed of the robot articulations.

According to Eq. (26)

$$U_n = \lambda \text{sign}\left(k_p e(t) + k_i \int e(t)dt + k_d \dot{e}(t)\right)$$

To calculate U_{eq}, it is necessary that $\dot{S}(t) = 0$

$$k_p \dot{e}(t) + k_i e(t) + k_d \ddot{e}(t) = 0.$$
$$\ddot{e}(t) = k_d^{-1}\left(k_p \dot{e}(t) + k_i e(t)\right)$$

with

$$\ddot{e}(t) = \ddot{q}_d - \ddot{q}$$

$$\ddot{q} = M(q)^{-1}\left(U_{eq} - H(q,\dot{q})\dot{q} - G(q) - F(\dot{q})\right) = \ddot{q}_d + k_d^{-1}\left(k_p \dot{e}(t) + k_i e(t)\right)$$

$$U_{eq} = M(q)\left(\ddot{q}_d + k_d^{-1}\left(k_p \dot{e}(t) + k_i e(t)\right)\right) + H(q,\dot{q})\dot{q} + G(q) + F(\dot{q})$$

Finally, the PID-SMC torque presented as in [18], with the demonstration of the Lyapunov stability condition, becomes

$$U = M(q)\left(\ddot{q}_d + k_d^{-1}\left(k_p \dot{e}(t) + k_i e(t)\right)\right) + H(q,\dot{q})\dot{q} + G(q) + F(\dot{q})$$
$$+ M(q)k_d^{-1} \lambda \text{sign}\left(k_p e(t) + k_i \int_0^{tf} e(t)dt + k_d \dot{e}(t)\right) \tag{29}$$

5 Simulation and Results

The main goal of this work is to optimize the parameters of SMC with PID surface for the trajectory control of 2DOF robot manipulator by the minimization of ITAE and ISTE objective functions mentioned as

$$J_1 = \text{ITAE} = \int_{t1}^{tf} |e(t)|dt \tag{30}$$

$$J_2 = \text{ISTE} = \int_{t1}^{tf} e(t)^2 dt \tag{31}$$

The parameters of the robot that have been taken in application are $m_1 = 10$ kg, $m_2 = 5$ kg, $l_1 = 1$ m, $l_2 = 0.5$ m, and the gravity $g = 9.8$ m/s^2. First, we apply the

Table 1 Cost function for the first and second articulation without disturbance

Cost functions	ITAE		ISTE	
	First	Second	First	Second
SMC_PID_ALO	9.669×10^{-5}	3.896×10^{-5}	5.519×10^{-11}	3.951×10^{-7}
SMC_PID_GWO	82.2×10^{-5}	300.9×10^{-5}	2.095×10^{-8}	8.118×10^{-8}

Table 2 SMCPID parameters for the first and second articulation

ISTE	K_{p1}	K_{i1}	K_{d1}	λ_1	K_{p2}	K_{i2}	K_{d2}	λ_2
SMC_PID_ALO	500	45.22	1.18	58.5	500	16.5	17.135	34.83
SMC_PID_GWO	500	49.89	15.75	43.01	603	37	46	48
ITAE	K_{p1}	K_{i1}	K_{d1}	λ_1	K_{p2}	K_{i2}	K_{d2}	λ_2
SMC_PID_ALO	500	10.97	1	2.86	500	7.33	1	1
SMC_PID_GWO	500	65.91	12.72	59.8	507	63.79	79	40

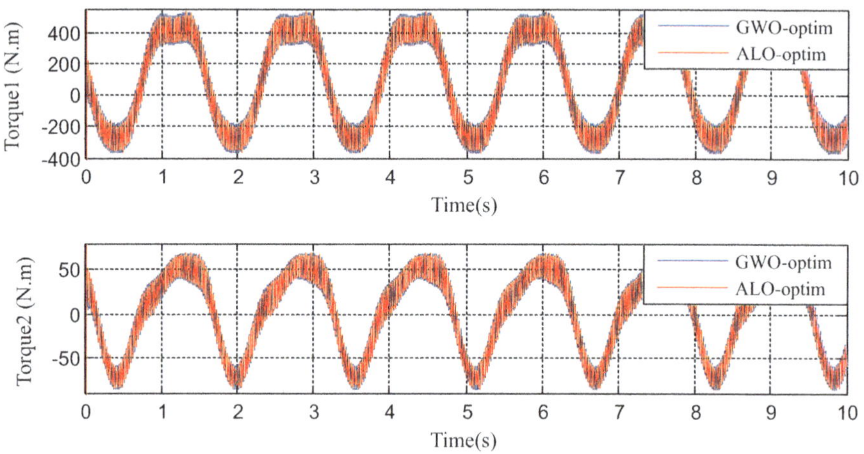

Fig. 1 Control torque of link 1 and link 2 with (sign) function ITAE criteria

algorithms described above to tune SMC controller. The objective function values for different optimization algorithms obtained with ITAE and ISTE criteria, defined in Eqs. (30) and (31), respectively, have been shown in Table 1.

It can be seen from the Table 1 that ALO algorithm gives minimum to the objective function compared with those of GWO, which means that ALO algorithm gives the best optimum that has minimum objective function better then GWO algorithm. The corresponding optimum parameters of PIDSMC were recapitulated in Table 2.

Figure 1 shows the control input applied to the first and second articulations obtained so far by both optimization algorithms. In order to avoid the chattering

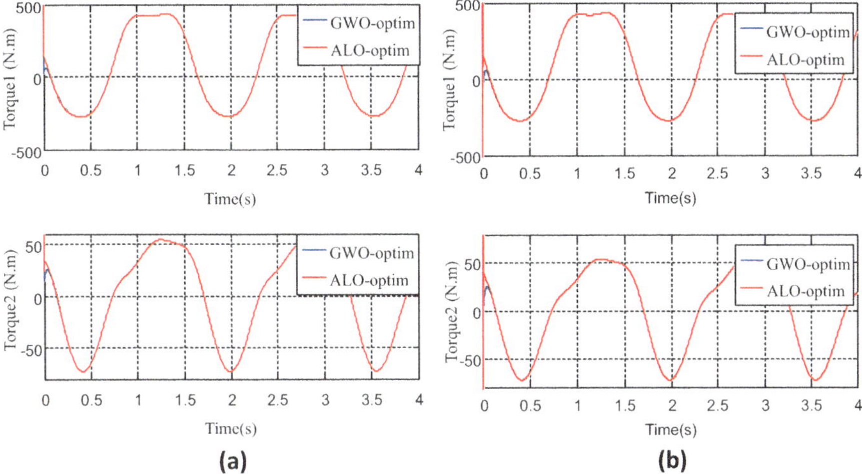

Fig. 2 Control torque signal of SMCPID controller with (tan*h*) function: (**a**) ISTE criteria and (**b**) ITAE criteria

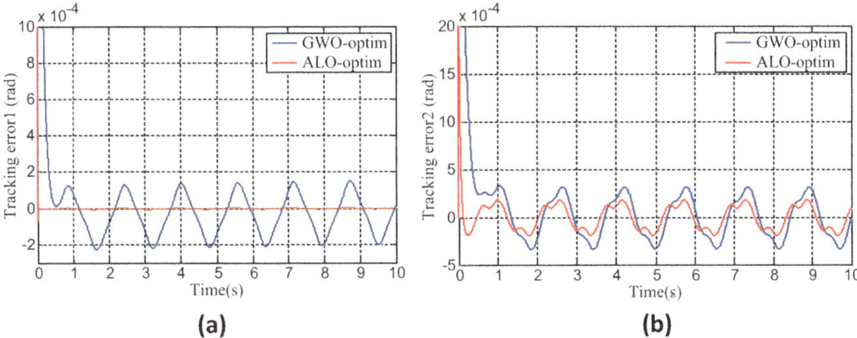

Fig. 3 Tracking error of both articulations with ISTE criteria and (tan*h*) function

effect of the "sign" function, used in Eq. (29), the latter is replaced by the "tan*h*" (hyperbolic tangent) function. It can be seen from Fig. 2 that the resulting torque was almost close for both chosen criteria and the two optimization algorithms.

Figures 3 and 4 show the error of the robot manipulator to track the desired trajectory by the minimization of ISTE and ITAE criteria, respectively. We can see from the figures that ALO algorithm, which has smaller cost function, outperforms GWO algorithm even if we change the objective function. The convergence curve of the used functions was represented in Fig. 5. The corresponding position of robot manipulator controlled by SMCPID controller optimized by the two algorithms was shown in Fig. 6.

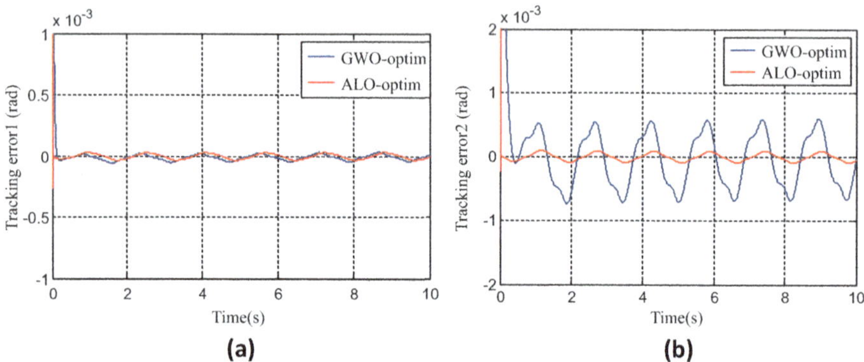

Fig. 4 Tracking error of both articulations with ITAE criteria and (tan$h$) function

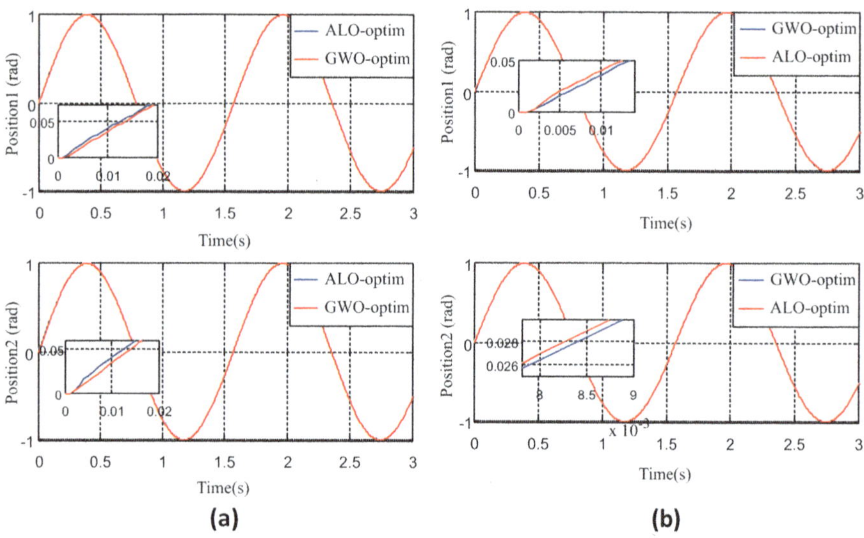

Fig. 5 Convergence curve of cost function: (a) ITAE criteria and (b) ISTE criteria

6 Conclusion

In this paper the optimization of the SMC with PID surface was realized with new techniques of optimization called ALO and GWO algorithms; the ALO presents more robustness in trajectory tracking control of 2DOF robot manipulator, regarding the convergence curve of the cost function, even if we change the objective function. From the observations of the simulations, we can realize the benefits of using evolutionary algorithms to tune the controller parameters than the traditional methods, especially when the system is highly nonlinear or in presence of distur-bances where an online optimization is recommended.

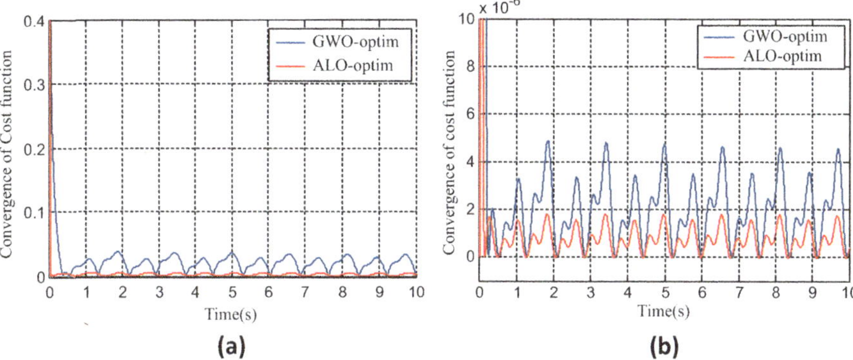

Fig. 6 Position of two articulations controlled by SMCPID controller with (tan*h*) function: (**a**) ITAE criteria and (**b**) ISTE criteria

References

1. Adhikary, N., & Mahanta, C. (2018). Sliding mode control of position commanded robot manipulators. *Control Engineering Practice, 81*, 183–198.
2. Jung, S. (2018). Improvement of tracking control of a sliding mode controller for robot manipulators by a neural network. *International Journal of Control, Automation and Systems, 16*(2), 937–943.
3. Yoo, D. (2009). A comparison of sliding mode and integral sliding mode controls for robot manipulators. *Transactions of the Korean Institute of Electrical Engineers, 58*(1), 168–172.
4. Liu, R., & Li, S. (2014). Optimal integral sliding mode control scheme based on pseudo spectral method for robotic manipulators. *International Journal of Control, 87*(6), 1131–1140.
5. Zhao, Y., Sheng, Y., & Liu, X. (2014). A novel finite time sliding mode control for robotic manipulators. *IFAC Proceedings, 47*, 7336–7341.
6. Azlan, N. Z., & Osman, J. H. S. (2006). Proportional integral sliding mode control of hydraulic robot manipulators with chattering elimination. In: First International Conference on Industrial and Information Systems, Sri Lanka.
7. Das, M., & Mahanta, C. (2014). Optimal second order sliding mode control for nonlinear uncertain systems. *ISA Transactions, 53*, 1191–1198.
8. Adelhedi, F., Jribi, A., Bouteraa, Y., & Derbe, N. (2015). Adaptive sliding mode control design of a SCARA robot manipulator system under parametric variations. *Journal of Engineering Science and Technology Review, 8*(5), 117–123.
9. Jiang, S., Zhao, J., Xie, F., Fu, J., Wang, X., & Li, Z. (2008). A novel adaptive sliding mode control for manipulator with external disturbance. In: 37th Chinese Control Conference (CCC).
10. Yi, S., & Zhai, J. (2019). Adaptive second-order fast nonsingular terminal sliding mode control for robotic manipulators. *ISA Transactions, 90*, 41–51.
11. Jing, C., Xu, H., & Niu, X. (2019). Adaptive sliding mode disturbance rejection control with prescribed performance for robotic manipulators. *ISA Transactions, 91*, 41–51.
12. Mirshekaran, M., Piltan, F., Esmaeili, Z., Khajeaian, T., & Kazeminasab, M. (2013). Design sliding mode modified fuzzy linear controller with application to flexible robot manipulator. *International Journal of Modern Education and Computer Science, 10*, 53–63.
13. Tran, M.D., & Kang, H.J. (2015). Adaptive fuzzy PID sliding mode controller of uncertain robotic manipulator. In: International Conference on Intelligent Computing: Intelligent Computing Theories and Methodologies, ICIC, pp. 92–103.

14. Ataei, M., & Shafiei, S. E. (2008). Sliding mode PID-controller design for robot manipulators by using fuzzy tuning approach. In: 27th Chinese Control Conference.
15. Nejatbakhsh Esfahani, H., & Azimirad, V. (2013). A new fuzzy sliding mode controller with PID sliding surface for underwater manipulators. *International Journal of Mechatronics, Electrical and Computer Technology, 3*(9), 224–249.
16. Kharabian, B., Bolandi, H., Ehyaei, A. F., Mousavi Mashhadi, S. K., & Smailzadeh, S. M. (2017). Adaptive tuning of sliding mode-PID control in free floating space manipulator by sliding cloud theory. *American Journal of Mechanical and Industrial Engineering, 2*(2), 64–71.
17. Vijay, M., & Jena, D. (2018). PSO based backstepping sliding mode controller and observer for robot manipulators. In: International Conference on Power, Instrumentation, Control and Computing (PICC).
18. Vijay, M., & Jena, D. (2017). PSO based neuro fuzzy sliding mode control for a robot manipulator. *Journal of Electrical Systems and Information Technology, 4*(1), 243–256.
19. Mokeddem, D., & Draidi, H. (2018). Optimization of PID sliding surface using Antlion optimizer. International Symposium on Modelling and Implementation of Complex Systems MISC, pp. 133–145.
20. Mirjalili, S., Mirjalili, S. M., & Lewis, A. (2014). Grey wolf optimizer. *Advances in Engineering Software, 69*, 46–61.
21. Mirjalili, S. (2015). The ant lion optimizer. *Advances in Engineering Software, 83*, 80–98.

Error Analysis of Nonlinear WENO Schemes Using Modified Equation

Tamer Kasem

Abstract A theoretical study of the nonlinear weighted essentially non-oscillatory (WENO) method is presented. Single- and multistep explicit numerical time integration algorithms are covered. The main idea is adopting the modified equation method. The necessary lengthy derivations are achieved using a computer algebra system. Accurate theoretical estimates of error norms are derived. The introduced theoretical results are validated via quantitative comparison with numerical experiments.

Keywords WENO · Nonlinearity · Modified equation · Error norm

1 Introduction

About two decades ago, the nonlinear fifth-order accurate, weighted essentially non-oscillatory method (WENO) was developed [1]. WENO method enjoys two advantages: resolution of complex smooth features due to its high order and avoiding spurious oscillations near high gradients [2, 3]. WENO can be regarded as a nonlinear improvement of the linear fifth-order upwind (UW5) discretization. The relatively wide five-point stencil of UW5 is subdivided into three small sub-stencils. A smoothness indicator is calculated for each sub-stencil, to detect those which would induce spurious oscillations. Problematic sub-stencils with high smoothness indicators values are given lower weights and vice versa. The weighting step renders WENO the nonlinear feature.

A wide variety of improved WENO versions are presented in [4]. However, studies of theoretical properties are rather rare. Specifically, reliable estimates of

T. Kasem (✉)
Faculty of Engineering, Department of Engineering Mathematics and Physics, Cairo University, Cairo, Egypt

On Secondment to Nile University, Sheikh Zayed, Giza, Egypt
e-mail: tkasem@nu.edu.eg

© Springer Nature Switzerland AG 2020
M. H. Farouk, M. A. Hassanein (eds.), *Recent Advances in Engineering Mathematics and Physics*, https://doi.org/10.1007/978-3-030-39841-1_3

error norms for standard benchmark problems have never been provided. A primary step upon developing all improved versions is coding the original WENO algorithm for validation and clarifying advantages. Instead of comparing with theoretical error norms, results were validated through convergence studies. Error norms of consecutive numerical experiments are obtained on a series of refined grids to estimate the order of approximation [5, 6]. However, the results of this procedure are strongly dependent on WENO numerical parameters [6]. Consequently, providing theoretical estimates for WENO error norms would be useful for code development and validation.

In the current work, the modified equation (ME) method is used [7]. The discrete equation is transformed into a *modified* partial differential equation, describing the *true* solution obtained numerically. An important advantage of ME is being generally applicable to nonlinear discretizations, including WENO. Theoretical error estimates are obtained based on the solutions of the nonlinear ME. Derivation of ME is based on Taylor series expansion. Successive differentiations and substitutions into the discrete equation are required. The generally tedious task is further complicated due to nonlinearity and multistep time integration. The computer algebra system Maxima [8] is employed to generate and simplify the ME. The theoretical results are validated against various numerical experiments.

2 WENO Discretization

It is required to discretize the following:

$$\frac{\partial u}{\partial t} + \frac{\partial f(u)}{\partial x} = 0 \tag{1}$$

The main task is to obtain a high-order finite difference discretization of the space derivative $\frac{\partial f(u)}{\partial x}$, while avoiding spurious oscillations due to Gibbs phenomenon. The upwinded version will be described in detail. The downwinded version is obtained similarly in a straightforward way [9]. The space derivative of f at x_i can be generally approximated as

$$\left.\frac{\partial f}{\partial x}\right|_i \approx \frac{f_{i+1/2} - f_{i-1/2}}{h} \tag{2}$$

where $f_{i+1/2}$ is the *flux* at point $x_{i+1/2}$ and $h = x_i - x_{i-1} = L/N$ is the constant grid spacing, L is the domain width, and N is the total number of subdivisions.

Substituting $f_{i-1/2} = \bar{f}_{i-1/2}$ or $f_{i-1/2} = \tilde{f}_{i-1/2}$ in Eq. (2) results in the fifth-order upwinded (UW5) or the weighted essentially non-oscillatory (WENO) space discretizations, respectively. On the next few lines, $\bar{f}_{i-1/2}$ and $\tilde{f}_{i-1/2}$ algorithms are explained.

Consider a third-order approximation of $\partial f/\partial x$:

$$\left.\frac{\partial f}{\partial x}\right|_i = \frac{\hat{f}^j_{i+1/2} - \hat{f}^j_{i-1/2}}{h} + O(h^3). \tag{3}$$

Three choices for $\hat{f}^j_{i-1/2}$ are possible based on the chosen sub-stencil j, i.e.:

$$\hat{f}^j_{i-1/2} = \begin{cases} \hat{f}^0_{i-1/2} = \dfrac{1}{3}f_{i-3} - \dfrac{7}{6}f_{i-2} + \dfrac{11}{6}f_{i-1} \\[2mm] \hat{f}^1_{i-1/2} = -\dfrac{1}{6}f_{i-2} + \dfrac{5}{6}f_{i-1} + \dfrac{1}{3}f_i \\[2mm] \hat{f}^2_{i-1/2} = \dfrac{1}{3}f_{i-1} + \dfrac{5}{6}f_i - \dfrac{1}{6}f_{i+1} \end{cases} \tag{4}$$

In smooth regions, the three choices can be combined to get the linear *upwinded* fifth-order accurate (UW5) approximation for $\partial f(u)/\partial x$, using the following constants [1, 9]:

$$d_0 = 0.1, d_1 = 0.6, d_2 = 0.3. \tag{5}$$

The fifth-order accurate numerical flux $\bar{f}_{i-1/2}$ can be calculated as

$$\bar{f}_{i-1/2} = \sum_{j=0}^{j=2} d_j \hat{f}^j_{i-1/2}. \tag{6}$$

In order to avoid spurious oscillations near high gradients, (Eq. 6) is modified as

$$\tilde{f}_{i-1/2} = \sum_{j=0}^{j=2} \omega_j \hat{f}^j_{i-1/2}. \tag{7}$$

The weights ω_j should be higher for stencils that yield smooth solutions and vice versa. Smoothness indicators (SI) are used to calculate ω_j. For each subinterval j, the SI is termed as β_j. Let us assume for the moment that β_j are given. The weights ω_j are calculated as

$$\alpha_j = \frac{d_j}{\left(\epsilon + \beta_j\right)^2}. \tag{8}$$

$$\omega_j = \frac{\alpha_j}{\sum_{s=0}^{s=2} \alpha_s}. \tag{9}$$

Here is a very small arbitrary parameter introduced to avoid division by zero; α_j is an intermediate weight used to calculate ω_j. A rather important result of [6] is that the original value of $\epsilon = 10^{-6}$ used by [1] would have a strong effect on the error norm calculated by numerical experiments. The value adopted in the current work is $\epsilon = 10^{-40}$, since its effect on error norms is negligible [6]. Normalization is achieved by (Eq. 9).

The $L2$ norm squared was used to calculate β_j in [1] using the following formula:

$$\beta_j = \sum_{m=1}^{2} \int_{x_{i-1/2}}^{x_{i+1/2}} h^{2m-1} \left(q_j^m \right)^2 dx. \tag{10}$$

Here q_j^m is the mth derivative of the polynomial used to evaluate $\hat{f}_{i-1/2}^{j}$. It can be proved that β_j is equal to

$$\beta_j = \begin{cases} \beta_0 = \dfrac{13}{12}\left(f_{i-3} - 2f_{i-2} + f_{i-1} \right)^2 + \dfrac{1}{4}\left(f_{i-3} - 4f_{i-2} + 3f_{i-1} \right)^2 \\[2mm] \beta_1 = \dfrac{13}{12}\left(f_{i-2} - 2f_{i-1} + f_i \right)^2 + \dfrac{1}{4}\left(f_{i-2} - f_i \right)^2 \\[2mm] \beta_2 = \dfrac{13}{12}\left(f_{i-1} - 2f_i + f_{i+1} \right)^2 + \dfrac{1}{4}\left(3f_{i-1} - 4f_i + f_{i+1} \right)^2 \end{cases} \tag{11}$$

3   Analysis

3.1   Governing Equation

Similar to [10, 11], the analysis will focus on the linear flux $f(u) = u$ used in Eq. (1)

$$\frac{\partial u}{\partial t} + \frac{\partial u}{\partial x} = 0 \tag{12}$$

Here the domain D is defined as $\{(x,t) \in \mathscr{R}^2 : 0 \le x \le L, 0 \le t < \infty\}$. The problem is studied subject to the space periodic boundary condition:

$$u(0,t) = u(L,t) \tag{13}$$

Considering the initial condition $u(x,t = 0) = F(x)$, and using the method of characteristics, the exact solution $u_e(x,t)$ is obtained:

$$u_e(x,t) = F(x - t) \tag{14}$$

It should be noted that, although Eq. (12) is linear, WENO discretization is *nonlinear*.

3.2 Time Discretization

The first-order explicit Runge–Kutta (RK1) solution of Eq. (1) is written as

$$u^{n+1} = u^n + \Delta L(u^n) \tag{15}$$

Here $\Delta = t^{n+1} - t^n$ is the time step; $L(u)$ is space discretization operator used to approximate $-\frac{\partial f(u)}{\partial x}$.

The second-order total variation diminishing Runge–Kutta (TVDRK2) explicit solution of Eq. (1) is written as

$$
\begin{aligned}
u^{(1)} &= u^n + \Delta L(u^n) \\
u^{n+1} &= \frac{1}{2} u^n + \frac{1}{2} u^{(1)} + \frac{1}{2} \Delta L\left(u^{(1)}\right)
\end{aligned}
\tag{16}
$$

The third-order total variation diminishing Runge–Kutta (TVDRK3) explicit solution of Eq. (1) is written as

$$
\begin{aligned}
u^{(1)} &= u^n + \Delta L(u^n) \\
u^{(2)} &= \frac{3}{4} u^n + \frac{1}{4} u^{(1)} \Delta L\left(u^{(1)}\right) \\
u^{n+1} &= \frac{1}{3} u^n + \frac{2}{3} u^{(2)} + \frac{2}{3} \Delta L\left(u^{(2)}\right)
\end{aligned}
\tag{17}
$$

Here, $u^{(1)}$ and $u^{(2)}$ are intermediate solution vectors, and u^n, u^{n+1} are solution vectors at t^n, t^{n+1}, respectively.

3.3 Modified Equation Analysis

Derivation of Modified Equations: The procedure of deriving modified equations (ME) will be revised. Using the first $M + 1$ terms of the Taylor expansion, the following expression is obtained:

$$
\begin{aligned}
f_{i+j} = f(x_i + jh) &= \sum_{m=0}^{M} \frac{(jh)^m}{m!} \left.\frac{\partial^m f(x)}{\partial x^m}\right|_{x=x_i} \\
&= f(x_i) + (jh)\left(\left.f(x)_x\right|_{x=x_i} \right) + \frac{(jh)^2 \left(\left.f(x)_{xx}\right|_{x=x_i} \right)}{2} \\
&\quad + \frac{(jh)^3 \left(\left.f(x)_{xxx}\right|_{x=x_i} \right)}{6} + \frac{(jh)^4 \left(\left.f(x)_{xxxx}\right|_{x=x_i} \right)}{24} + \text{H.O.T.}
\end{aligned}
\tag{18}
$$

Here M is the order of the ME to be derived, and H.O.T. stands for higher-order terms. Substituting Eq. (18) into space discretizations and neglecting H.O.T., an operator $L(U)$ is approximated as a series in terms of space derivatives. For instance, substituting Eq. (18) into Eqs. (2), (4), and (6), an approximate series presentation for UW5 space discretization is obtained:

$$L_{\text{UW5}}(u) \approx -u_x + \frac{h^2(u_{xxxxx})}{60} - \frac{h^6(u_{xxxxxx})}{140} + \frac{h^7(u_{xxxxxxx})}{240} + \text{H.O.T.} \tag{19}$$

To reduce clutter the symbol, x_i is replaced by x and the subscript i is dropped. In order to analyze the time integration scheme, the following Taylor expansion is used:

$$
\begin{aligned}
u_i^{n+1} = u(x_i, t^n + \Delta) &= \sum_{m=0}^{M} \frac{\Delta^m}{m!} \left. \frac{\partial^m u}{\partial t^m} \right|_{t=t^n} \\
&= u(x_i, t^n) + \Delta \left(u(x_i, t)_t \big|_{t=t^n} \right) + \frac{\Delta^2 \left(u(x_i, t)_{tt} \big|_{t=t^n} \right)}{2} \\
&\quad + \frac{\Delta^3 \left(u(x_i, t)_{ttt} \big|_{t=t^n} \right)}{6} + \frac{\Delta^4 \left(u(x_i, t)_{tttt} \big|_{t=t^n} \right)}{24} + \text{H.O.T.}
\end{aligned}
\tag{20}
$$

Equation 20 should be used to derive approximate series in terms of time derivatives evaluated at t^n. For instance, substituting Eq. (20) into Eq. (15) along with Eq. (19), the UW5-RK1 procedure is approximated by a series in terms of space and time derivatives:

$$
\begin{aligned}
&\frac{\Delta(u_{tt})}{2} + \frac{\Delta^2(u_{ttt})}{6} + \frac{\Delta^3(u_{tttt})}{24} + \frac{\Delta^4(u_{ttttt})}{120} + \frac{\Delta^5(u_{tttttt})}{720} + \frac{\Delta^6(u_{ttttttt})}{5040} \\
&- \frac{h^5(u_{xxxxx})}{60} + \frac{h^6(u_{xxxxxx})}{140} + u_t + u_x = 0
\end{aligned}
\tag{21}
$$

The superscript of n is dropped for convenience. The last step is repeated for each extra intermediate step. For instance, the step should be repeated twice and thrice for Eqs. (16) and (17), respectively. Finally, all time derivatives (except u_t) are eliminated by successive differentiation of Eq. (21). The procedure is fully detailed at [7]. Pseudo-code is provided in Appendix 1. For instance, the final form of the ME for UW5-RK1 is

$$
\begin{aligned}
&\frac{h^6(u_{xxxxxxx})\lambda^6}{7} + \frac{h^5(u_{xxxxxx})\lambda^5}{6} + \frac{h^4(u_{xxxxx})\lambda^4}{5} + \frac{h^3(u_{xxxx})\lambda^3}{4} \\
&+ \frac{h^2(u_{xxx})\lambda^2}{3} - \frac{h^6(u_{xxxxxx})\lambda}{60} + \frac{h(u_{xx})\lambda}{2} + \frac{h^6(u_{xxxxxx})}{140} \\
&- \frac{h^5(u_{xxxxx})}{60} + u_x + u_t = 0
\end{aligned}
\tag{22}
$$

Solutions for a Sinusoidal Initial Condition: Consider the initial condition

$$u(x, t = 0) = A \sin(kx) \tag{23}$$

Solving Eq. (22) to satisfy Eq. (23) is straightforward. Consider the trial solution

$$u(x, t) = Ae^{\alpha t} \sin(kx - \omega t) \tag{24}$$

Substitution of Eq. (24) in Eq. (22) and canceling the common factor $Ae^{\alpha t}$, we get

$$
\begin{aligned}
&\left(a + \frac{\theta^6}{60h} - \frac{\theta^2 \lambda}{2h} + \frac{\theta^4 \lambda^3}{4h} - \frac{\theta^6 \lambda^5}{6h} \right) \sin(kx - \omega t) \\
&+ \left(+\omega + \frac{\theta}{h} - \frac{\theta^7}{140h} + \frac{\theta^7 \lambda}{60h} - \frac{\theta^3 \lambda^2}{3h} + \frac{\theta^5 \lambda^4}{5h} - \frac{\theta^7 \lambda^6}{7h} \right) cos(kx - \omega t) = 0
\end{aligned} \tag{25}
$$

A necessary condition for satisfaction of Eq. 25 by arbitrary values of $(x,t) \in D$ is to equate the coefficients of $\cos(kx - \omega t)$ and $\sin(kx - \omega t)$ to zeros. As a result, closed-form expressions are obtained for α and ω:

$$\alpha = \frac{-\theta^6 + 30\theta^2 \lambda - 15\theta^4 \lambda^3 + 10\theta^6 \lambda^5}{60h} \tag{26}$$

$$\omega = -\frac{-420\theta + 3\theta^7 - 7\theta^7 \lambda + 140\theta^3 \lambda^2 - 84\theta^5 \lambda^4 + 60\theta^7 \lambda^6}{420h} \tag{27}$$

The symbol θ is defined as $\theta = kh$. The reader is reminded that both the initial condition and the resulting solution have the same wave number k. This is expected for linear problems.

4 Results

4.1 Theoretical Results

The results of applying ME to WENO are presented for single- and multistep algorithms.

Single-Step First-Order Runge–Kutta: Substituting Eq. (18) into Eqs. (2), (4), and (7), the following operator is obtained:

$$
\begin{aligned}
L_w(u) = {}& -u_x - \frac{h^5 (u_{xx})(u_{xxx})^2}{10(u_x)^2} - \frac{h^6 (u_{xx})^2 (u_{xxx})^2}{10(u_x)^3} + \frac{h^6 (u_{xxx})^3}{20(u_x)^2} \\
& + \frac{h^5 (u_{xxx})(u_{xxxx})}{5(u_x)} + \frac{19h^6 (u_{xx})(u_{xxx})(u_{xxxx})}{100(u_x)^2} - \frac{9h^6 (u_{xxxx})^2}{100(u_x)} \\
& - \frac{9h^6 (u_{xxx})(u_{xxxxx})}{100(u_x)} + \frac{h^5 (u_{xxxxxx})}{60} - \frac{h^6 (u_{xxxxxxx})}{140}
\end{aligned} \tag{28}
$$

Further substitution into Eq. (15) following the same steps done in Sect. 3.3, the following equation truncated at $M = 6$ is obtained:

$$\frac{103h^6(u_{xxxxxx})\lambda^6}{720} + \frac{h^5(u_{xxxxx})\lambda^5}{6} + \frac{h^4(u_{xxxx})\lambda^4}{5} + \frac{h^3(u_{xxx})\lambda^3}{4}$$

$$+ \frac{h^2(u_{xx})\lambda^2}{3} - \frac{h^6(u_{xxxxxx})\lambda}{60} + \frac{h^6(u_{xx})(u_{xxxx})\lambda}{5(u_x)} - \frac{h^6(u_{xxx})^2\lambda}{5(u_x)}$$

$$+ \frac{2h^6(u_{xx})(u_{xxx})(u_{xxxx})\lambda}{5(u_x)^2} + \frac{h^6(u_{xx})^3\lambda}{10(u_x)^2} - \frac{h^6(u_{xx})^2(u_{xxx})^2\lambda}{5(u_x)^3} + \frac{h(u_{xx})\lambda}{2}$$

$$+ \frac{h^6(u_{xxxxxx})}{140} - \frac{h^5(u_{xxxxx})}{60} + \frac{9h^6(u_{xxx})(u_{xxxx})}{100(u_x)} + \frac{9h^6(u_{xxx})^2}{100(u_x)}$$

$$- \frac{19h^6(u_{xx})(u_{xxx})(u_{xxxx})}{100(u_x)^2} - \frac{h^5(u_{xxx})(u_{xxxx})}{5(u_x)} - \frac{h^6(u_{xxx})^3}{20(u_x)^2} + \frac{h^6(u_{xx})^2(u_{xxx})^2}{10(u_x)^3}$$

$$+ \frac{h^5(u_{xx})(u_{xxx})^2}{10(u_x)^2} + u_x + u_t = 0 \tag{29}$$

The nonlinearity in Eq. (29) appears starting from $O(h^5)$ terms. The numerators of the rational nonlinear $O(h^5)$ terms contain u_{xxx} that cancels out with the u_x in the denominator. Similarly, nonlinear $O(h^6)$ terms simplify to be linear in $sin(kx - \omega t)$ and $cos(kx - \omega t)$. Fortunately, the same procedure of Sect. 3.3 can be applied with success Eq. (29) to get

$$\left(\frac{7\theta^5 k}{60} + \alpha - \frac{\theta k\lambda}{2} + \frac{\theta^3 k\lambda^3}{4} - \frac{\theta^5 k\lambda^5}{6}\right) \sin(kx - \omega t) +$$

$$\left(k - \frac{33\theta^6 k}{700} + \frac{7\theta^6 k\lambda}{60} - \frac{\theta^2 k\lambda^2}{3} + \frac{\theta^4 k\lambda^4}{5} - \frac{103\theta^6 k\lambda^6}{720} - \omega\right) \cos(kx - \omega t) = 0 \tag{30}$$

$$\alpha = \frac{-7\theta^5 k + 30\theta k\lambda - 15\theta^3 k\lambda^3 + 10\theta^5 k\lambda^5}{60} \tag{31}$$

$$\omega = -\frac{-25200k + 1188\theta^6 k - 2940\theta^6 k\lambda + 8400\theta^2 k\lambda^2 - 5040\theta^4 k\lambda^4 + 3605\theta^6 k\lambda^6}{25200} \tag{32}$$

Two-Step Second-Order TVD Runge–Kutta: The numerical algorithm obtained by applying WENO space discretization with RKTVD2 will be termed as WENO-RK2. The same steps explained in Sect. 4.1 can be applied to obtain a ME for WENO-RK2. The resulting equation is

$$\frac{13h^6(u_{xxxxxx})\lambda^6}{720} + \frac{9h^6(u_{xxx})(u_{xxxxx})}{100(u_x)} + \frac{9h^6(u_{xxxx})^2}{100(u_x)}$$
$$-\frac{19h^6(u_{xx})(u_{xxx})(u_{xxxx})}{100(u_x)^2} - \frac{h^6(u_{xxx})^3}{20(u_x)^2} + \frac{h^6(u_{xx})^2(u_{xxx})^2}{10(u_x)^3} - \frac{h^5(u_{xxx})(u_{xxxx})}{5(u_x)}$$
$$+\frac{h^5(u_{xx})(u_{xxx})^2}{10(u_x)^2} - \frac{h^4(u_{xxxxx})\lambda^4}{20} - \frac{h^3(u_{xxxx})\lambda^3}{8} - \frac{h^2(u_{xxx})\lambda^2}{6}$$
$$+\frac{h^6(u_{xxxxxxx})}{140} - \frac{h^5(u_{xxxxxx})}{60} + u_x + u_t = 0 \tag{33}$$

$$\alpha = \frac{-14h^5k^6 + 15h^3k^4\lambda^3}{120} \tag{34}$$

$$\omega = -\frac{-25200k + 1188h^6k^7 - 4200h^2k^3\lambda^2 + 1260h^4k^5\lambda^4 + 455h^6k^7\lambda^6}{25200} \tag{35}$$

Three-Step Third-Order TVD Runge–Kutta: Substituting Eq. (28) into Eq. (17), we get the $O(h^6)$ nonlinear ME

$$\frac{h^6(u_{xxxxxx})\lambda^6}{240} + \frac{h^6(u_{xxxxxxx})}{140} + \frac{9h^6(u_{xxxx})^2}{100(u_x)} - \frac{19h^6(u_{xx})(u_{xxx})(u_{xxxx})}{100(u_x)^2}$$
$$-\frac{h^6(u_{xxx})^3}{20(u_x)^2} + \frac{h^6(u_{xx})^2(u_{xxx})^2}{10(u_x)^3} + \frac{9h^6(u_{xxx})(u_{xxxxx})}{100(u_x)} - \frac{h^5(u_{xxx})(u_{xxxx})}{5(u_x)}$$
$$+\frac{h^5(u_{xx})(u_{xxx})^2}{10(u_x)^2} - \frac{h^5(u_{xxxxxx})\lambda^5}{72} - \frac{h^5(u_{xxxxx})}{60} + \frac{h^4(u_{xxxxx})\lambda^4}{30}$$
$$+\frac{h^3(u_{xxxx})\lambda^3}{24} + u_x + u_t = 0 \tag{36}$$

$$\alpha = \frac{-42h^5k^6 - 15h^3k^4\lambda^3 + 5h^5k^6\lambda^5}{360} \tag{37}$$

$$\omega = \frac{-8400k + 396h^6k^7 - 280h^4k^5\lambda^4 + 35h^6k^7\lambda^6}{8400} \tag{38}$$

The same procedure of Sect. 3.3 can be applied with to success Eq. (36) to get

$$u_6^{RK3}(x,t) = Ae^{\left(-\frac{7h^5k^6}{60} - \frac{h^3k^4\lambda^3}{24} + \frac{h^5k^6\lambda^5}{72}\right)t} \sin\left(k(x-t) + \left(\frac{h^6k^7\lambda^6}{240} - \frac{h^4k^5\lambda^4}{30} + \frac{33h^6k^7}{700}\right)t\right) \tag{39}$$

4.2 Numerical Results

Spatial and temporal error variations are clarified as follows. The *numerical* error $e_N(x_i, t^n)$ obtained directly from the computations is defined as

$$e_N(x_i, t^n) = u(x_i, t^n) - u_e(x_i, t^n) \tag{40}$$

On the other hand, $e_6(x_i, t^n)$ is the *theoretical* error calculated based on the $O(h^6)$ ME:

$$e_6(x_i, t^n) = u_6(x_i, t^n) - u_e(x_i, t^n) \tag{41}$$

Similarly, the error norms are defined as

$$L_1(N) = \frac{\sum_{i=1}^{N} |e_N(x_i, t^n)|}{N}, \tag{42}$$

$$L_1^6(N) = \frac{\sum_{i=1}^{N} |e_6(x_i, t^n)|}{N}, \tag{43}$$

Numerical solutions based on RK1, TVDRK2, and TVDRK3 algorithms are presented. A single harmonic initial condition of the form $u(x, t = 0) = \sin(kx)$ is adopted. The error norms for RK1 are provided in Table 1. Two values of $k = 7\pi$, 10π ($\theta = 0.2199, 0.31416$) were adopted. A clear agreement between $L_1(200)$ and $L_1^6(200)$ (based on ME) is found. The error time growth is well predicted.

Second-order TVDRK2 and third-order TVDRK3 error norms are provided in Tables 2 and 3, respectively. Two values of $k = 6\pi$, 8π ($\theta = 0.1885, 0.2513$) were

Table 1 Error norms calculated numerically and theoretically for WENO-RK1

t_{max}	$k = 7\pi, \theta = 0.2199$		$k = 10\pi, \theta = 0.31416$	
	$L_1(200)$	$L_1^6(200)$	$L_1(200)$	$L_1^6(200)$
2	2.958×10^{-2}	2.964×10^{-2}	5.154×10^{-2}	5.012×10^{-2}
4	6.054×10^{-2}	6.066×10^{-2}	1.086×10^{-1}	1.042×10^{-1}
6	9.295×10^{-2}	9.314×10^{-2}	1.707×10^{-1}	1.626×10^{-1}

$L = 2.0, \lambda = 0.01, h = 1/100$

Table 2 Error norms calculated numerically and theoretically for WENO-RK2

t_{max}	$k = 6\pi, \theta = 0.1885$		$k = 8\pi, \theta = 0.2513$	
	$L_1(200)$	$L_1^6(200)$	$L_1(200)$	$L_1^6(200)$
2	7.081×10^{-4}	7.286×10^{-4}	3.666×10^{-3}	3.734×10^{-3}
4	1.411×10^{-3}	1.456×10^{-3}	7.099×10^{-3}	7.446×10^{-3}
6	2.110×10^{-3}	2.184×10^{-3}	1.041×10^{-2}	1.114×10^{-2}

$L = 2.0, \lambda = 0.05, h = 1/100$

Table 3 Error norms calculated numerically and theoretically for WENO-RK3

t_{max}	$k = 6\pi$, $\theta = 0.1885$		$k = 8\pi$, $\theta = 0.2513$	
	$L_1(200)$	$L_1^6(200)$	$L_1(200)$	$L_1^6(200)$
2	6.759×10^{-4}	6.658×10^{-4}	3.709×10^{-3}	3.739×10^{-3}
4	1.354×10^{-3}	1.331×10^{-3}	7.138×10^{-3}	7.455×10^{-3}
6	2.019×10^{-3}	1.995×10^{-3}	1.044×10^{-2}	1.115×10^{-2}

$L = 2.0$, $\lambda = 0.05$, $h = 1/100$

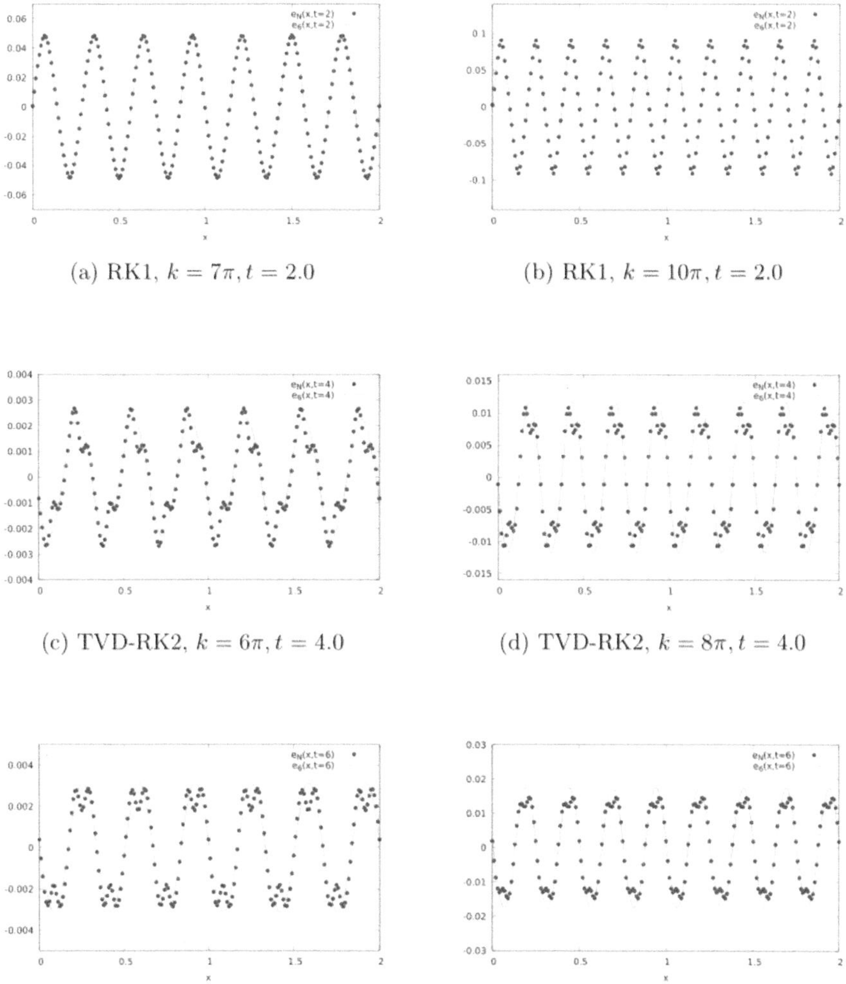

(a) RK1, $k = 7\pi, t = 2.0$

(b) RK1, $k = 10\pi, t = 2.0$

(c) TVD-RK2, $k = 6\pi, t = 4.0$

(d) TVD-RK2, $k = 8\pi, t = 4.0$

(e) TVD-RK3, $k = 6\pi, t = 6.0$

(f) TVD-RK3, $k = 8\pi, t = 6.0$

Fig. 1 Spatial error distributions, various time algorithms, and initial conditions. (**a**) RK1, $k = 7\pi$, $t = 2.0$, (**b**) RK1, $k = 10\pi$, $t = 2.0$, (**c**) TVD-RK2, $k = 6\pi$, $t = 4.0$, (**d**) TVD-RK2, $k = 8\pi$, $t = 4.0$, (**e**) TVD-RK3, $k = 6\pi$, $t = 6.0$, and (**f**) TVD-RK3, $k = 8\pi$, $t = 6.0$

adopted. The error norms of TVDRK3 are lower than those of TVDRK2, which is expected since errors should decrease as integration order increases. For further validation, the error spatial distributions are plotted for various cases in Fig. 1. The reader is reminded to the fact that the theoretical formulas do not include ε. Hence the error formulas should be valid regardless of the value of used in the code.

5 Conclusions

The nonlinear weighted essentially non-oscillatory WENO method was studied using the modified equation (ME) method. Nonlinearity was reflected in the ME in terms of rational terms. However, the rational terms were reduced into linear terms for single harmonic initial conditions. Hence, theoretical solutions were obtained using a symbolic manipulator (Maxima). The procedure was applied for single- and multistep time discretizations. Numerical errors were obtained theoretically (ME solutions) and experimentally (numerical computations). Agreement between both results was illustrated, in terms of error norms and spatial error distributions. Hence, the theoretical formulas can be used to get reliable error estimates. These results should play a key role in validation and future development of numerical WENO codes.

Appendix 1: Elimination of Time Derivatives

Consider the equation

$$\sum_{m=1}^{M} h^m \sum_{n=0}^{m} a_{m,n} \frac{\partial^m u}{\partial x^{m-n} \partial t^n} = 0 \tag{44}$$

Here $a_{0,1} = a_{1,0} = 1$. It is required to eliminate all time derivatives and mixed derivatives of order 2 or higher, i.e., $\frac{\partial^m u}{\partial x^{m-n} \partial t^n}$, such that $(m, n) \in (\{2, 3, ..., M\} \times \{1, 2, ..., M\})$. A pseudo-code is given below.

Step 1: Prepare a list L of ordered pairs $(m,n) = [(0,2),(1,1),$

$$(0, 3), (1, 2), (2, 1),$$

$$\cdots$$

$$(0, M), \ldots, (M-1, 1)]$$

Step 2: Apply successive differentiation and elimination

$$\text{modified}_{eq} = \text{Eq. }(44)$$

$$\text{for m} = 1, M$$

$$n_{\max} = (m/2)(m+3)$$

$$n_{\min} = n_{\max} - m$$

$$\text{modified}_{eq1} = \text{Truncate Eq. }(44) \text{ to } O\!\left(h^{M-m+1}\right)$$

$$\text{for } n = n_{\min}, n_{\max}$$

$$m_a = L[n][1]$$

$$n_a = L[n][2] - 1$$

$$\text{modified}_{eq2} = \frac{\partial^{m_a+n_a}}{\partial x^{m_a}\,\partial t^{n_a}}\left(\text{modified}_{eq1}\right)$$

$$\text{cof} = \text{coefficient of } \frac{\partial^{m_a+n_a+1} u}{\partial x^{m_a}\,\partial t^{n_a+1}} \text{ in modified}_{eq}$$

$$\text{modified}_{eq} = \text{modified}_{eq} - \text{cof} \times \text{modified}_{eq2}$$

$$\text{return modified}_{eq}.$$

References

1. Jiang, G.-S., & Shu, C.-W. (1996). Efficient implementation of weighted ENO schemes. *Journal of Computational Physics, 126*, 202–228.
2. Ekaterinaris, J. (2005). High-order accurate, low numerical diffusion methods for aerodynamics. *Progress in Aerospace Sciences, 41*, 192–300.
3. Shu, C. W. (2009). High order weighted essentially nonoscillatory schemes for convection dominated problems. *SIAM Review, 51*, 82–126.
4. Book of Abstracts, International Conference on Spectral and High-Order Methods, Rio De Janeiro, Brazil. (2016). Retrieved from http://icosahom2016.swge.inf.br/BOA20160620.pdf.
5. Borges, R., Carmona, M., Costa, B., & Don, W. S. (2008). An improved weighted essentially non-oscillatory scheme for hyperbolic conservation laws. *Journal of Computational Physics, 227*, 3191–3211.
6. Henrick, A., Aslam, T., & Powers, J. (2005). Mapped weighted essentially non-oscillatory schemes: Achieving optimal order near critical points. *Journal of Computational Physics, 207*, 542–567.
7. Warming, R. F., & Hyett, B. J. (1974). The modified equation approach to the stability and accuracy analysis of finite-difference methods. *Journal of Computational Physics, 14*, 159–179.
8. Maxima.sourceforge.net. Maxima, a Computer Algebra System. Version 5.34.1. (2014). Retrieved from http://maxima.sourceforge.net/.
9. Shu, C. W. (1998). Essentially non-oscillatory and weighted essentially non-oscillatory schemes for hyperbolic conservation laws. *Lecture Notes in Mathematics, 1697*, 325–432.

10. Motamed, M., Macdonald, C. B., & Ruuth, S. J. (2011). On the linear stability of the fifth-order WENO discretization. *Journal of Scientific Computing, 47*, 127–149.
11. Wang, R., & Spiteri, R. J. (2007). Linear instability of the fifth-order WENO method. *SIAM Journal on Numerical Analysis, 45*, 1871–1901.

Enhanced Modified-Polygon Method for Point-in-Polygon Problem

Mostafa El-Salamony and Amr Guaily

Abstract The present contribution is an extension to our previously published algorithm, modified-polygon method (MPM), in which we modified the algorithm to account for the case of slender bodies which was a drawback in the original work. The modification is $O(N)$ in time. The modified algorithm is tested against existing techniques with an apparent degree of success.

Keywords Point in polygon · Modified area · Arbitrary polygon · Slender bodies

1 Introduction

In geographic information systems (GIS), in order to describe exactly the topological relation between spatial objects, the node-arc-polygon model functions are used as one of the essential representations. Point-in-polygon problem [1] is used to determine whether or not a certain location is situated in a certain region [2].

In the level-set method [3], in order to capture the moving interface, it is crucial to initially determine the location of the interface with respect to the domain of interest. Hence, one shall solve the PIP problem for each node. One important application is the modeling of fuel combustion inside rocket engine. Initially, the fuel grain has a known shape, then burning starts and the fuel grain shape is burnt out. To trace the grain distortion, PIP shall be solved [3].

M. El-Salamony (✉)
Department of Aeronautics and Astronautics, College of Engineering, Peking University, Beijing, China
e-mail: elsalamony.mostafa@pku.edu.cn

A. Guaily
Smart Engineering Systems Research Center (SESC), Nile University, Giza, Egypt

Faculty of Engineering, Engineering Mathematics and Physics Department, Cairo University, Cairo, Egypt
e-mail: aguaily@nu.edu.eg

© Springer Nature Switzerland AG 2020
M. H. Farouk, M. A. Hassanein (eds.), *Recent Advances in Engineering Mathematics and Physics*, https://doi.org/10.1007/978-3-030-39847-7_4

In robot motion and planning, the computer generates a map of the obstacles around the robot, find a position for the robot which is suitable for the desired action, construct a path for moving the robot to that position, and translate that plan to the robot's actuators. Hence, the process of determining the robot position and the obstacles places contains PIP problem [4].

Computer vision and length measurements have many applications [5], starting from vision-based measurement, to measure the deformation in bodies [6]. Also, this can be applied to unmanned aerial vehicles (UAVs) and its autonomous landing process. The UAV shall detect if a certain position is suitable for landing and measure how far it is, then trace its borders by detecting the intersections of the surrounding obstacles; hence, the UAV can decide whether it will land or not. This can be applied on landing decision for a helicopter [7] or a fixed-wing aircraft over a conventional runway [8] or a ship [9].

This article is structured as follows. Previous methods are stated in Sect. 2. Section 3 reviews the original method and evaluated in Sect. 4. The proposed modification for slender bodies is elucidated in Sect. 5, followed by the conclusion in Sect. 6.

2 Literature Survey

One of the primary methods to solve the PIP problem and the most popular method is the ray-crossing (ray intersection) method which was derived by Nordbeck and Rystedt [10] in 1967. Although it can deal with all kinds of polygons, it fails to detect if the investigated point lies on the polygon circumference or if the ray intersects a polygon vertex. It is executed in $O(n)$ in time where n is the number of the polygon boundaries. Hence, this method is stable from numerical viewpoint [11].

An important upgrade to this algorithm was developed in 1994 [12] where a modified version was proposed to calculate the ray intersection method based on the vector representation and determinants.

The second method is the area method [10]. Its basic idea is to connect the investigated point to all vertices and make triangles of them, then calculate the area of each triangle individually and add them. If the total area is the same as the initial polygon area, then the point is inside the polygons. If the area increases, then the point is outside the polygon. On one hand, it is easy to calculate the areas of a polygon, but on the other hand, it is not applicable for concave polygons because fictitious areas are generated.

In SWATH method [13], the polygon is subdivided into swaths (trapezoids), and the evaluation time is of order $O(n \log n)$ [14, 15]. Then by using binary search, the trapezoid is located for the investigated point in time $O(\log n)$; then the ray intersecting method is performed on the polygon sides of the trapezoid. This method is preferred when many points are to be tested on the same polygon, but the preprocessing is a drawback because of the high computational effort.

Sum of angles method [16, 17] is O(n) in time and used widely, but it suffers from two main disadvantages: it is slow because calculating the angles is computationally expensive, and it is sensitive to the truncation and rounding errors [18]. Thus, it is less suitable for polygons with large number of vertices.

Wedge method [16] can deal with convex polygons efficiently. In the first step, the polygon is divided into n wedges using a random point inside the polygon, which is O(n) in time. Then by utilizing binary search, the wedge that includes the point of interest is found out in O(log n) in time. Finally, by using cross-product operator, the investigated point relative location is found out in O(1) time.

Coded coordinate system (CCS) method is proposed by Chen and Townsend [19]. A coordinate system is created on the tested point, and the polygon vertices are held out against it and will be positioned on the coordinate axes or in one of its quadrants. The location of these vertices is matched to one of five possibilities. The following step is to use the ray-crossing method by horizontal testing ray that starts from the point of interest to negative infinity. Hence, only the edge that has vertices belonging to the second and third quadrants is to be examined. The authors [19] stated that this method can be made faster by approximately 25% if compared with the ray-crossing algorithm. But on the other hand, this method suffers from the same disadvantages of the ray-crossing method.

In sign-of-offset method, the polygon side on which the point of interest is located can be detected by computing the cross-product of vectors consisting of a combination of the point and polygon edges. If the cross-product sign is the same for all polygon edges, it is concluded that the point is inside the polygon. This algorithm has O(n) in time, but the main disadvantage is that it cannot be utilized for concave polygons [12].

Method of thin regular slices (TRS) [20] is an upgrade of the wedge method. In this method, the polygon is subdivided into a large number of slices/slabs. To attain the minimum time required, two polygon edges at most should be placed at each side of the slice. Since polygon sides in a convex polygon create two monotone chains, this preprocessing demands O(n) in time. The slice that includes the point can be found out using a simple formula, which is O(1) in time.

In the triangle-based method [21], the polygon is disintegrated into smaller triangles, which is O(n) in time, and each triangle is defined by a vertex located at the origin, and the other two vertices are located on a polygon edge. These triangles are categorized based on their orientation direction: positive or negative. Next, a ray starting from the origin and includes the point of interest is tested by all of the triangles. This method calculates how many intersections are found with the positively and negatively oriented triangles. If the outcome has a negative sign, the conclusion is that the point of interest is located outside the polygon. The authors [21] reported that this algorithm can be faster than the ray-crossing method.

Skala [21], in 1996, suggested a method using dual representation. It requires preprocessing, which has O(n) in time. Then the restriction test needs constant time O(1). One drawback of this method is that it can only be utilized for convex polygons without holes.

Huangand and Shih [22], in 1997, proposed the grid method which is appropriate for raster-based cases. The polygon is depicted by a group of cells or pixels. The pixel color at a certain position can be used to detect if the polygon includes the point of interest. This algorithm can be executed in $O(n)$ time. The authors [22] declared that this algorithm is not appropriate for vector-based representation of polygons.

Zalik and Kolingerova [17], in 2001, developed an algorithm based on grid indication. This algorithm has two steps. Initially, a grid of uniform-sized cells is constructed, and the polygon is placed on that grid. Then heuristic approach is utilized for cell dimensioning. The grid cells are labeled as being outside, inside, or over the polygon edge. An adapted flood-fill algorithm is utilized for cell categorization. Next, the points are to be examined independently. If the point of interest lies in an inner or outer cell, the outcome is reported directly with no further processing. In case that the cell encloses the polygon edge, it is feasible to find out the local point location.

In the same year, Hormann and Agathos [23] investigated in detail the basic concepts of ray-crossing and angle summation methods, and they found that they are mathematically alike. Thus, they derived a new algorithm and optimized it thoroughly. It is fast but wrongly detects the point location with respect to a polygon in case that the polygon lies between two peaks (e.g., star-shaped polygon), and also to detect if the point lies on a vertex, it must compare if the coordinates of the point match one of the vertices' coordinates. This procedure needs $O(n)$ in time.

R-tree search method is proposed by Zhou et al. [24] in 2013. This method is based on classifying each point in a treelike hierarchy; then search for the point in this tree. The main disadvantage is that the polygon must be represented in the R-tree indices, which creates limitations on drawing arbitrary polygons. Another drawback is that it has to be evaluated to each single point in the domain, which means huge execution time especially if the domain contains many (thousands) points.

Suwardi et al. [25], in 2015, proposed an algorithm to deal with a polygon based on the Boolean overlay. This method is based on counting all points in the polygon and the point of interest, then classifying them using color coding. Finally, the output is obtained by visual detection of the color of the output figure/map. Later, Khatun and Sharma [26] modified this algorithm to include the condition of point on boundary by adding one more "If condition" to check if the point lies on the body. This algorithm needs preprocessing to detect the shape and position of the point and the polygon, which can be extracted from the GIS, but this limits the application of this method.

Using ray intersection method, Zhang and Wang [27], in 2017, presented an algorithm to solve the PIP problem in 3D spherical coordinates directly instead of transforming the spherical polygon into a set of planar ones by projections which is a time-consuming process.

El-Salamony and Guaily proposed the modified-polygon method [28] which is based on the very basic definition of the polygon size with the drawback of producing misleading results for the case of slender bodies due to the self-intersection. The aim of the present work is to suggest a remedy for the problem of slender bodies while keeping the superiority of the original technique.

3 The Original Modified-Polygon Method (MPM)

The concept of this method is developed from the way of detection if a point divides a line internally or externally; if the point divides the line internally, the total distance between the line endings passing through the point is the same as the line length. In case that the point divides the line segment externally, the total distance of the line passing from the line endings through the point will be longer than the original length.

Extending this concept to an arbitrary point and a side of a polygon in 2D space, the polygon area is to be calculated rather than the length. The line passing from one end point of the polygon side to the other end point passing through the point of interest shall have longer length except if the point lies over the line. Also, this side will be split into two: one from the start point to the point of interest and another from that point to the end point. This new deformation will change the polygon initial area. If the point lies outside the polygon, the area of the modified polygon will expand and vice versa. In case that the point is located over the polygon circumference, the area will remain the same. Hence, by comparing the areas of the modified and the initial polygons, the point position with respect to the polygon can be determined precisely regardless of the polygon shape or any other constraint. The polygon area can be calculated as

$$A = \frac{1}{2} \left| \sum_{i=1}^{N} x_1 * y_{1+1} - y_1 * x_{i+1} \right| \tag{1}$$

where x_i and y_i are the vertices' coordinates. Points x_{N+1} and y_{N+1} are evaluated as x_1 and y_1. The absolute value is to be taken to cancel the effect of the direction around the polygon, i.e., CW or CCW.

To decide which line is to be connected to the point, a simple search algorithm is followed. Firstly, measure the distance between the point "P" and all the vertices of the polygon. The nearest vertex "A" will be one of the two ends of the line that shall be modified to include the point. The time required for this process is O(n). The next step is to choose which line to be modified: AB or AC; see Figs. 1 and 2. The choice here is arbitrary except for one case: if the line connecting between the point and the other end (say, line PB) intersects with the other polygon side (line AC) as shown in Figs. 1 and 2. Hence, an intersection check must be done. The time required for this process is O(1). The flowchart of this algorithm is demonstrated in Fig. 3.

Fig. 1 Intersection exists between lines PB and AC

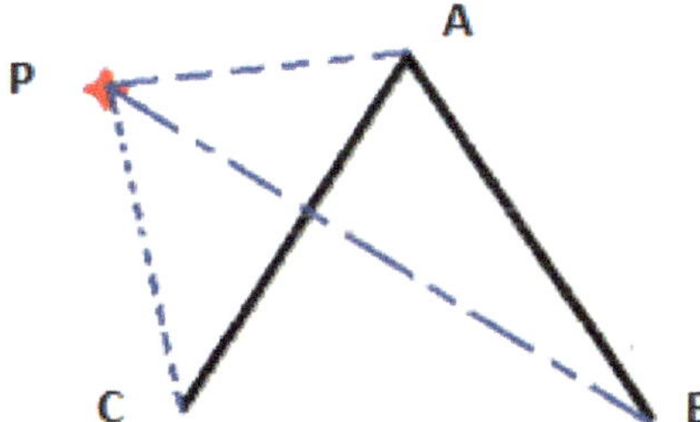

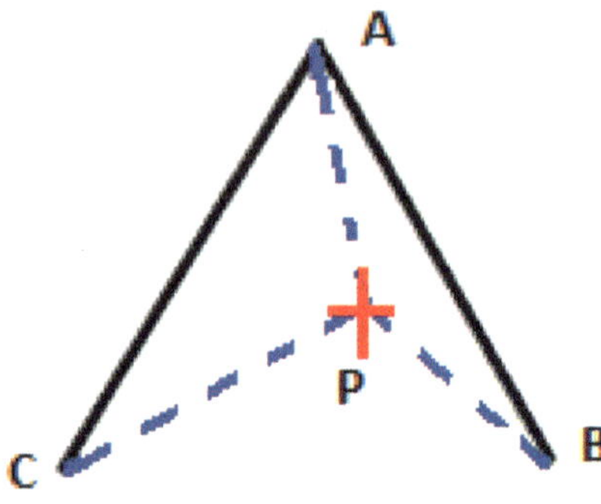

Fig. 2 No intersection exists between lines PB and AC

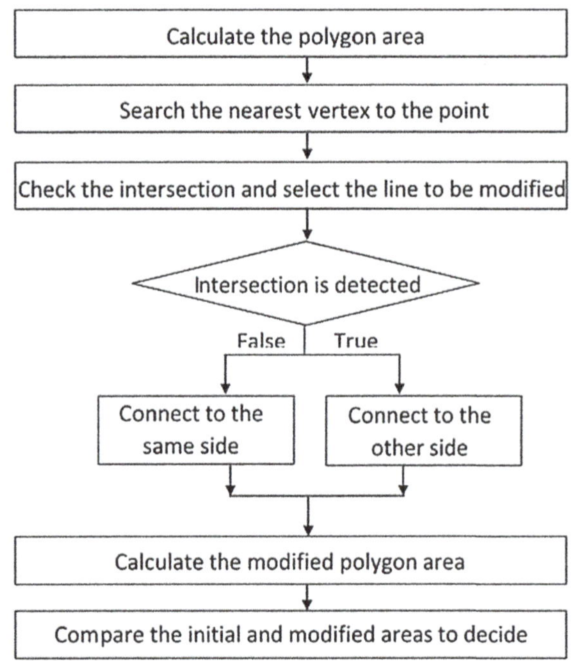

Fig. 3 Flowchart of the proposed algorithm

3.1 The Modified-Polygon Method (MPM) Vs. the Area Method

To understand the differences between the proposed "modified-polygon" method and the area method [10], one shall compare them together from the viewpoint of the basic principle used. The area method is based on connecting the point of interest to all of the polygon sides and construct triangles, then to sum up the areas of each triangle and compare the initial and new areas. If the point is located inside the shape, the total area will stay the same, but the area will enlarge if the point lies outside. The leading drawback is that if the polygon shape is concave (e.g., star), when connecting the triangles, they may overlap and intersect with the initial polygon itself, which generates untrue and self-intersecting areas, and this method will lose its functionality.

In the modified-polygon method [28], the point of interest is implanted in the polygon (i.e., the polygon will deform to pass through this point); then the whole area of this modified polygon is to be evaluated directly. By comparing the modified and initial areas, one of three possibilities are expected: (1) the modified area is larger, which means that the point is located outside the polygon; (2) the modified area is smaller, which means that the point is located inside the polygon; and (3) the modified area is the same, which means that the point is located precisely over the circumference of the polygon. The fictitious areas and the self-intersections cannot happen here because of the check that is done in step 4 in the flowchart (Fig. 3) except for slender bodies. This case is studied in detail, and a solution is proposed in Sect. 5. One more advantage for this method over the area method and the majority of the other methods is that it can determine if the point lies over the polygon.

4 Evaluation of the Proposed Method

4.1 Linearity of the Running Time

To evaluate the method, a challenging test case is investigated by this method and others. The five-pointed star is one of the most challenging problems, and hence it is used as a test case because it contains both convex and concave regions. The comparison is done with three other methods which are (1) ray-crossing [10], (2) sum of angles [16], and (3) algorithm number 6 of [24] since it is relatively fast. The calculating time is stated in Table 1. The average time is computed from averaging elapsed time of calculating 200 times. The reported calculations are obtained using MATLAB on an Intel i5 processor and 4 GB RAM.

As informed from Table 1, the angle summation method is the most computationally expensive method, then comes the ray-crossing and algorithm 6, and the proposed algorithm is the fastest. By investigating the methodology of these algorithms, it can be informed that in ray-crossing, an intersection check occurs between a ray passing from the investigated point and cutting the polygon. This needs to check the intersection problem n times. For algorithm 6, it is fast since it primarily depends on logic operators, but in order to check if the point lies over the polygon, it is necessary to test all of the vertices' coordinates. This test has $O(n)$ in time.

Table 1 Computational time comparison

Method	Average time	Minimum time
Present work (MPM)	3.69e–4	1.94e–4
Ray intersection	9.67e–4	6.11e–4
Sum of angles (SoA)	34.0e–4	25.0e–4
Algorithm number 6 [24]	6.39e–4	3.10e–4

The key point behind the speed of the modified-polygon method is that it calculates the areas using Eq. (1) two times which is $O(n)$ in time also, but it needs no more processing except a single check for line intersection which is O (1). In order to examine if the point is located over the circumference of the polygon, the other methods use logic operators which are $O(n)$ in time. On the other hand, the proposed method uses single logic operator only. Besides, this method is the most accurate among the other methods because it depends on the very basic definition of the polygon size.

4.2 Numerical Check for the Elapsed Time

For the purpose of investigating the behavior of the proposed method with the variation of the number of points, the method is tested on a quadrant of N points, and the minimum and average time consumed are computed. The averaging process is made over 2000 times, and N is varied from 100 to 10,000. The results show that this method has linear dependence on time as shown in Fig. 4.

4.3 Test Case with Multi-Perforated Shape

In order to stand on the strength points of the proposed method, a more challenging test case, shown in Fig. 5, is examined, which is used in solid propellant rocket

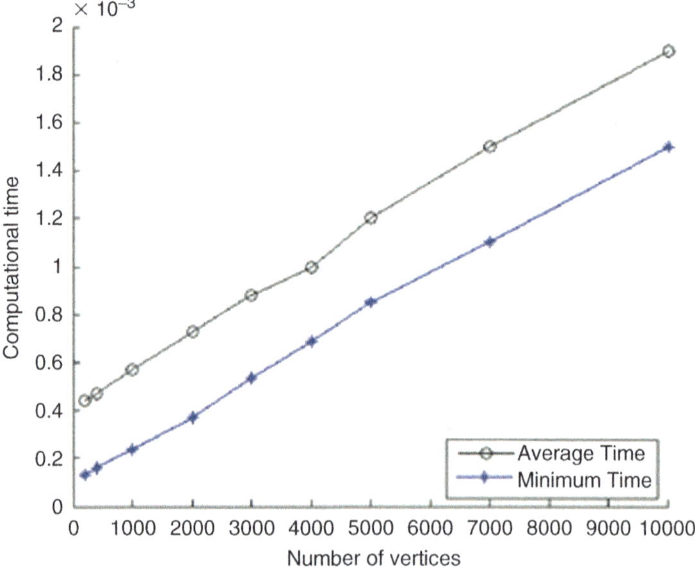

Fig. 4 Processing time versus number of vertices

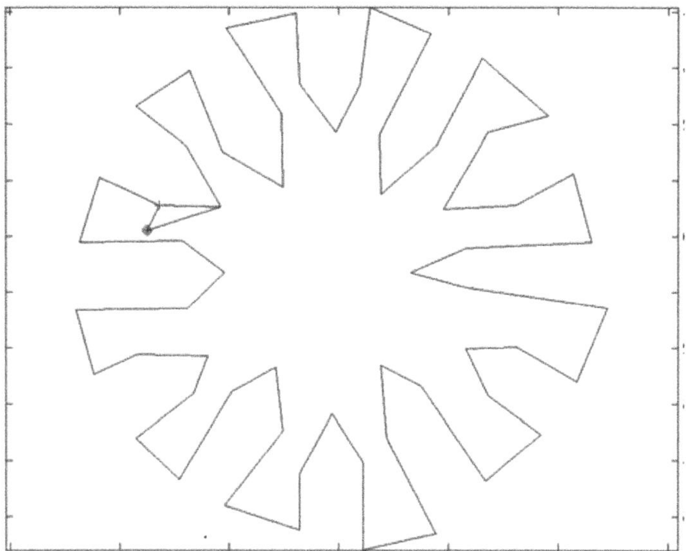

Fig. 5 Complex shape: the test point is located inside

engines. In case that the point is located outside the shape, area of the shape in black will increase. As shown in Figs. 5, 6, and 7, the method detected properly that the point is located inside, outside, or over the polygon, respectively. It is notable that algorithm 6 [24] fails to detect the point position relative to the polygon in the second case, Fig. 6, because the point is surrounded by the polygon from both sides.

5 Treatment of Slender Shapes and Self-Intersection

Consider a triangle ABC with relatively large obtuse angle B, and a point P needing to be investigated lies downside angle B and the opposite side AC as shown in Fig. 8. Using the aforementioned algorithm, the point will be connected to vertex B as it is the nearest one. Then the algorithm will insert the investigated point between AB and BC, but in this particular case, both choices are wrong simply because the line BP is already intersecting the original shape. Hence, the correct choice is to connect P through AC. This problem can appear in any polygon with slender dimensions.

To solve this issue, the point is to be connected to the nearest polygon side. To do that, first connect P to the nearest vertex; then check the intersection between this line (PB for the mentioned example) and the polygon sides. This intersection problem is $O(N)$ in time. If several intersections are discovered, the point must be connected to the side nearest to the point which corresponds to the nearest intersection point. Flowchart of the corrected method is illustrated in Fig. 9.

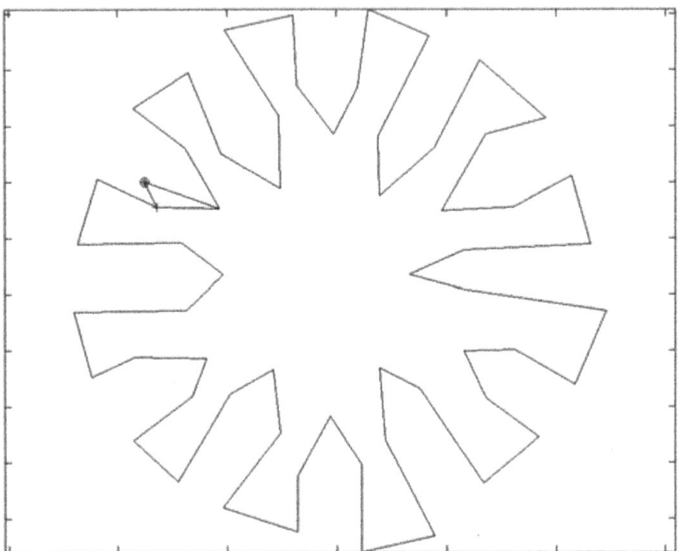

Fig. 6 Complex shape: the test point is located outside

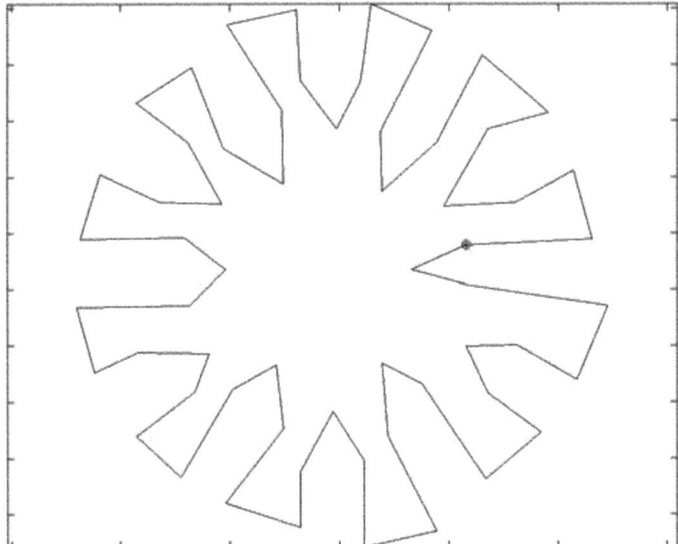

Fig. 7 Complex shape: the test point is located over the shape contour

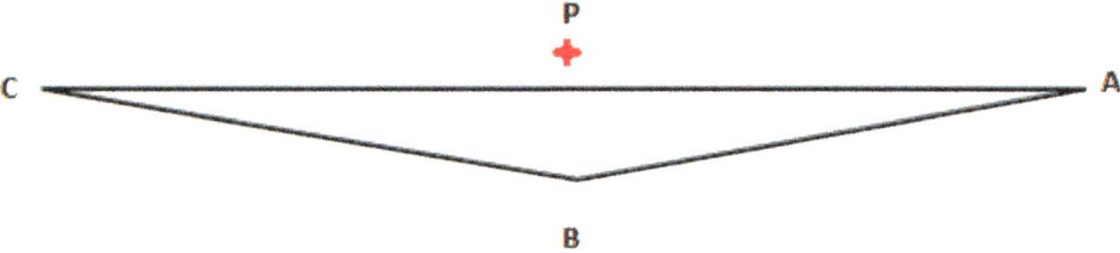

Fig. 8 Line PB is self-intersecting with the initial shape

Fig. 9 Flowchart of the modified algorithm

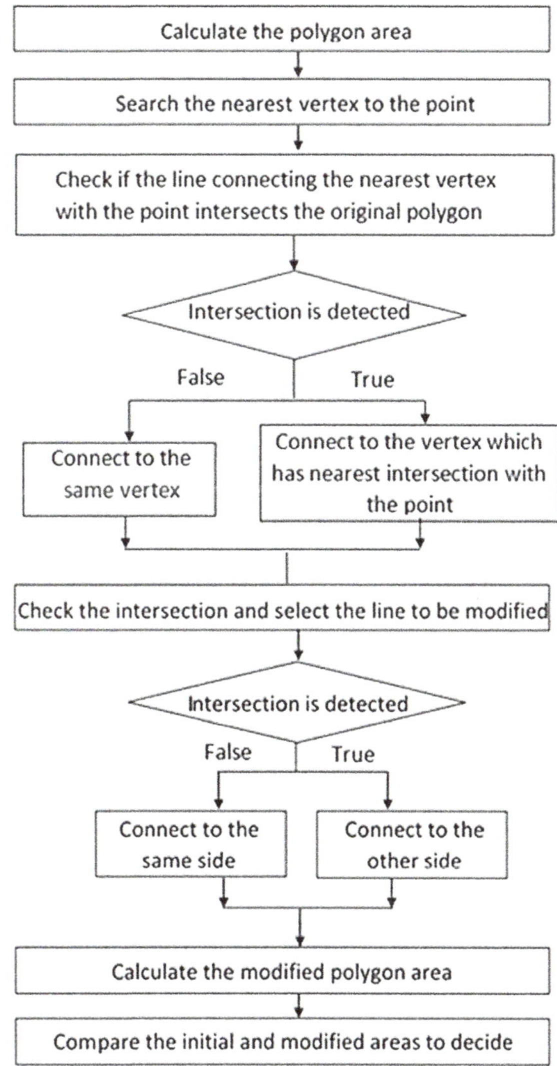

More complex example is shown in Fig. 10, and the method successfully classified the point as external point as illustrated in Fig. 11, where the nearest vertex is marked in red and the bypassed side is colored by magenta. This correction will make the time consumption in the same order of the ray intersection method.

6 Conclusion

The modified polygon method (MPM) is modified to account for slender bodies. The proposed method has many advantages, the most important of which are that (1) it deals with concave and convex polygons, (2) it is faster when compared with the existing methods, (3) it depends on the basic definition of the polygon size (line/area/volume) conditional to the spatial dimension of the considered shape, and (4) time consumption of this method varies linearly with the number of vertices of the polygon. This method is tested against different shapes, and the results are compared with some of the existing methods with a great degree of success showing the superiority of the proposed algorithm. The correction for slender bodies is successfully tested and proved to be efficient.

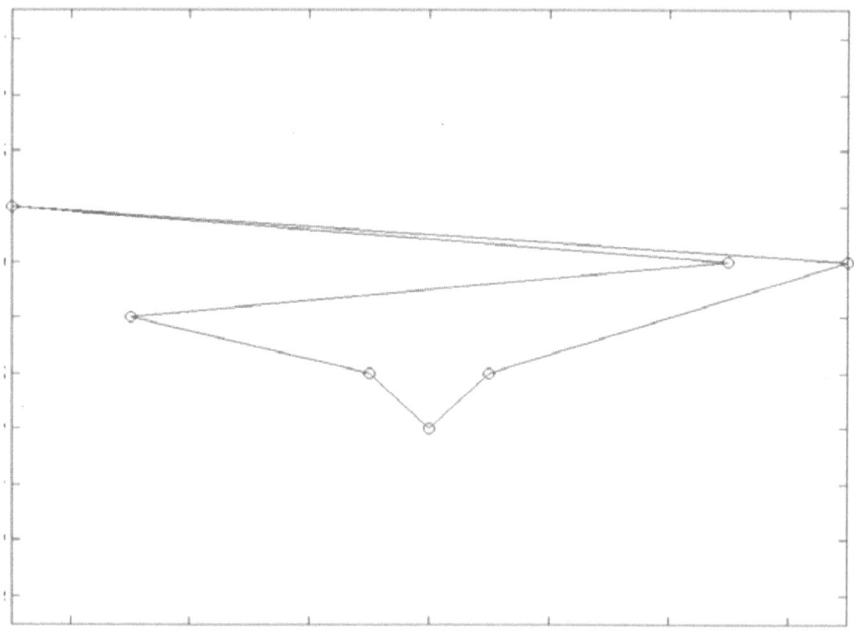

Fig. 10 Complex geometry for the proposed method

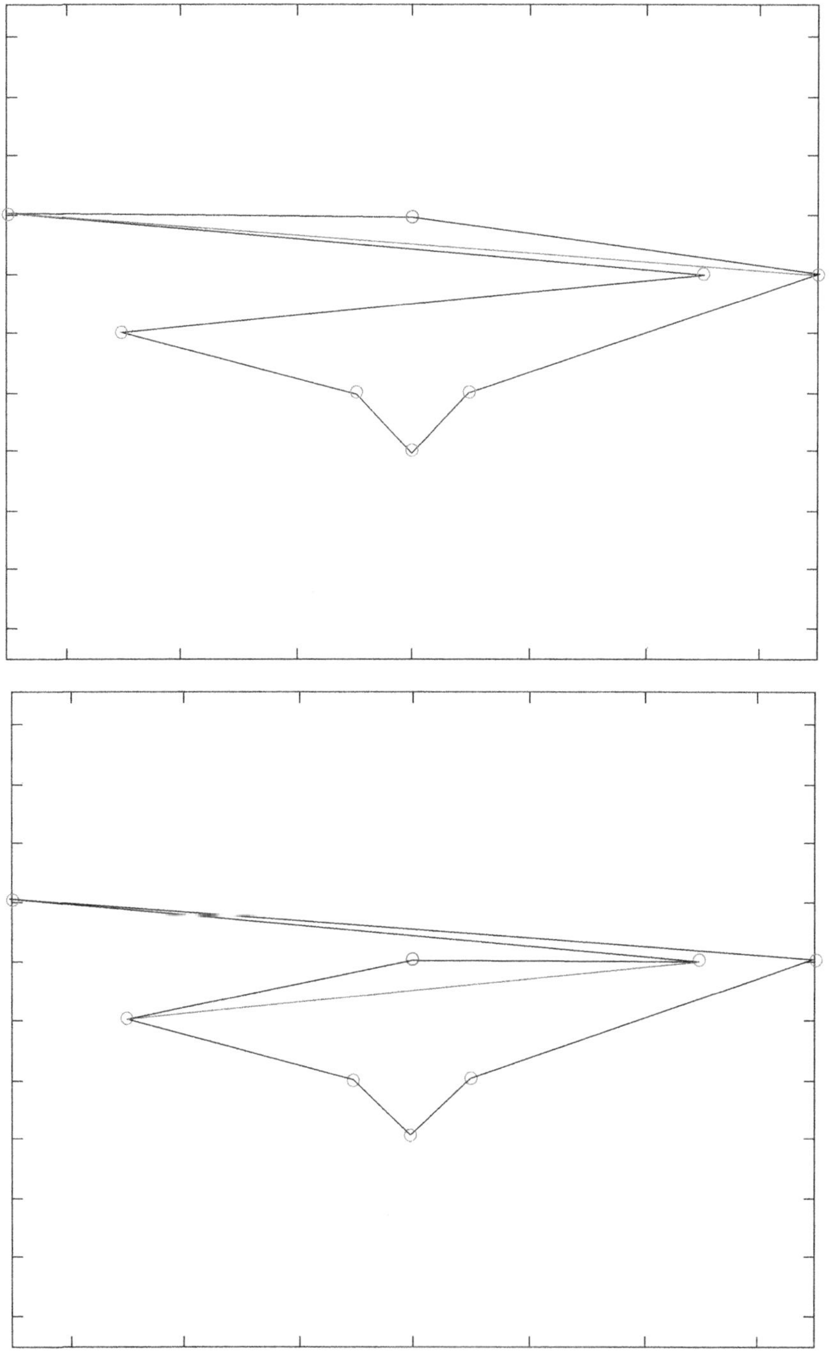

Fig. 11 The proposed method detected the position of the point with respect to the complex geometry

References

1. Foley, J. D., Van, F. D., Van Dam, A., Feiner, S. K., Hughes, J. F., Hughes, J., & Angel, E. (1996). *Computer graphics: Principles and practice* (Vol. 12110). Boston: Addison-Wesley.
2. Rigaux, P., Scholl, M., & Voisard, A. (2003). Spatial databases with application to GIS. *SIGMOD Record, 32*(4), 111.
3. Osher, S., Fedkiw, R., & Piechor, K. (2004). *Level set methods and dynamic implicit surfaces.* Cham: Springer.
4. Del Pobil, A. P., & Serna, M. A. (1995). *Spatial representation and motion planning* (Vol. 1014). Berlin: Springer Science & Business Media.
5. Shirmohammadi, S., & Ferrero, A. (2014). Camera as the instrument: The rising trend of vision based measurement. *IEEE Instrumentation and Measurement Magazine, 17*(3), 41–47.
6. Luo, P. F., Chao, Y. J., Sutton, M. A., & Peters, W. H. (1993). Accurate measurement of three-dimensional deformations in deformable and rigid bodies using computer vision. *Experimental Mechanics, 33*(2), 123–132.
7. Johnson, A., Montgomery, J., & Matthies, L. (2005, April). Vision guided landing of an autonomous helicopter in hazardous terrain. In *Proceedings of the 2005 IEEE International Conference on Robotics and Automation* (pp. 3966–3971). New York: IEEE.
8. Shang, J., & Shi, Z. (2007). Vision-based runway recognition for UAV autonomous landing. *International Journal of Computer Science and Network Security, 7*(3), 112–117.
9. Xu, G., Zhang, Y., Ji, S., Cheng, Y., & Tian, Y. (2009). Research on computer vision-based for UAV autonomous landing on a ship. *Pattern Recognition Letters, 30*(6), 600–605.
10. Nordbeck, S., & Rystedt, B. (1967). Computer cartography point-in-polygon programs. *BIT Numerical Mathematics, 7*(1), 39–64.
11. Manber, U. (1989). *Introduction to algorithms: A creative approach.* Boston: Addison-Wesley.
12. Taylor, G. (1994). Point in polygon test. *Survey Review, 32*(254), 479–484.
13. Salomon, K. B. (1978). An efficient point-in-polygon algorithm. *Computers & Geosciences, 4*(2), 173–178.
14. Seidel, R. (1991). A simple and fast incremental randomized algorithm for computing trapezoidal decompositions and for triangulating polygons. *Computational Geometry, 1*(1), 51–64.
15. Žalik, B., & Clapworthy, G. J. (1999). A universal trapezoidation algorithm for planar polygons. *Computers & Graphics, 23*(3), 353–363.
16. Preparata, F. P., & Shamos, M. I. (1985). Convex hulls: Basic algorithms. In *Computational geometry* (pp. 95–149). New York, NY: Springer.
17. Žalik, B., & Kolingerova, I. (2001). A cell-based point-in-polygon algorithm suitable for large sets of points. *Computers & Geosciences, 27*(10), 1135–1145.
18. Chen, M., & Townsend, P. (1987). Efficient and consistent algorithms for determining the containment of points in polygons and polyhedra. *Eurographics, 87*, 423.
19. Skala, V. (1994). *Point-in-polygon with O (1) complexity. TR 68/94.* Plzeò: University of West Bohemia.
20. Feito, F., Torres, J. C., & Urena, A. (1995). Orientation, simplicity, and inclusion test for planar polygons. *Computers & Graphics, 19*(4), 595–600.
21. Skala, V. (1996). Line clipping in E2 with O (1) processing complexity. *Computers & Graphics, 20*(4), 523–530.
22. Huang, C. W., & Shih, T. Y. (1997). On the complexity of point-in-polygon algorithms. *Computers & Geosciences, 23*(1), 109–118.
23. Hormann, K., & Agathos, A. (2001). The point in polygon problem for arbitrary polygons. *Computational Geometry, 20*(3), 131–144.
24. Zhou, T., Wei, H., Zhang, H., Wang, Y., Zhu, Y., Guan, H., & Chen, H. (2013, November). Point-polygon topological relationship query using hierarchical indices. In *Proceedings of the 21st ACM SIGSPATIAL International Conference on Advances in Geographic Information Systems* (pp. 572–575). New York: ACM.

25. Suwardi, I. S., Lestari, D. P., & Satya, D. P. (2015, August). Handling arbitrary polygon query based on the boolean overlay on a geographical information system. In *2015 2nd International Conference on Advanced Informatics: Concepts, Theory and Applications (ICAICTA)* (pp. 1–4). New York: IEEE.
26. Khatun, F., & Sharma, P. (2016). Arbitrary polygon query handling algorithm on GIS based on three value logic-an approach. *International Journal of Computer Applications, 975*(8887).
27. Li, J., Zhang, H., & Wang, W. (2017, June). Fast and robust point-in-spherical-polygon tests using multilevel spherical grids. In *International workshop on next generation computer animation techniques* (pp. 56–66). Cham: Springer.
28. El-Salamony, M., & Guaily, A. (2019). Modified-polygon method for point-in-polygon problem. *8th European Conference for Aeronautics and Space Sciences, Madrid, Spain.*

Part II
Applications in Mechanical Engineering

Transient Temperature Profiles in Powder Beds During Additive Manufacturing by 3D Printing of Metal Powders: A Lattice Boltzmann Study

Mohammed A. Boraey

Abstract The present work aims at predicting the transient temperature profiles in the powder beds of metal powders during the 3D printing of metals using laser melting. The effects of the laser beam power, diameter, and speed on the developed temperature profiles are predicted using the lattice Boltzmann method (LBM). These profiles are needed to predict the behavior of the powder bed during the 3D-printing process which affects many manufacturing parameters like the production speed, the product quality, and the properties. The results show that a compromise has to be made between the laser beam speed, which directly affects the production speed, and the developed depth of the heated zone in the powder bed, which determines how many laser passes are needed for a specific product.

Keywords 3D printing of metals · Temperature profiles · The lattice Boltzmann method

1 Introduction

Additive manufacturing (AM) is becoming one of the most popular manufacturing techniques for the production of products and goods that have special configurations or which are difficult or even impossible to be manufactured using traditional manufacturing techniques [1–4]. The technique has recently been used in the manufacturing of replacement tissues and organs for patients with chronic diseases [5].

The popularity of the technique comes from its ability to use almost any metal (polymers can be used as well [6]) to manufacture a product in any shape. This gives it many advantages over other manufacturing techniques which need mass

M. A. Boraey (✉)
Mechanical Power Engineering Department, Zagazig University, Zagazig, Egypt

Smart Engineering Systems Research Center, Nile University, Giza, Egypt
e-mail: boraey@ualberta.ca

© Springer Nature Switzerland AG 2020

65

M. H. Farouk, M. A. Hassanein (eds.), *Recent Advances in Engineering Mathematics and Physics*, https://doi.org/10.1007/978-3-030-39847-7_5

production to function in an economic mode [7] or can only work with a limited set of materials.

On the other hand, AM faces many challenges as well. Among these is the limit on the speed of the laser beam which scans the powder bed to produce the final product. Although higher speeds are desired, these do not allow the penetration of the beam energy to larger depths which results in an increased number of passes to produce a product with a given thickness [1].

Increasing the laser power has been suggested and actually implemented in some cases to overcome this problem, but this comes at the price of increased cost, increased hazard due to the explosive nature of most of the metal powders, and limiting the process to metals with high vaporization temperatures [8, 9].

The aim of the present work is to introduce a computational approach toward the prediction of the AM process using laser melting in order to optimize the process parameters. A lattice Boltzmann model (LBM) is developed and used to calculate the transient temperature profiles in a metal powder bed for different values of the laser beam power, diameter, and speed to choose the most suitable configuration that results in the shortest process time.

2 Problem Description

The problem under consideration is the heating of a metal powder bed using a moving laser beam with a specific scanning velocity. Figure 1 shows a schematic diagram of the laser-melting process in additive manufacturing. A powder bed is composed of many layers of a metal powder, and a laser beam is scanning the bed in a specific direction with a given velocity. The beam power results in melting the powder bed in place and forming a solid part. The scanning is repeated according to the desired shape of the final product until all the powder is melted and the final product is produced.

The goal of the present work is to investigate how different process parameters will change the transient heat profiles in the powder bed which determines the number of required scanning passes and subsequently the time needed for manufacturing.

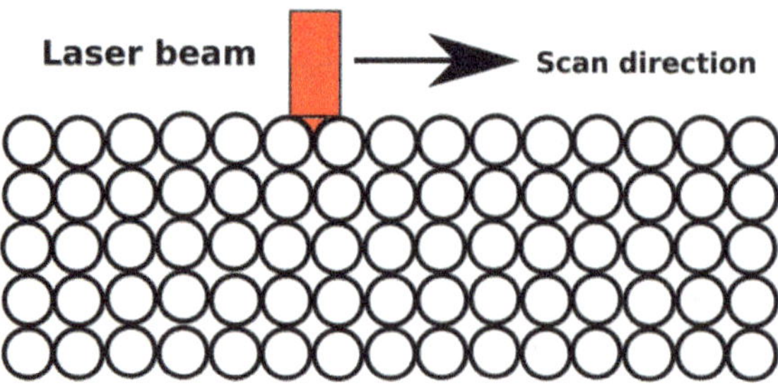

Fig. 1 Schematic of the 3D printing using laser-melting process

The details of the used numerical model and the tested cases are given in the following section.

3 Numerical Model

The numerical model developed uses the lattice Boltzmann method (LBM) [10] to resolve the transient temperature field inside the metal powder bed during the 3D-printing process using laser melting.

Figure 2 shows the computational domain and the used boundary conditions. The domain consists of a number of powder particle layers with adiabatic bottom and side boundaries. The top boundary is composed of three parts. The left part has already been heated by the laser beam. An adiabatic boundary condition is used for this part. The part under the laser beam (of diameter d_l) is a constant temperature boundary condition which is determined by the laser beam power as will be explained later. The unheated part of the top surface will be maintained at the initial temperature.

The transient temperature profiles are solved using the lattice Boltzmann method which solves a discretized version of the Boltzmann equation [11]:

$$g_i(x + c_i \Delta t, t + \Delta t) = g_i(x, t) - \frac{1}{\tau_t}[g_i(x, t) - g_i^{\text{eq}}(x, t)] \tag{1}$$

g is the temperature probability distribution function. τ_t is the relaxation time which is related to the thermal diffusivity α as follows [12]:

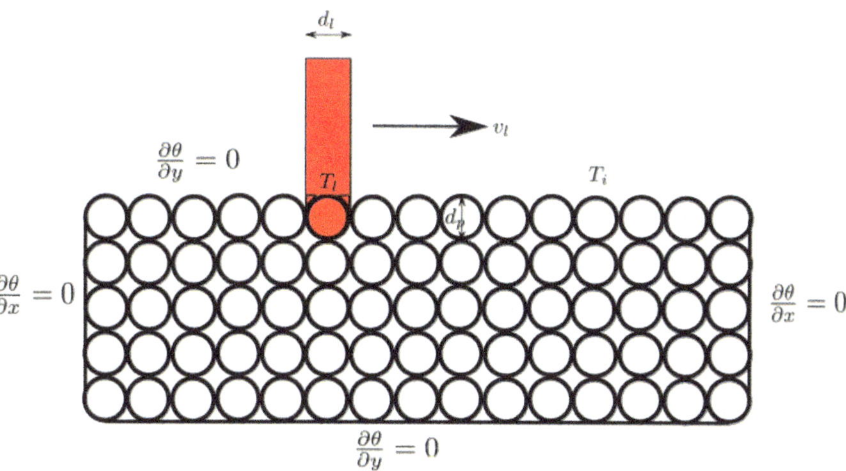

Fig. 2 The computational domain and boundary conditions

$$\alpha = c_s^2 \left[\frac{1}{\tau_f} - \frac{1}{2} \right] \tag{2}$$

$c_s = \sqrt{\frac{1}{3}\frac{\Delta x}{\Delta t}}$ is the lattice speed of sound.

The temperature is calculated as follows:

$$T = \sum_{i=0}^{8} g_i \tag{3}$$

For the used D2Q9 lattice configuration, the weights w_i and speeds $\mathbf{c}_i$ are given by [13]

$$w_i = \begin{cases} \dfrac{4}{9} & i = 0 \\[2mm] \dfrac{1}{9} & i = 1, 2, 3, 4 \\[2mm] \dfrac{1}{36} & i = 5, 6, 7, 8 \end{cases} \tag{4}$$

$$c_i = \begin{cases} (0,0) & i = 0 \\ (\pm c, 0), (0, \pm c) & i = 1, 2, 3, 4 \\ (\pm c, \pm c) & i = 5, 6, 7, 8 \end{cases} \tag{5}$$

The equilibrium distribution function is given by [14–16]

$$g_i^{\text{eq}}(x, t) = w_i T \left[1 + \frac{u.c_i}{c_s^2} + \frac{(u.c_i)^2}{2c_s^4} - \frac{u.u}{2c_s^2} \right] \tag{6}$$

4 Results and Discussion

In this section, the results of the numerical stimulation of the 3D-printing process are presented along with the discussion.

Three parameters were varied during the present study: the laser beam diameter d_l, velocity v_l, and power p_l. If the powder bed can be assumed as a semi-infinite solid exposed to an energy pulse (which is a good approximation for all the cases where the heating effect is not felt at the other end of the powder bed), then the surface temperature of the bed under the laser beam is directly proportional to the laser power [17]. In this case, the laser beam power can be represented by the surface temperature T_l.

Table 1 The used laser parameters

Parameter	Range
d_l/d_p	0.50–0.75
T_l/T_m	5.0–15.0
v_l/v_c	80–320

For each tested case, the temperature contours are plotted for the normalized temperature difference θ defined as

$$\theta = \frac{T - T_i}{T_m - T_i} \tag{7}$$

where T_i is the initial temperature of the powder bed and T_m is the melting temperature of the bed material. All regions in the powder bed with $\theta \geq 1$ belong to the molten pool of the bed.

Table 1 shows the range for each of the laser parameters which was varied during the present study. d_l and d_p are the laser beam and metal powder particle diameters, respectively. T_l and T_m are the surface temperature exposed to the laser beam and the metal melting temperature, respectively. v_l and v_c are the laser beam velocity and the characteristic speed for conduction in the metal powder bed, respectively.

For each of these cases, the simulation starts when the laser beam was heating the first powder particle and stopped when the beam has passed the eighth particle. Although the simulation could have been shorter due to the repetitive (i.e., cyclic) nature of the problem, a longer simulation time reveals the history of the laser heating process and the transient temperature profiles at different stages of the laser beam passing through the particles.

For each case, the temperature profiles and the maximum depth of the molten pool, L, are calculated. The molten pool is defined as the part of the powder bed with temperature equal to or greater than the metal melting temperature.

Figure 3 shows the temperature contours and molten pool size for few cases with different laser parameters. It is clear that as the laser power increases (i.e., higher T_l), the molten pool size increases as well. However, if this higher power is used at higher speeds, the beam does not have sufficient time to penetrate through the powder bed. Increasing the beam diameter also increases the molten pool size, but its effect is not as significant as the laser power.

To help quantify these effects, the maximum depth of the molten pool is calculated and plotted for all the simulated cases as a function of other laser parameters. The effect of the beam diameter is shown in Fig. 4.

The effect of the beam power is illustrated in Fig. 5 for the three power levels. The effect of the laser beam diameter d_l is less significant compared to the laser beam power level represented by T_l.

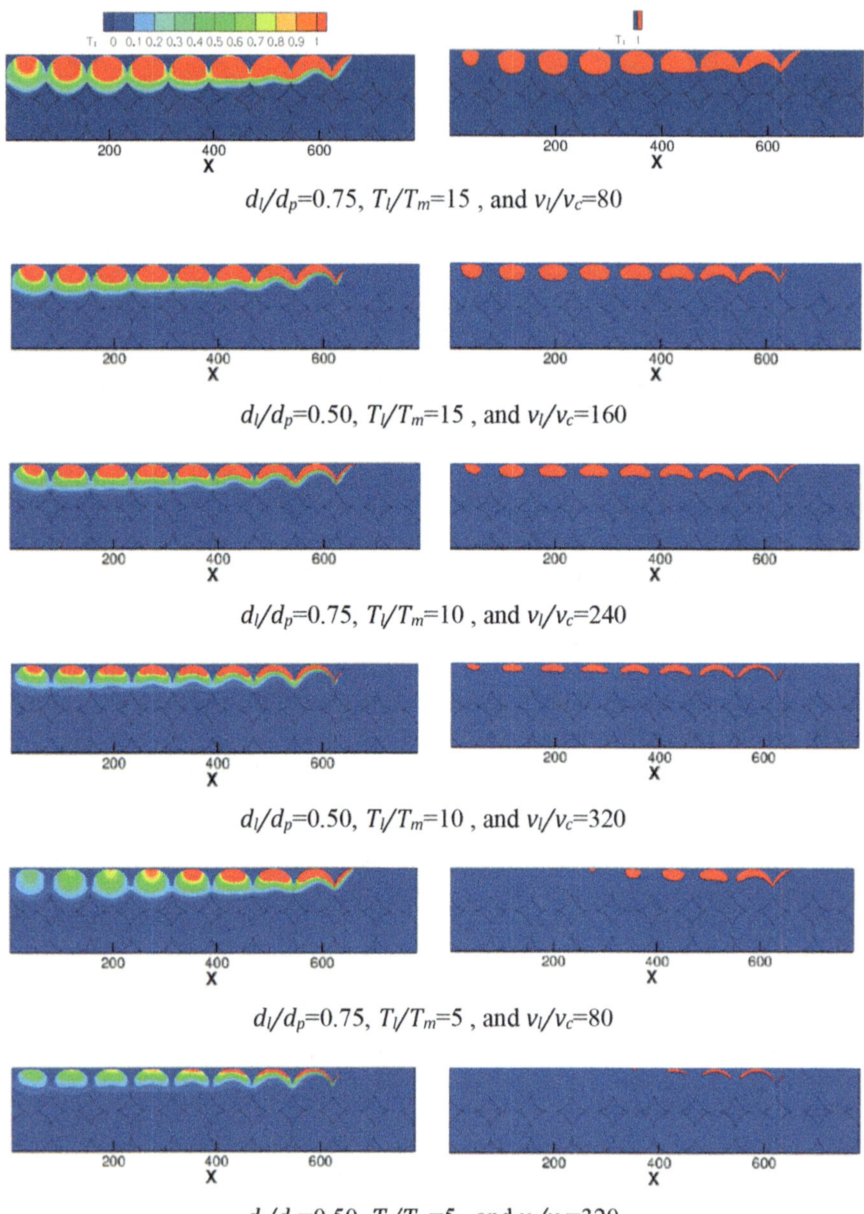

Fig. 3 Temperature contours (left) and molten pool size (right)

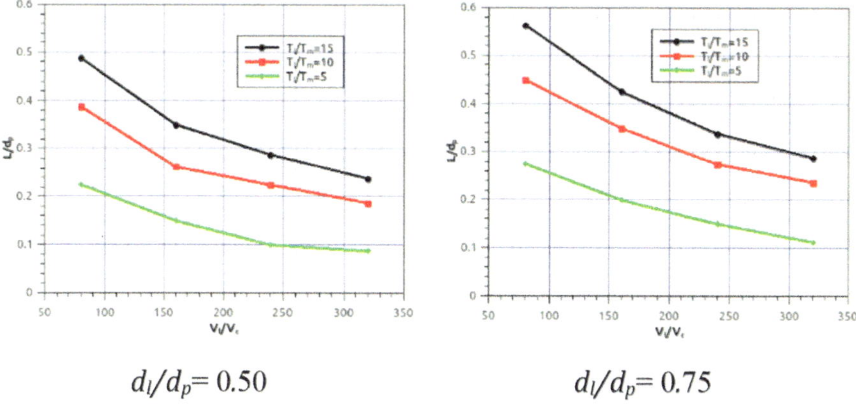

Fig. 4 The maximum depth of the molten pool, L, at different laser beam diameters

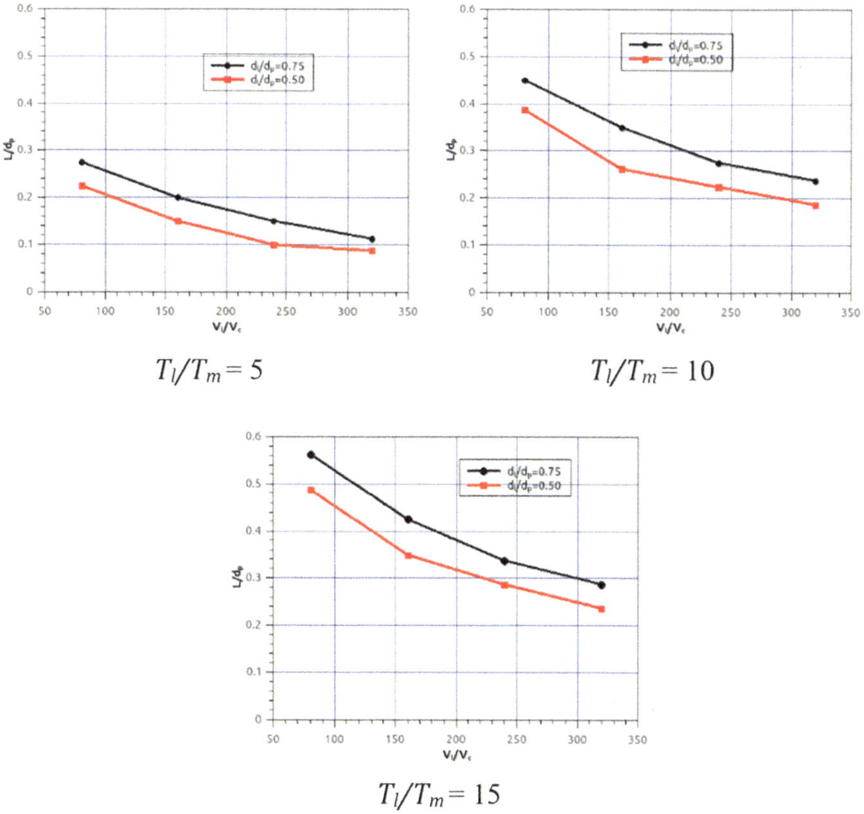

Fig. 5 The maximum depth of the molten pool, L, at different laser beam power levels

5 Conclusions

The transient temperature profiles and the molten pool size during the 3D-printing process of metal powders using laser melting are predicted using the lattice Boltzmann method (LBM). The effect of different laser beam parameters like the laser beam diameter, speed, and power on the temperature distribution inside the powder bed and on the size of the molten pool is investigated. The speed of the laser beam has the most significant effect with faster beams resulting in smaller molten pool size. The next significant factor is the laser power, while the least effect was due to the laser beam size.

Higher speeds for the laser beam result in shorter time for each laser beam pass but negatively affect the size of the molten pool. A compromise has to be made between the speed and pool size in order to reduce the total product manufacturing time.

References

1. Herzog, D., Seyda, V., Wycisk, E., & Emmelmann, C. (2016). Additive manufacturing of metals. *Acta Materialia, 117*, 371–392.
2. Wong, K. V., & Hernandez, A. (2012). A review of additive manufacturing. *ISRN Mechanical Engineering, 2012*, 208760.
3. Vaezi, M., Seitz, H., & Yang, S. (2013). A review on 3d micro-additive manufacturing technologies. *The International Journal of Advanced Manufacturing Technology, 67*(5–8), 1721–1754.
4. Gao, W., Zhang, Y., Ramanujan, D., Ramani, K., Chen, Y., Williams, C. B., Wang, C. C., Shin, Y. C., Zhang, S., & Zavattieri, P. D. (2015). The status, challenges, and future of additive manufacturing in engineering. *Computer-Aided Design, 69*, 65–89.
5. Melchels, F. P., Domingos, M. A., Klein, T. J., Malda, J., Bartolo, P. J., & Hutmacher, D. W. (2012). Additive manufacturing of tissues and organs. *Progress in Polymer Science, 37*(8), 1079–1104.
6. Eckel, Z. C., Zhou, C., Martin, J. H., Jacobsen, A. J., Carter, W. B., & Schaedler, T. A. (2016). Additive manufacturing of polymer-derived ceramics. *Science, 351*(6268), 58–62.
7. Atzeni, E., & Salmi, A. (2012). Economics of additive manufacturing for end-usable metal parts. *The International Journal of Advanced Manufacturing Technology, 62*(912), 1147–1155.
8. Gu, D., Meiners, W., Wissenbach, K., & Poprawe, R. (2012). Laser additive manufacturing of metallic components: Materials, processes and mechanisms. *International Materials Reviews, 57*(3), 133–164.
9. Murr, L. E., Gaytan, S. M., Ramirez, D. A., Martinez, E., Hernandez, J., Amato, K. N., Shindo, P. W., Medina, F. R., & Wicker, R. B. (2012). Metal fabrication by additive manufacturing using laser and electron beam melting technologies. *Journal of Materials Science and Technology, 28*(1), 1–14.
10. Mohamad, A. A. (2011). *Lattice Boltzmann Method. Fundamentals and Engineering Applications with Computer Codes*. Berlin: Springer.
11. Boraey, M. A. (2019). An asymptotically adaptive successive equilibrium relaxation approach for the accelerated convergence of the lattice Boltzmann method. *Applied Mathematics and Computation, 353*, 29–41.

12. Boraey, M. A. (2018). A hydro-kinematic approach for the design of compact corrugated plate interceptors for the de-oiling of produced water. *Chemical Engineering and Processing Process Intensification, 130*, 127–133.
13. Boraey, M. A., & Epstein, M. (2010). Lattice Boltzmann modeling of viscous elementary flows. *Advances in Applied Mathematics and Mechanics, 2*(4), 467–482.
14. Izquierdo, S., & Fueyo, N. (2009). Optimal preconditioning of lattice Boltzmann methods. *Journal of Computational Physics, 228*(17), 6479–6495.
15. Boraey, M. A. (2017). Simulation of the lid-driven cavity flow at Reynolds numbers between 100 and 1000 using the multi-relaxation-time lattice Boltzmann method. *Mugla Journal of Science and Technology, 3*(2), 110–115.
16. Boraey, M. A. (2019). Thermal optimization of square pin-fins in crossflow using the lattice Boltzmann method with quadratic thermal equilibrium. *Physica A: Statistical Mechanics and its Applications, 532*, 121880.
17. Cengel, Y. A., & Ghajar, A. J. (2014). *Heat and mass transfer: Fundamentals and applications.* New York: McGraw-Hill Education.

Numerical and Experimental Validation of an Unbalanced Oil Vane Pump Using RANS Approach

Ahmed H. El-Hennawi, Muhammed Eltahan, Mohammed Magooda, and Karim Moharm

Abstract Positive displacement pumps have a wide range of usage due to its compact size and efficiency as well. Vane pump is a positive displacement pump with many applications such as keeping the oil supply of the engines in the automotive and the fuel pumping in aviation industries. Pump testing is a direct technique to know its performance; however, it costs a lot to see the behavior of the fluid during the pump operation experimentally. Computational fluid dynamics (CFD) programs were developed to be another method in pump optimization as a result of its low cost and high speed. In this chapter, a validation process for a four-vane unbalanced pump model will be conducted, by comparing the results of the numerical model with the experimental results. The three-dimensional model was built up and simulated using ANSYS (Fluent). An experiment was held on a four-vane pump test rig, at the Mechanical Power Department of Ain Shams University. The results of the simulated model have been compared with the pump experimental results and gave a good agreement, with a maximum deviation of <6% in the flow rate.

Keywords RANS approach · Vane pump · Computational fluid dynamics · Model validation

A. H. El-Hennawi (✉)
Department of Mechanical Engineering, Ain Shams University, Cairo, Egypt

Mechanical Engineering Department, Technische Hochschule Luebeck, Luebeck, Germany
e-mail: ahmed.elhennawi@stud.th-luebeck.de

M. Eltahan
Aerospace Engineering Department, Cairo University, Cairo, Egypt

Institute of Bio-Geosciences (IBG-3, Agrosphere), Forschungszentrum Juelich GmbH, Juelich, Germany

Centre for High-Performance Scientific Computing, Geoverbund ABC/J, Juelich, Germany

M. Magooda
Aerospace Engineering Department, Cairo University, Cairo, Egypt

K. Moharm
Electrical Engineering Department, Alexandria University, Alex, Egypt

© Springer Nature Switzerland AG 2020

M. H. Farouk, M. A. Hassanein (eds.), *Recent Advances in Engineering Mathematics and Physics*, https://doi.org/10.1007/978-3-030-39847-7_6

1 Introduction

Vane pumps depend basically on volumetric displacement during their work. Many numerical and experimental studies have been exerted on the vane pumps in order to value their performance and enhance the design [1–3]. A three-dimensional CFD analysis of a novel vane pump power split transmission (VPPST) is studied by Frosina et al. [4]. The VPPST is a double-acting vane pump with two inlet chambers and two outlet chambers where the rotor and the ring contour are concentric. This study analyzed the output shaft power at different rotational pump speeds and pressures. Moreover, the influence of different gap heights between the vane tips and the ring has been analyzed. This study illustrated that by increasing the gap, the leakage and the output power decreases.

Pressure pulsation is one of the factors that can affect the pump by creating noise inside the pump. Adding to that, it produces low-pressure locations which lead to cavitation. Using metering grooves like the v-notch at the pump inlet is a good way to damp the pressure variations [5]. Pump cavitation is another factor that plays an important role in the pump's performance and lifetime [6, 7]. Cavitation simulation was performed using a code that has been developed to capture the gap shape between both the vane and stator with moving mesh by Zhang and Xu [8]. They investigated the influences of different vane tip radiuses on the cavitation area and the flow rate of the vane pump. They examined the cavitation size due to different radii of the vane tip. Moreover, Zhang and Xu found that the time-averaged flow rate at the outlet of the pump was a bit larger in the case of a small tip radius over a period.

The flow volume changes continuously through the pump rotation. As a result, the solver must be able to update the mesh every time step, as each mesh node displacement is composed of solid body rotation and a radial translation [7].

The purpose of this chapter is to validate an unbalanced positive displacement vane pump model by constructing a three-dimensional model, performing numerical simulations on it, and comparing its results with the experimental results. The Reynolds-averaged Navier–Stokes simulation (RANS) approach [9] has been used for flow modeling. The shear stress transport (SST) $k - \omega$ turbulence model has been used in all simulations.

2 Test Reg and Experiment Setup

2.1 Pump Description

A four-vane pump with volumetric displacement equals 18.0745 cm^3/rev has been used in the experiment to get the output flow rate and output pressure, by varying the rotational speed and recording the results. The pump comprises two eccentric circles: the inner is the rotor, which contains the vanes, and the outer is the stator. The pump dimensions in mm are shown in Fig. 1.

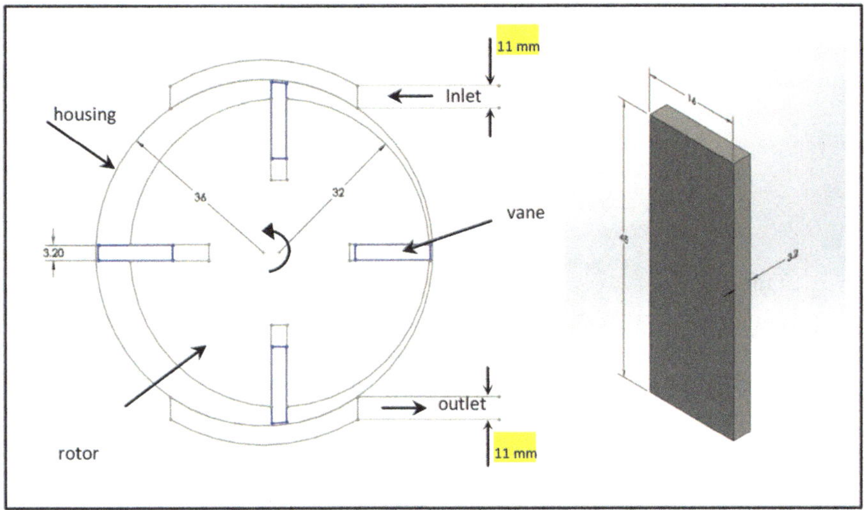

Fig. 1 The vane pump dimensions

2.2 Test Rig and Experiment Description

The test rig, which consists of an oil tank, electrical motor, vane pump, pressure gauge, frequency inverter, flow meter, and control valve, has been used in the experiment. An oil with density and dynamic viscosity of 891 kg/m^3 and 0.0163053 kg/ms, respectively, has been used. The measurement devices have been calibrated, and errors for data collections were specified on all components and have been described below:

- Flow meter error is $\pm1.5\%$ of measured flow rate. Maximum flow rate which can be measured is 2.5 m^3/h.
- Tachometer error for measuring rpm is ±5 rpm or 0.5% for tested speed.
- Pressure gauge error is 0.1% of full scale (10 bar), which is 1 kpa maximum for measurement. The test rig is shown in Fig. 2.

Different rotational speeds with different output pressures have been examined to get the output flow rate to determine the characteristic map in terms of pressure versus flow rate, as shown in Fig. 3.

Figure 3 shows the results of the experiment for three rotational speeds 775 rpm, 850 rpm, and 975 rpm, respectively, and as it is obvious, the pump's output flow rate is a function of the rotational speed. As a result, the flow rate increases by increasing speed. Moreover, the slope of the curve decreases due to the leakage decrease.

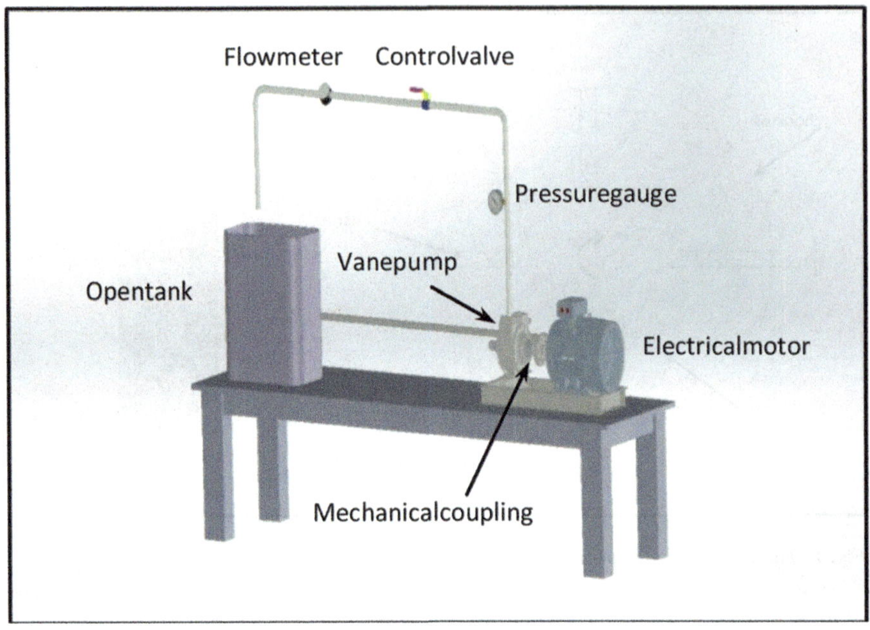

Fig. 2 The pump test rig

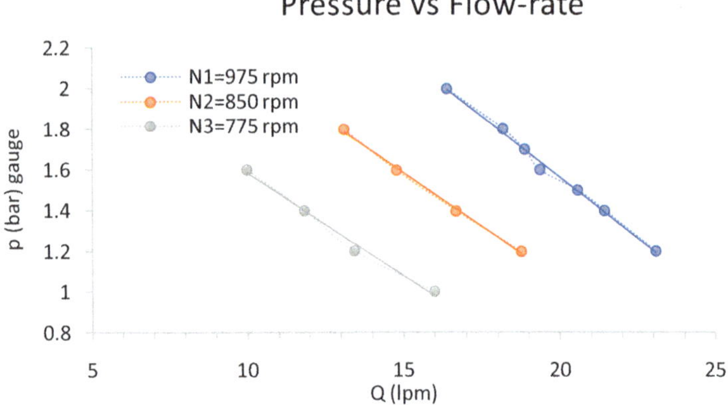

Fig. 3 Experimental results: pressure versus flow rate

3   Three-Dimensional CFD Model

3.1   Numerical Model and Mesh Generation

A numerical model has been generated to perform the flow simulations using ANSYS workbench (design modeler) version 16.0 [10]. The geometry includes two zones, stationary zone (inlet and outlet pipes) and moving deformation zone (rotating chambers). The model has been meshed with fully structured grids (hexahedron) to discretize the governing equations. A grid independence test has been performed to the model starting with very coarse mesh ending by a very fine mesh to assure that the mesh number doesn't affect the simulation results. The model with a grids number of 1,000,000 elements is chosen to perform all simulations with it. The three-dimensional model and the meshed model are shown in Fig. 4.

3.2   Boundary Conditions

A set of boundary conditions which have been used in the experiment are defined in Fluent solver to the pump model as follows:

(a) A total pressure of 0 bar-gauge (atmospheric pressure), turbulent intensity equals 5%, and hydraulic diameter of 11 mm are defined at the pump inlet.
(b) A set of static pressure starting with 1, 1.2, 1.4, 1.6, 1.8, and 2 bar-gauge, turbulent intensity equals 5%, and hydraulic diameter of 11 mm are defined at the pump outlet.
(c) The working fluid which is used in the experiment is oil with density of 891 kg/m^3 and dynamic viscosity of 0.0163053 pa·s.

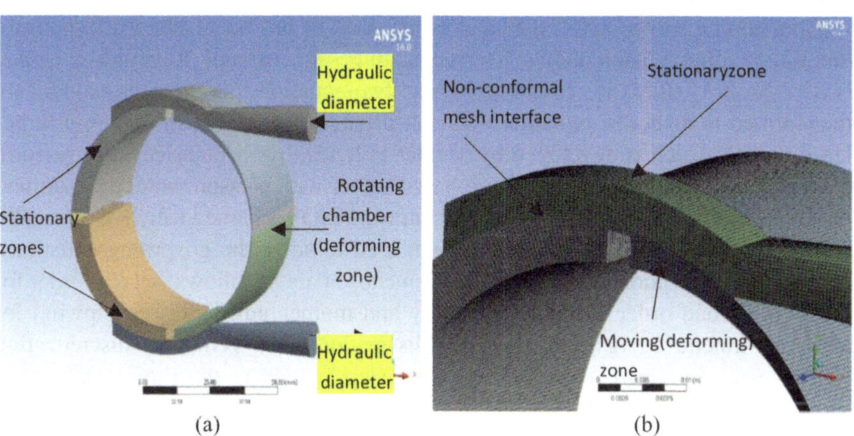

Fig. 4   (**a**) The three-dimensional model; (**b**) the meshed model

(d) Three rotational speeds are used in numerical simulations of 775, 850, and 975 rpm, equal to that used in the experimental data.

The turbulent intensity is calculated according to the formula, [10]:

$$I = 0.16 \ \mathrm{Re}_{d_h}^{-\frac{1}{8}} \tag{1}$$

where I is the turbulent intensity, d_h is the hydraulic diameter, and Re_{d_h} is the Reynolds number as a function of hydraulic diameter of 11 mm, velocity corresponding to 975 rpm, density of 891 kg/m^3, and dynamic viscosity of 0.0163053 pa·s.

4   Governing Equations and Solver Details

The Reynolds-averaged Navier–Stokes (RANS) approach was used for flow simulation. The continuity and momentum conservation equations for Newtonian fluids are given as follows [11]:

$$\frac{\partial}{\partial t}(\rho) + \nabla \cdot (\rho U) = 0 \tag{2}$$

$$\frac{\partial}{\partial t}(\rho U) + \nabla \cdot (\rho UU) = -\nabla P + \nabla \cdot \left(T + T^{\mathrm{Turb}}\right) \tag{3}$$

where ρ is density, U is velocity vector, P is pressure, μ is dynamic viscosity, and T^{Turb} is turbulent stress tensor. The convective term (ρUU) represents a differential balance of momentum, on a control volume. The right-hand side of Eq. (3) represents the summation of external forces: pressure and shear stresses.

Commercial solver ANSYS Fluent (version 16.0) has been selected to perform the finite volume simulations. Transient, three-dimensional Reynolds-averaged Navier–Stokes (RANS) equations plus two-equation turbulence model are transformed to algebraic equations. The shear stress transport (SST) $k - \omega$ turbulence model, which is developed by Menter [12, 13], has been selected to perform the numerical simulations. A pressure-based solver with pressure-velocity coupling, Pressure-Implicit with Splitting of Operators (PISO) scheme, which is normally chosen for transient calculations, has been used to solve the governing equations. A spatial discretization data which has been chosen are as follows: second order for pressure, second order upwind for density and momentum, first order upwind for turbulent kinetic energy, and specific dissipation rate. Temporal discretization scheme was examined under various time steps which is the time step test, as shown in Fig. 5.

Three complete revolutions were computed in all simulations in order to assure obtaining the repeatable pattern, with 1° time step which is found to be sufficient to

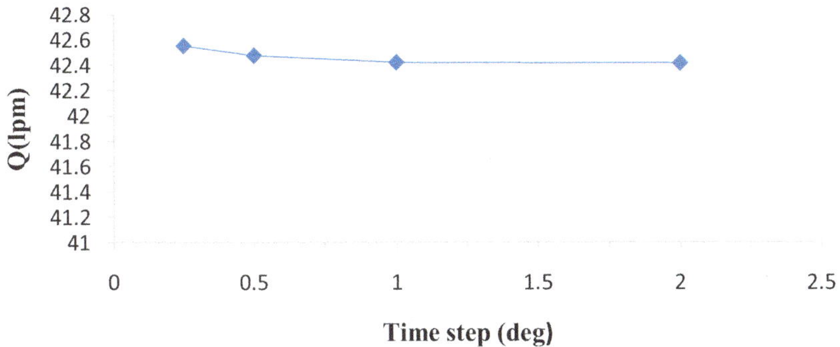

Fig. 5 Time step test

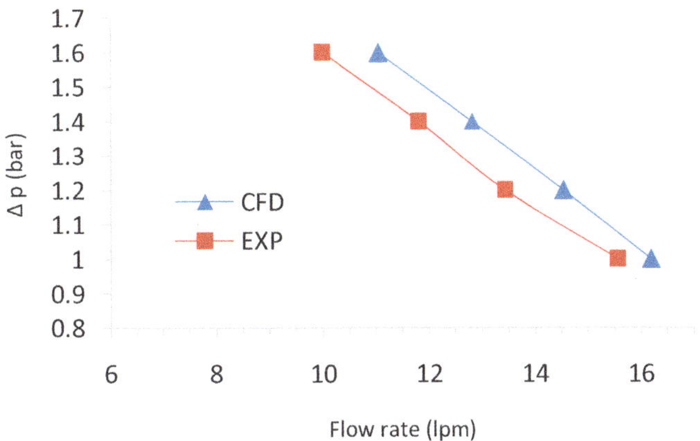

Fig. 6 Experimental versus numerical results at 775 rpm

resolve time-dependent features and to get the real behavior of transient changes besides maintaining the numerical solver stability.

5 Numerical Simulation Results and Discussion

5.1 Validation of the Numerical Model

Several numerical simulations have been performed using three rotational speeds of 775, 850, and 975 rpm. The characteristic maps of these three speeds in terms of output pressure versus flow rate are shown in Figs. 6, 7, and 8, respectively.

The values of the flow rate from numerical modeling at different pressures were compared with the experimental data at certain speed 975 rpm, as presented in Fig. 8. It is apparent that the numerical results have a good agreement and are very similar to

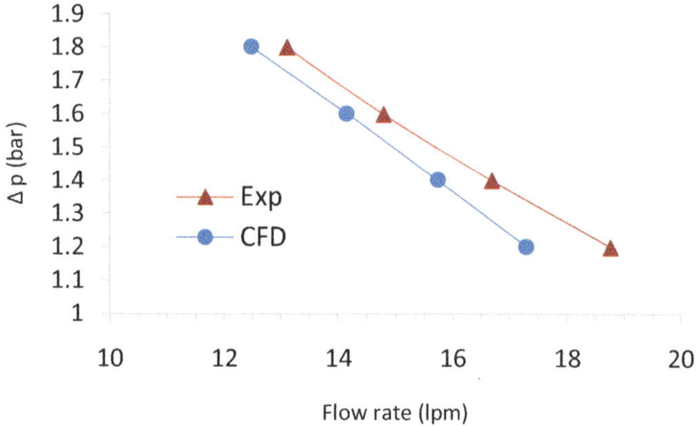

Fig. 7 Experimental versus numerical results at 850 rpm

Fig. 8 Experimental versus numerical results at 975 rpm

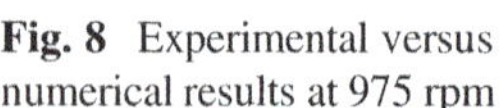

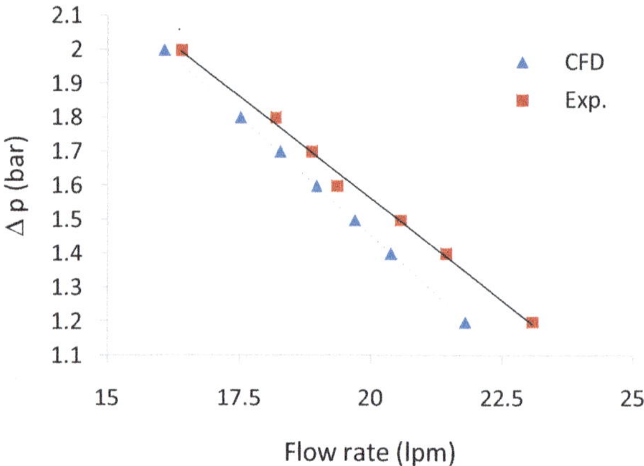

the experimental results, with maximum deviation <6% in the flow rate at the same pressure head. As a result, this confirms the validity of the numerical model.

Moreover, the pressure contours, density contours, and velocity vectors inside the chambers, inlet, and outlet ports, at rotational speed of 975 and output pressure equals to 2 bar, are shown in Figs. 9, 10, and 11, respectively.

Figures 9 and 10 illustrate the pressure and density distribution inside the pump at a crank angle of $720°$, and they give a good insight of the low and high (pressure and density) places, which affect the performance of the pump. Figure 11 shows the velocity distribution inside the pump. The velocity increases obviously at the gap heights where the volume between the vane tip and the cam ring decreases.

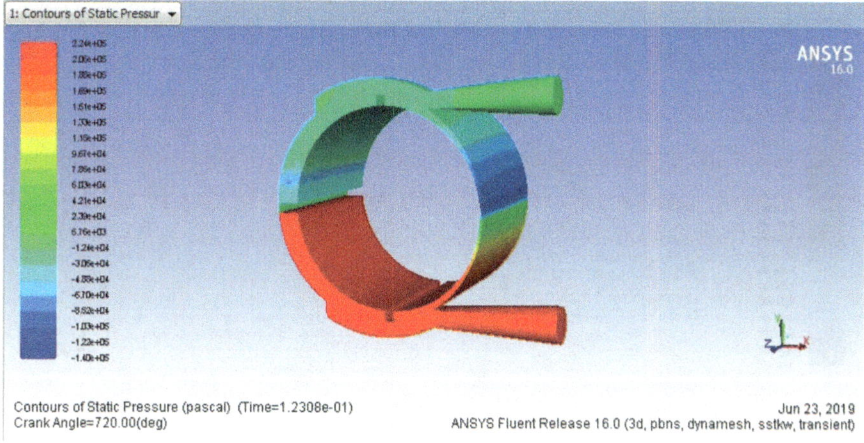

Fig. 9 Pressure contours inside the vane pump at 975 rpm

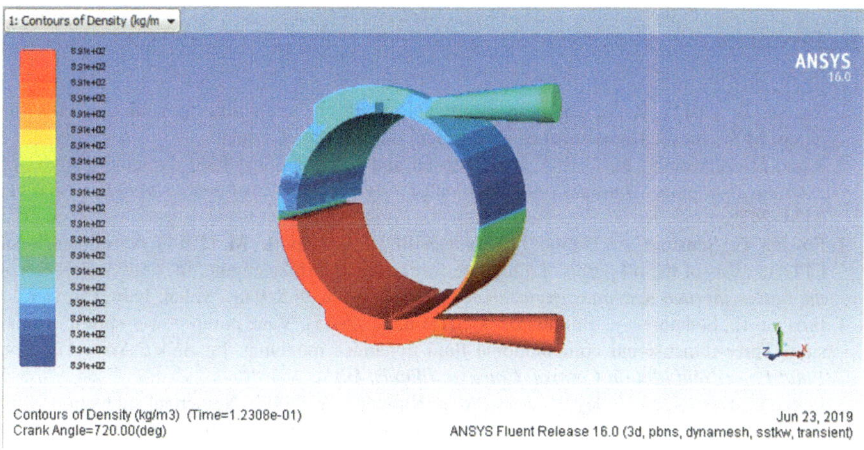

Fig. 10 Density contours inside the vane pump at 975 rpm

6 Conclusion

In the present chapter, a validation process to a positive displacement vane pump has been performed using Reynolds-averaged Navier–Stokes simulation (RANS) approach. A three-dimensional CFD model has been generated using ANSYS Fluent 16.0, and several numerical simulations have been done. The numerical results have been compared with the experiment results which was conducted in the lab. The numerical results were close to the experimental results with less than 6% error in the flow rate. In the future work, a CFD parametric study will be performed on the vane pump model.

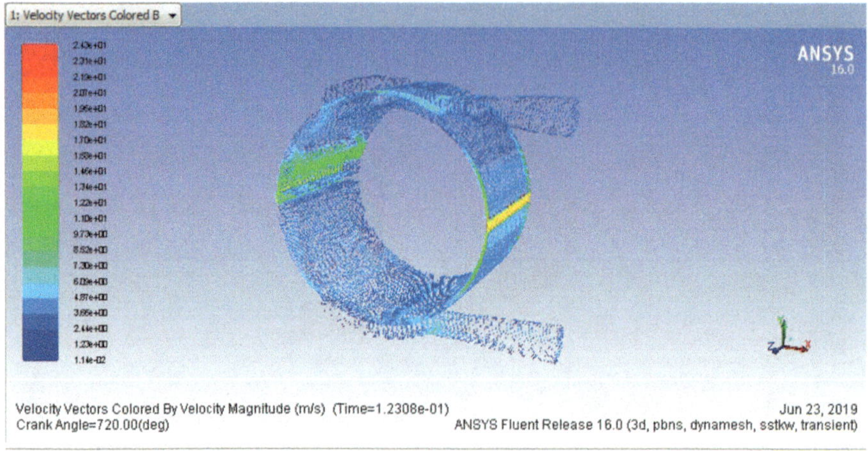

Fig. 11 Velocity vectors inside the vane pump at 975 rpm

References

1. Rancic, S. (2014). Reduction of pressure pulsations on automotive transmission oil vane pump. M.Sc. thesis, Department of Mechanical Engineering, Canada.
2. Rana, D., & Kumar, N. (2014). Experimental and computational fluid dynamic analysis of external gear pump. *International Journal of Engineering Development and Research, 2*, 2474–2478.
3. Frosina, E., Senatore, A., Buono, D., Manganelli, M., & Olivitte, M. (2014). A tridimensional CFD analysis of the oil pump of a high performance motorbike engine. In *68th conference of the Italian thermal machines engineering association* (pp. 938–948). Milan, Italy: Elsevier.
4. Frosina, E., Senatore, A., Buono, D., & Stelson, K. (2015). Vane pump power split transmission: Three-dimensional computational fluid dynamics modeling. In: *ASME Symposium on Fluid Power and Motion Control, Chicago, Illinois, USA.*
5. Lessa, L., Kessler, M., Foley, L., Johri, A., & Khondge, A. (2007). Numerical simulation of oil flow and noise in a power steering pump. In: *3rd European Automotive CFD Conference, Frankfurt, Germany.* pp. 129–137.
6. Hain-Würtenberger, S. (2007). Simulation of cavitating flow in vane pumps. In: *3rd European Automotive CFD Conference, Frankfurt, Germany.* pp. 285–293.
7. Brusiani, F., Bianchi, G., Costa, M., Squarcini, R., & Gasperini, M. (2009). Evaluation of air/cavitation interaction inside a vane pump. In: *4th European Automotive Simulation Conference, Munich, Germany.*
8. Zhang, Q., & Xu, X. (2014). Numerical simulation on cavitation in a vane pump with moving mesh. In: *ICCM Conference, Cambridge, England.*
9. Gad-el-Hak, I., Hussin, A., Hamed, A., & Mahmoud, N. (2017). 3D numerical modeling of zeotropic mixtures and pure working fluids in an ORC turbo-expander. *International Journal Turbomachinery Propulsion and Power, 2*(1), 2.
10. ANSYS FLUENT 14.5. (2010). *Documentation: User Guide. ANSYS Inc., Technical Report, Nov. 2010.*

11. Versteeg, K., & Malalasekera, W. (2007). *An introduction to computational fluid dynamics: The finite volume method*. London: Pearson Education.
12. Menter, R. (1993). Zonal two equation k-ω turbulence models for aerodynamic flows. *AIAA Journal Paper*, 93–2906.
13. Menter, R. (1994). Two-equation eddy-viscosity turbulence models for engineering applications. *AIAA Journal, 32*(8), 1598–1605.

Numerical Study of an Unbalanced Oil Vane Pump Using Shear Stress Transport (SST) $k - \omega$ Turbulence Model

Ahmed El-Hennawi, Muhammed Eltahan, Mohammed Magooda, and Karim Moharm

Abstract Vane pumps have several applications in the automotive and aviation industries. Many factors affect pump performance. So, modifying the pump design plays an important role in pump performance and efficiency. Computational fluid dynamics (CFD) techniques became an adequate tool to design and optimize the pump as they can save time and cost. In this chapter, a parametric study for an unbalanced vane pump model will be shown, by studying the effect of two parameters which are the model vanes number and the gap height between the vane tip and the pump casing. A three-dimensional model was built and simulated using ANSYS (Fluent). The shear stress transport (SST) $k - \omega$ turbulence model was used in all simulations. Three different gap heights and three different vanes number models were simulated. The model with gap height of 0.1 mm gave the best flow rate and output power. Moreover, three models with four, six, and eight rotor vanes were modeled. The characteristic map was determined, and the eight-vane model gave the best output flow rate.

Keywords SST $k - \omega$ turbulence model · Vane pump · Computational fluid dynamics · RANS approach

---

A. El-Hennawi (✉)
Department of Mechanical Engineering, Ain Shams University, Cairo, Egypt

Mechanical Engineering Department, Technische Hochschule Luebeck, Luebeck, Germany
e-mail: ahmed.elhennawi@stud.th-luebeck.de

M. Eltahan
Aerospace Engineering Department, Cairo University, Cairo, Egypt

Institute of Bio-Geosciences (IBG-3, Agrosphere), Forschungszentrum Juelich GmbH, Juelich, Germany

Centre for High-Performance Scientific Computing, Geoverbund ABC/J, Juelich, Germany

M. Magooda
Aerospace Engineering Department, Cairo University, Cairo, Egypt

K. Moharm
Electrical Engineering Department, Alexandria University, Alex, Egypt

© Springer Nature Switzerland AG 2020

M. H. Farouk, M. A. Hassanein (eds.), *Recent Advances in Engineering Mathematics and Physics*, https://doi.org/10.1007/978-3-030-39847-7_7

1 Introduction

Unbalanced vane pumps usually comprise of two eccentric circles: the inner is the rotor which the vanes extrude from, and the outer is the casing. During the pump rotation, the fluid confines between the rotor, stator, and vanes from the suction part (large volume) to the output part (small volume), causing the pumping action.

Quite a lot of numerical studies were performed on the vane pumps, but the most basic challenge is that the flow volume changes continuously over time. Therefore, the numerical solver must be able to deform and generate the mesh every time step. Pump performance and optimization are important issues for engineers. Consequently, modification of the pump even by adjusting the inlet and outlet ports [1–3], or by minimizing the pressure pulsation [4], will increase the flow rate and hence the efficiency.

Another important issue that affects the pump's performance and lifetime is the cavitation. It could happen when the fluid pressure becomes less than the vapor pressure at the working conditions. During experiments it is hard to detect the cavitation in the pump; however, the CFD tools have the ability to determine the cavitation size and position. Several additional studies paid more attention to the cavitation simulation and its effect on the pump, to avoid its occurrence and prevent the pump body from surface erosion [5–7].

The purpose of this chapter is to perform a numerical study for an unbalanced oil vane pump with CFD three-dimensional code using ANSYS Fluent version 16.0, which can generate the mesh for different vanes number models and different gap heights between the vane tip and stator in order to illustrate their effect on the flow rate and, hence, the efficiency. The Reynolds-averaged Navier–Stokes (RANS) simulation approach [8] was used for flow modeling. The shear stress transport (SST) $k - \omega$ turbulence model was used in all simulations.

2 Computational Method

2.1 Numerical Model and Dynamic Mesh

A detailed three-dimensional model was generated in order to perform the fluid simulations using ANSYS Workbench (design modeler) version 16.0. The geometry includes two zones, stationary zone (inlet and outlet pipes) and moving deformation zone (rotating chambers). The model dimensions are shown in [9]. The three-dimensional model is shown in Fig. 1. The model was meshed with fully structured grid (hexahedron) with (1,000,000 elements). The model mesh and zones are illustrated in Fig. 2.

ANSYS Fluent dynamic mesh capability was used to model the change-in-time shape of rotor chambers. The motion of the mesh nodes and the vanes is described by a user-defined function (UDF). The mesh motion is a function of chamber

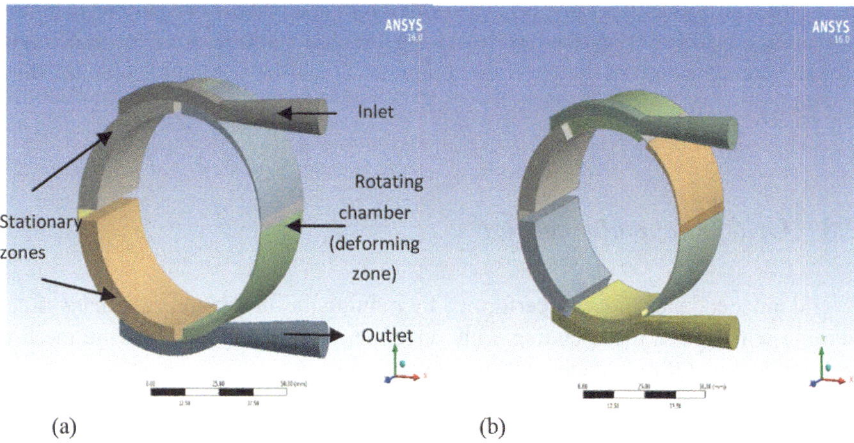

(a)   (b)

Fig. 1 The pump model; (**a**) four-vane model and (**b**) six-vane model

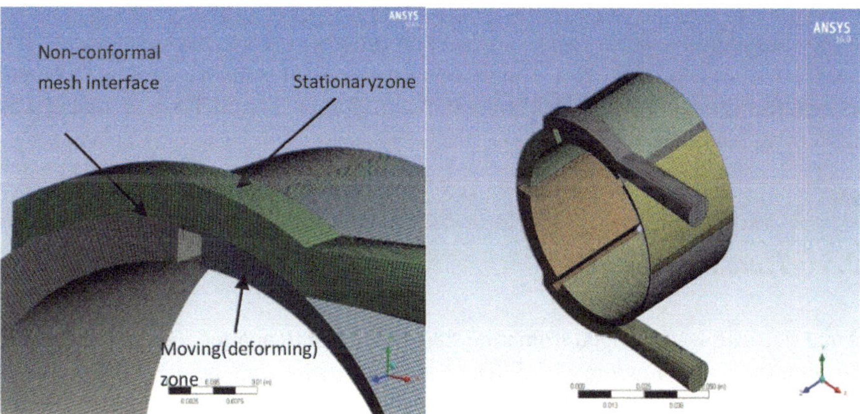

Fig. 2 The model with structured mesh

eccentricity, revolution speed, and the rotor case shape [5]. As a result of the existence of two different zones, one is stationary and the other is deforming, it is necessary to use a moving–sliding methodology, whereby the stationary and moving zones are meshed separately. Each zone connects to the other via a nonconformal mesh interface, which is updated at each time step.

2.2 Numerical Tests

Numerical tests have been conducted to ensure that the numerical model has minimal error. These tests are a grid independence test, turbulence model test, and time step test. Numerical tests were performed for six complete revolutions to assure obtaining a repeatable pattern. All following simulations were done for three revolutions when it is noticed that there is no change in the flow rate pattern after the third revolution.

2.3 Grid Independence Test

A grid independence test was performed by running the simulation case many times starting with coarse mesh ending with very fine mesh. It is noted that the mesh is independent at 1,000,000 elements as shown in Fig. 3.

2.4 Turbulence Model Test

Many turbulence models were simulated in order to choose the best model. The shear stress transport (SST) $k - \omega$ turbulence model's results were found to be the closest to the experimental results, as shown in Fig. 4. The SST $k - \omega$ model was used in all simulations.

2.5 Time Step Test

Time step size was changed from time step equivalent to quarter degree of the pump rotation to $2°$, as shown in Fig. 5. The test results show that the output flow rate is nearly constant. Therefore, in the current study, the time step size equivalent to $1°$ of the pump rotation was selected in all numerical simulations.

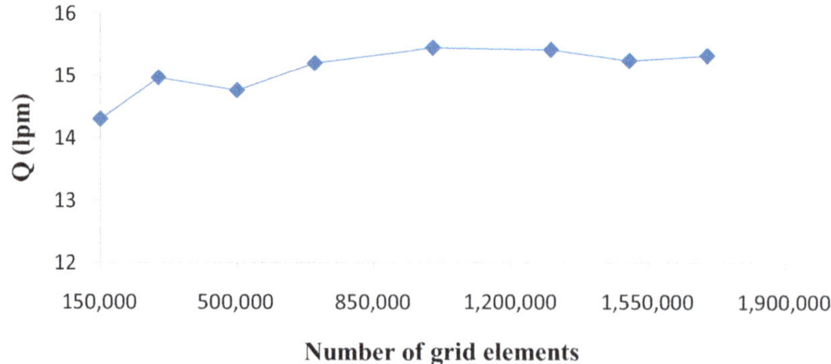

Fig. 3 Grid independence test

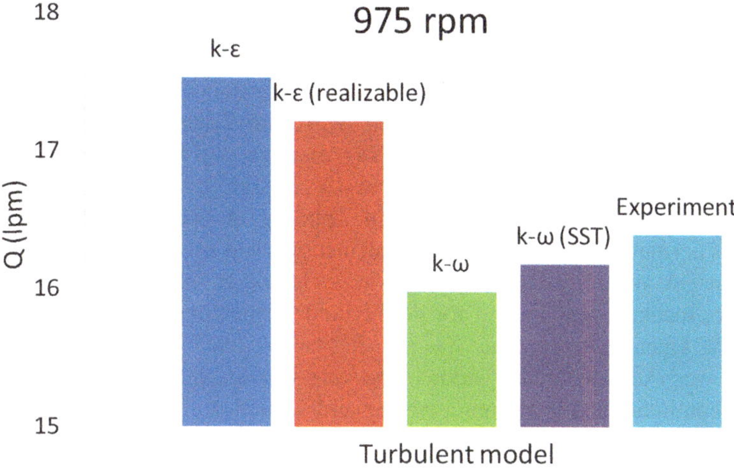

Fig. 4 Turbulence model test

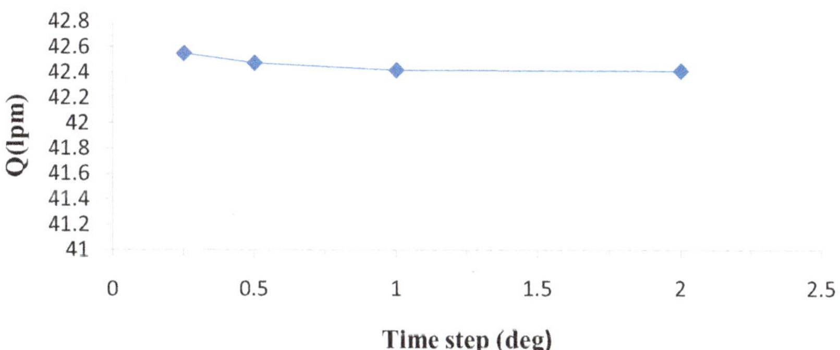

Fig. 5 Time step test

2.6 Boundary Conditions

Boundary conditions were set as follows: 0-gauge total pressure at inlet, set of static pressure starting with 2 bar ending with 8 bar, turbulence intensity 5%, hydraulic diameter 11 mm and oil density, and dynamic viscosity 891 kg/m^3, 0.0163053 kg/ms, respectively. Two rotational speeds were simulated, 1500 and 2500 rpm.

The turbulent intensity is calculated according to the formula [10]:

$$I = 0.16 \ \mathrm{Re}_{d_h}^{-\frac{1}{8}} \tag{1}$$

where I is the turbulent intensity, d_h is the hydraulic diameter, and Re is the Reynolds number.

2.7   Solver Details and Governing Equations

ANSYS Fluent solver 16.0 was used to apply the control volume discretization method on the control volumes formed with grid. Transient, three-dimensional Reynolds-averaged Navier–Stokes (RANS) equations plus two-equation turbulence model is transformed to algebraic equations. A pressure-based solver with pressure–velocity coupling, pressure-implicit with splitting of operators (PISO) scheme, which is normally chosen for transient calculations, was used to solve the governing equations. A spatial discretization data was chosen as follows: second order for pressure, second-order upwind for density and momentum, first-order upwind for turbulent kinetic energy, and specific dissipation rate. Temporal discretization scheme was examined under various time steps which is the time step test. Navier–Stokes continuity and momentum conservation equations for Newtonian fluids are given as follows [11]:

$$\frac{\partial}{\partial t}(\rho) + \nabla \cdot (\rho U) = 0 \tag{2}$$

$$\frac{\partial}{\partial t}(\rho U) + \nabla \cdot (\rho U U) = -\nabla P + \nabla \cdot \left(T + T^{\mathrm{Turb}}\right) \tag{3}$$

where ρ is density, U is velocity vector, P is pressure, μ is dynamic viscosity, and T^{Turb} is turbulent stress tensor. The convective term $(\rho U U)$ represents a differential balance of momentum, on a control volume. The right-hand side of Eq. (3) represents the summation of external forces: pressure and shear stresses.

2.8   SST k – ω Turbulence Model

The shear stress transport (SST) $k - \omega$ turbulence model, which is developed by Menter [12, 13], was selected to perform the numerical simulations. It is a hybrid two-equation model that combines the advantages of both $k - \varepsilon$ and $k - \omega$ models. The SST $k - \omega$ model can be switched to a $k - \varepsilon$ model in the free stream avoiding the common $k - \omega$ problem and its oversensitivity to the inlet free-stream turbulence properties and performs much better than $k - \varepsilon$ model for boundary layer flows.

The eddy viscosity model is used for the Reynolds stress term to close the equation set [6]:

$$R_{ij} = -\rho \overline{u_i' u_j'} = 2\mu_t S_{ij} - \frac{2}{3}\left(\mu_t \frac{\partial u_k}{\partial x_k} + \rho k\right)\delta_{ij} \tag{4}$$

$$S_{ij} = \frac{1}{2}\left(\frac{\partial u_i}{\partial x_j} + \frac{\partial u_j}{\partial x_i}\right)$$

where S_{ij} is the deformation rate tensor, δ_{ij} is the Kronecker operator, k is a turbulent kinetic energy, and μ_t is the turbulent viscosity.

Both the turbulence kinetic energy k and the specific dissipation rate ω are obtained from the following transport equations [6]:

$$\frac{\partial}{\partial_t}(\rho k) + \frac{\partial}{\partial x_i}(\rho u_i k) = \tau_{ij}\frac{\partial U_i}{\partial x_j} - \beta^* k\rho\omega + \frac{\partial}{\partial x_j}\left[\left(\mu + \frac{\mu_t}{\sigma_k}\right)\frac{\partial k}{\partial x_j}\right] \tag{5}$$

$$\frac{\partial}{\partial_t}(\rho\omega) + \frac{\partial}{\partial x_i}(\rho u_i \omega) = \frac{\gamma}{\nu_t}\tau_{ij}\frac{\partial U_i}{\partial x_j} - \beta\rho\omega^2 + \frac{\partial}{\partial x_j}\left[\left(\mu + \frac{\mu_t}{\sigma_\omega}\right)\frac{\partial\omega}{\partial x_j}\right] + $$
$$2\rho(1 - F_1)\sigma_{\omega 2}\frac{1}{\omega}\frac{\partial k}{\partial x_j}\frac{\partial\omega}{\partial x_j} \tag{6}$$

$$\mu_t = \rho\frac{k}{\omega}$$

where σ_k, $\beta*$, γ, β, and F_1 are the empirical constants obtained from theoretical analysis combined with experimental data.

3 Results and Discussion

In the current chapter, the analysis was focused on two aspects: the effect of the gap-height between the vane tip and the outer casing on the flow rate and the effect of a number of rotor vanes models on flow rate and the output power. The results were analyzed at different pressures and rotational speeds of 1500 and 2500 rpm.

3.1 Effect of Gap Height

Numerical simulations were performed for four-vane model to determine the pump characteristic map, in terms of pressure versus flow rate curves at the outlet, three gap heights (0.5, 0.3, and 0.1 mm) for two rotational pump speeds (1500 and 2500 rpm), as shown in Fig. 6.

Figure 6a, b shows the effect of three gap heights on output flow rate, where there is a significant increase between the gaps of 0.5, 0.3, and 0.1 mm. As shown in Fig. 6, it is obvious that the gap height affects the flow rate. Consequently, the more increase in the gap height, the lower outlet flow rate. In addition, the slope of the curves decreases by decreasing the gap height which leads to less leakage in the flow but more friction between the vane tip and the outer casing.

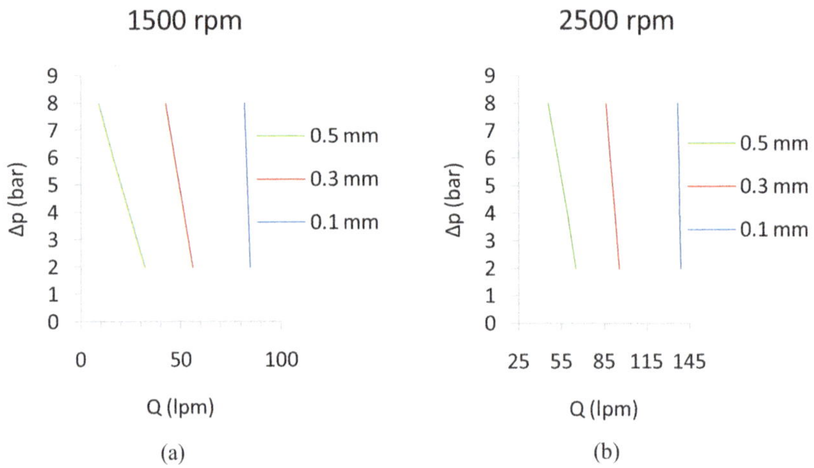

Fig. 6 Pressure vs. flow rate; (**a**) 1500 rpm and (**b**) 2500 rpm

Moreover, the pressure in the pump chambers decreases by increasing the gap height as shown in Fig. 7.

Using the numerical model, the output power has been analyzed too using Eq. (7). Figure 8 presents the numerical results per one revolution obtained at two speeds as a function of the shaft position at eight bars:

$$P_{\text{out}} = \Delta p * Q_{\text{out}} \tag{7}$$

Figure 8 demonstrates the great influence of the gap height on the output power, showing that the ripples and the value of the power decrease by increasing the gap height which affects directly on the total performance.

3.2 Effect of Number of Rotor Vanes

Another factor was taken into consideration in simulations which is the number of vanes. The pump characteristic map, in terms of pressure vs. flow rate curves at the outlet, for three models having four, six, and eight vanes at a gap height of 0.3 mm was modeled and examined for two rotational pump speeds (1500 and 2500 rpm) as shown in Fig. 9.

Figure 9a, b shows the output flow rate at different rotational pump speeds and different vanes number. It is apparent that by increasing the number of vanes, the flow rate increases. In (a) and (b), the effect of the number of vanes is obvious in flow rate increasing between four-, six-, and eight-vane models in the output flow rate. The curve slope in Fig. 9 reduces by increasing the number of vanes, which prevents

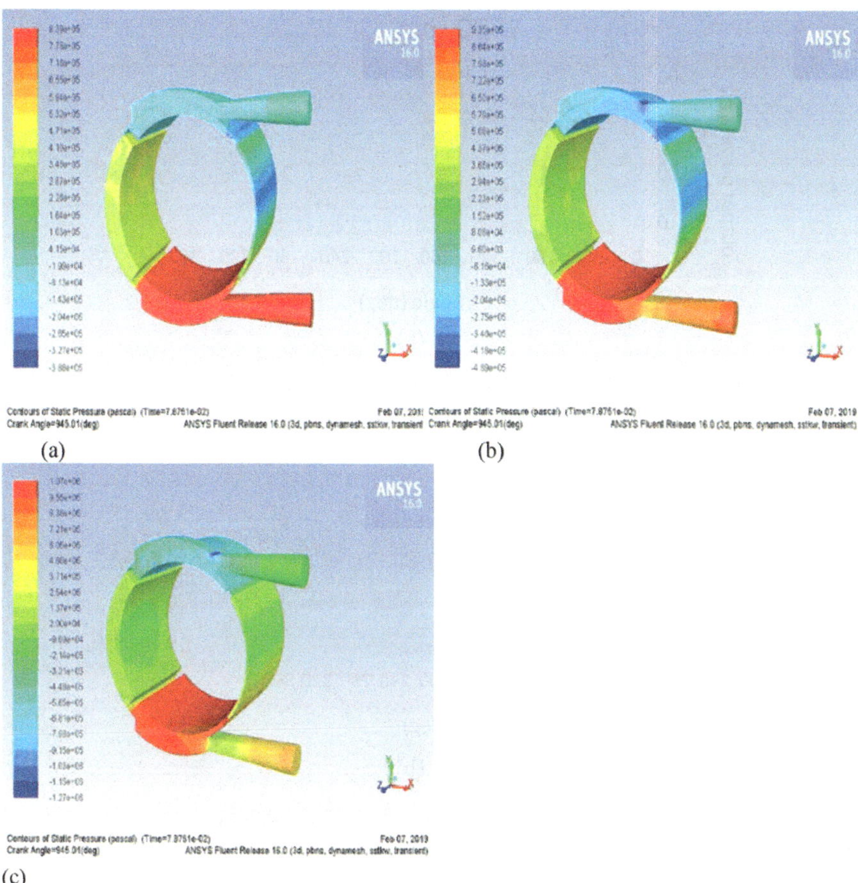

Fig. 7 Pressure in pump chambers at crank angle of 225°, 8-bar outlet, and 2500 rpm (**a**) 0.5 mm gap, (**b**) 0.3 mm gap, and (**c**) 0.1 mm gap

the flow of inner leakage. The pressure contours in the pump chambers in the three models at 2500 rpm and 8 bar also is illustrated in Fig. 10.

4 Conclusion

In this work, a parametric study for an unbalanced vane pump using CFD modeling was performed. The study analyzed the effect of the gap height and its effect on the flow rate and therefore the output power at different rotational pump speeds and pressures by varying three gap heights between the vane tip and the outer casing. The model with gap height of 0.1 mm gave the best performance and flow rate. Moreover, three models with four, six, and eight vanes were modeled. The characteristic

                                                                A. El-Hennawi et al.

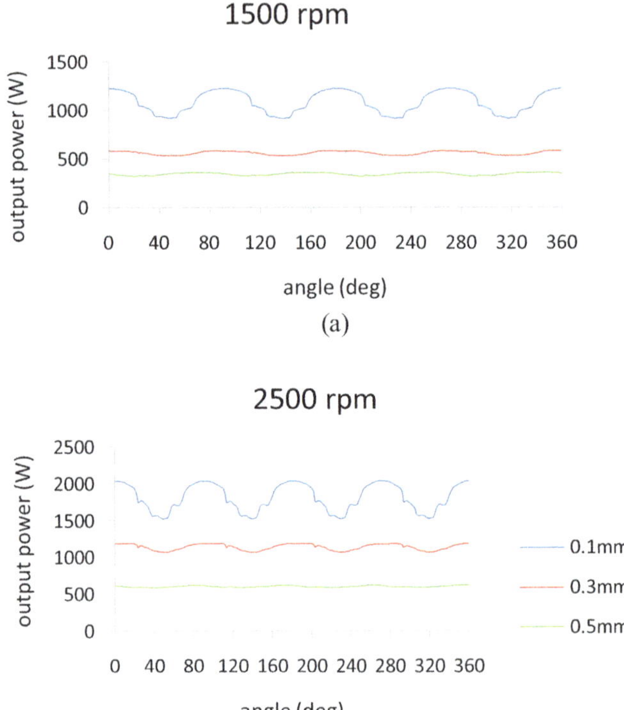

Fig. 8 Output power at different gap heights: (**a**) 1500 rpm and (**b**) 2500 rpm

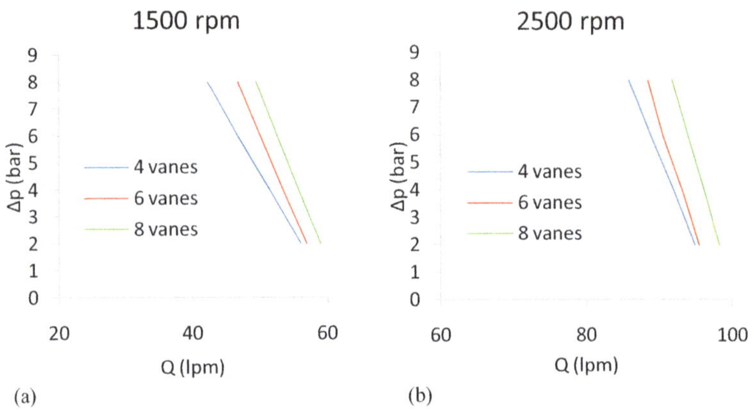

Fig. 9 Pressure vs. flow rate; (**a**) 1500 rpm and (**b**) 2500 rpm

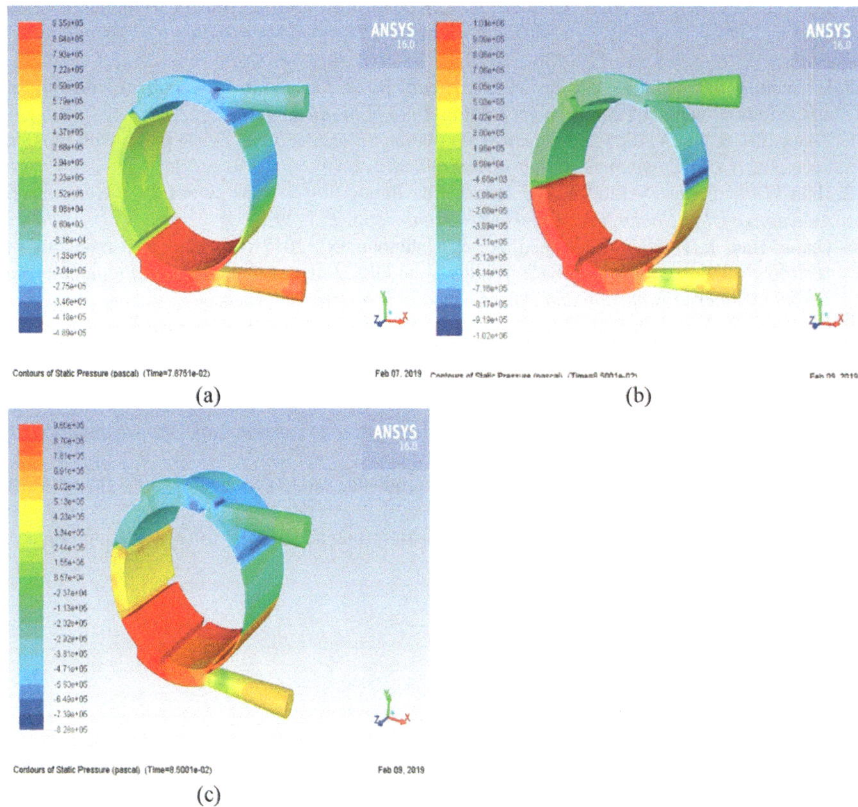

Fig. 10 Pressure in pump chambers at crank angle of 225°, 8-bar outlet, and 2500 rpm (**a**) four-vane model, (**b**) six-vane model, and (**c**) eight-vane model

map was plotted at different speeds and pressures, and the eight-vane model gave the best output flow rate.

References

1. Rancic, S. (2014). Reduction of pressure pulsations on automotive transmission oil vane pump. M.Sc. thesis, Department of Mechanical Engineering, Canada
2. Rana, D., & Kumar, N. (2014). Experimental and computational fluid dynamic analysis of external gear pump. *International Journal of Engineering Development and Research, 2,* 2474–2478.
3. Frosina, E., Senatore, A., Buono, D., Manganelli, M., & Olivitte, M. (2014). A tridimensional CFD analysis of the oil pump of a high performance motorbike engine. In *68th Conference of the Italian Thermal Machines Engineering Association* (pp. 938–948). Milan, Italy: Elsevier.

4. Lessa, L., Kessler, M., Foley, L., Johri, A., & Khondge, A. (2007). Numerical simulation of oil flow and noise in a power steering pump. In: *3rd European Automotive CFD Conference, Frankfurt, Germany*. pp. 129–137.
5. Brusiani, F., Bianchi, G., Costa, M., Squarcini, R., & Gasperini, M. (2009). Evaluation of air/cavitation interaction inside a vane pump. In: *4th, Munich, Germany*.
6. Zhang, Q., & Xu, X. (2014). Numerical simulation on cavitation in a vane pump with moving mesh. In: *ICCM Conference, Cambridge, England*.
7. Hain-Würtenberger, S. (2007). Simulation of cavitating flow in vane pumps. In: *3rd European Automotive CFD Conference, Frankfurt, Germany*. pp. 285–293.
8. Gad-el-Hak, I., Hussin, A., Hamed, A., & Mahmoud, N. (2017). 3D numerical modeling of zeotropic mixtures and pure working fluids in an ORC turbo-expander. *International Journal Turbomachinery Propulsion and Power, 2*(1), 2.
9. El-Hennawi, A., Eltahan, M., Magooda, M., & Moharm, K. (2019). *Numerical and experimental validation of an unbalanced oil vane pump using RANS approach*. Submitted.
10. ANSYS FLUENT 14.5. (2010). *Documentation: User Guide. ANSYS Inc., Technical Report, Nov. 2010*.
11. Versteeg, K., & Malalasekera, W. (2007). *An introduction to computational fluid dynamics: The finite volume method*. Pearson Education.
12. Menter, R. (1993). Zonal two equation $k - \omega$ turbulence models for aerodynamic flows. *AIAA Journal Paper*, 93–2906.
13. Menter, R. (1994). Two-equation Eddy-viscosity turbulence models for engineering applications. *AIAA Journal, 32*(8), 1598–1605.

Part III
Solid-State Electronics

Transport Properties of Ferromagnetic Silicene Superlattice-Based Nanostructure

Ahmed Saeed Abdelrazek Bayoumi and Adel Helmy Phillips

Abstract A ferromagnetic silicene superlattice field effect transistor (FET) study is presented considering the properties of spin transport with induced ac-field at various frequencies and applied magnetic field. The valley-resolved conductance and spin-resolved conductance are deduced using transfer matrix method and Landuaer–Butticker equation. The spin polarization is formulated using spin-resolved conductance of both spin alignments, while valley polarization is formulated using valley-resolved conductance of both K and K' points. The obtained results show sharp resonant peaks with specific widths for both valley- and spin-dependent conductance. These resonant sharp peaks might be due to the effect of applied ac-field and magnetic field. This research might have scientific impact in design and understanding silicene superlattice-based nanodevices and spin filters by tailoring the investigated parameters.

Keywords Spin and valley polarizations · Silicene superlattice · Induced ac-field · Electric field · EuO

1 Introduction

The utilization of electron spin and electron charge in combination with photons causes the emergence of a new field named as spintronics [1–6]. The rise of spintronics improves some old function in addition to the creation of new features in data processing, storage, and transfer in other words in the field of information

A. S. A. Bayoumi (✉)
Faculty of Engineering, Department of Engineering Physics and Mathematics, Kafr-Elsheikh University, Kafr-Elsheikh, Egypt
e-mail: ahmed.bayoumi@eng.kfs.edu.eg

A. H. Phillips
Faculty of Engineering, Department of Engineering Physics and Mathematics, Ain-Shams University, Cairo, Egypt
e-mail: adel_phillips@eng.asu.edu.eg

© Springer Nature Switzerland AG 2020
M. H. Farouk, M. A. Hassanein (eds.), *Recent Advances in Engineering Mathematics and Physics*, https://doi.org/10.1007/978-3-030-39847-7_8

technology (IT). Also, spintronics devices are a promising avenue for quantum computing and quantum communication [7–10].

The rise of graphene [11] increases the research in field of two-dimensional (2D) Dirac materials especially hexagonal lattice structures such as phosphorene [12], transition metal dichalcogenides [13, 14], and silicene [15, 16].

Due to the atomic arrangement of sp^3/sp^2-like hybridized silicene atoms [16, 17], about 0.46 Å distance separates two sublattice atom planes, which enhances spin–orbit coupling (SOC) leading to tunable energy gap spin–valley dependent [16, 18]. Silicene has a strong intrinsic SOC and a buckled structure. A gap at Dirac points could be opened due to strong SOC [16, 19, 20] which results in a correlation between valley and spin degrees of freedom. The control of band gap could be done by normal external electric field because of the buckled structure of silicene [20–23]. Silicene has valley degrees of freedom K and K' which are not equivalent, and silicene is unique in this [16]. K and K' in the first Brillouin zone enable silicene to give an extra degree of freedom with spin and charge for quantum computation and information [24]. Recently, the importance of data processing and encoding make spin- and valleytronics promising candidates because of their degrees of freedom spin, valley, and charge [25–27]. Spin-diffusion time of silicene is ($t_s = 1$ ns at 85 K and 500 ns at 60 K) and spin coherence length is ($\ell_s = 10$, 350, 2000 µm at room temperature) which are longer besides large SOC gap (1.55 meV) [16, 19, 28]. So accordingly, silicene is a good candidate for spintronics applications. In recent years spin–valley polarization of ferromagnetic silicene superlattices has been intensively explored [29–36].

This chapter is dedicated to explore the valley polarization and spin polarization transport behavior of monolayer ferromagnetic silicene superlattice under influence of induced ac-field and applied magnetic field. It is made up of four parts; the introduction was first of all; the theoretical model was presented in the next section and its formalism. The numerical results for single-layer silicene superlattice transport characteristics are presented in Sect. 3. Finally, the argument is given in Sect. 4 with a summary.

2 Theoretical Model

The proposed spintronic FET (Fig. 1) is N superlattice period that consists of $(2N-1)$ monolayer normal silicene strips with $2N$ monolayer ferromagnetic silicene strips interlocking. Ferromagnetic and normal silicene regions have equal lengths d. A SiO_2 substrate with metallic leads for source and drain was used in this investigation.

The spin–valley polarization transport under the influence of magnetic field and induced ac-field is carried out. The oscillating ac-field could introduce photon-assisted conduction channels, which can be modified by the gate voltage to position it in that nanostructure conduction window [37–40]. Zeeman splitting that was introduced because of magnetic field in addition to photon-assisted conduction channels makes the transport distinct for both possibilities of valley and spin

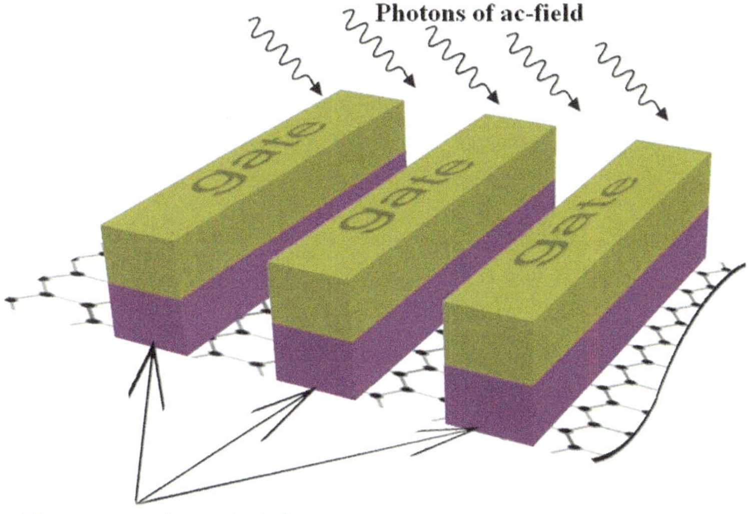

Fig. 1 A schematic diagram of proposed device

alignments and then photon-assisted spin and valley-resolved polarized conduction results.

For the proposed nanostructure, the Dirac equation is

$$H\Psi = E\Psi \tag{1}$$

The Dirac Hamiltonian is

$$H = \hbar v_F \left(k_x \tau_x - \eta k_y \tau_y \right) - (\eta \sigma \lambda_{\mathrm{SO}} - \Delta_z)\tau_z + eV_{\mathrm{ac}} \cos \omega t - \sigma_z h_o + U \tag{2}$$

where the parameter U is

$$U = V_d + eV_{sd} + eV_g + \frac{1}{2} g\mu_B B\sigma_z \tag{3}$$

where v_F is the Fermi velocity, τ_x, τ_y, τ_z are the Pauli matrices in the sublattice pseudospin space, $\eta(\eta') = +1(-1)$ corresponds to the $K(K')$ valley, $\sigma = +1(-1)$ is for spin up and spin down configuration, λ_{SO} is the intrinsic spin–orbit coupling, and Δ_z represents on-site potential difference between A and B sublattices. The on-site potential difference (Δ_z) could be tuned by normal electric field which creates opportunities for tunable energy gap applications; the exchange energy is represented by h_o in ferromagnetic regions; $\hbar$ represents reduced Planck's constant; induced ac-field peak and frequency are $\mathbf{V_{ac}}$ and, ω; barrier height, V_b; bias and gate voltages are V_{sd} and V_g; and Lande g-factor, Bohr magneton, and magnetic field strength are g, μ_B, and B. The following eigenfunctions are obtained by solving Eq. (1).

The left normal silicene eigenfunction is

$$\Psi_{NS}^{(\ell)} = \sum_{n=1}^{\infty}$$
$$\times \left[e^{i(k_x x + k_y y)} \begin{pmatrix} \hbar v_F k_F e^{i\eta\theta} \\ E_{NS} \end{pmatrix} + r_{\eta\sigma} e^{i(-k_x x + k_y y)} \begin{pmatrix} -\hbar v_F k_F e^{-i\eta\theta} \\ E_{NS} \end{pmatrix} \right] J_n \left(\frac{eV_{ac}}{n\hbar\omega} \right) e^{-in\omega t} \tag{4}$$

where k_F is expressed as $k_F = \frac{\sqrt{E_F^2 - \lambda_{SO}^2}}{\hbar v_F}$, E_F is Fermi energy, θ is the incident angle which is expressed as $\theta = \tan^{-1}(k_y/k_x)$, and r is the reflection coefficient. The parameter E_{NS} is given by

$$E_{NS} = E + \eta\sigma\lambda_{SO} - \Delta_z \tag{5}$$

The eigenfunction in the normal silicene in the superlattice region is

$$\Psi_{NSL} = \sum_{n=1}^{\infty}$$
$$\times \left[a_{\eta\sigma} \begin{pmatrix} \hbar v_F k_+ . e^{i\eta\phi} \\ E_{NS} \end{pmatrix} e^{ik_x x} + b_{\eta\sigma} \begin{pmatrix} -\hbar v_F k_- e^{-i\eta\phi} \\ E_{NS} \end{pmatrix} e^{-ik_x x} \right] e^{ik_y y} . J_n \left(\frac{eV_{ac}}{n\hbar\omega} \right) . e^{-in\omega t} \tag{6}$$

where ϕ is the incident angle on the interface between the normal and ferromagnetic silicene where $\phi = \tan^{-1}(k_y/k_x')$.

The eigenfunctions in the ferromagnetic silicene in the superlattice regions are

$$\Psi_{FSL} = \sum_{n=1}^{\infty}$$
$$\times \left[c_{\eta\sigma} \begin{pmatrix} \hbar v_F k_+' . e^{i\eta\phi} \\ E_{FS} \end{pmatrix} . e^{ik_x' x} + d_{\eta\sigma} \begin{pmatrix} -\hbar v_F k_-' . e^{i\eta\phi} \\ E_{FS} \end{pmatrix} . e^{-ik_x' x} \right] e^{ik_y y} . J_n \left(\frac{eV_{ac}}{n\hbar\omega} \right) . e^{-in\omega t} \tag{7}$$

The eigenfunction in the right normal silicene is

$$\Psi_{NS}^{(r)} = \sum_{n=1}^{\infty} t_{\eta\sigma}.e^{i\left(k_x x + k_y y\right)} \cdot \begin{pmatrix} \hbar v_F k_F.e^{i\eta\theta} \\ E_{NS} \end{pmatrix} . J_n\left(\frac{eV_{ac}}{n\hbar\omega}\right).e^{-in\omega t} \tag{8}$$

where $t_{\eta\sigma}$ is the transmission coefficient and the parameters E_{FS}, k_x, k_y, k_x', $k_{\pm}$, $k_{\pm}'$ are given as follows:

$$E_{FS} = E + \eta\sigma\lambda_{SO} - \Delta_z - U + \sigma_z h_o, \tag{9}$$

$$k_x = \frac{\sqrt{E^2 - (\eta\sigma\lambda_{SO} - \Delta_z)^2}}{\hbar v_F} \cdot \cos\theta, \tag{10}$$

$$k_y = \frac{\sqrt{E^2 - (\eta\sigma\lambda_{SO} - \Delta_z)^2}}{\hbar v_F} \cdot \sin\theta, \tag{11}$$

$$k_{\pm} = k_x \pm i\eta k_y \quad \text{and} \quad k_{\pm}' = k_x' \pm i\eta k_y, \tag{12}$$

and

$$k_x' = \frac{\sqrt{(E - U + \sigma_z h_o)^2 - (\eta\sigma\lambda_{SO} - \Delta_z)^2 - (\hbar v_F k_y)^2}}{\hbar v_F}.\cos\phi. \tag{13}$$

and $J_n\left(\frac{eV_{ac}}{n\hbar\omega}\right)$ is the nth order of the first kind of Bessel function [37–40]. In a silicene superlattice nanostructure, the presence of various minibands n has to be considered in the solutions of Eqs. (4), (6)–(8) and combines solutions with $\exp(-in\omega t)$ phase factor [37–40]. Using transfer matrix method, boundary conditions are applied at the interfaces [29–36] where we get the following equation:

$$\begin{pmatrix} t_{\eta\sigma} \\ 0 \end{pmatrix} = [R_1(d)]^{-N}.[R_2(d).T]^N \begin{pmatrix} 1 \\ r_{\eta\sigma} \end{pmatrix} \tag{14}$$

where

$$T = R_1^{-1}(d)M_1^{-1}M_2 R_2(d)M_2^{-1}M_1 \tag{15}$$

and

$$M_1 = \begin{pmatrix} \hbar v_F k_+ & -\hbar v_F k_- \\ E_{NS} & E_{NS} \end{pmatrix} \tag{16}$$

$$M_2 = \begin{pmatrix} \hbar v_F k_+' & -\hbar v_F k_-' \\ E_{FS} & E_{FS} \end{pmatrix} \tag{17}$$

$$R_1 = \begin{pmatrix} e^{ik_x d} & 0 \\ 0 & e^{-ik_x d} \end{pmatrix} \tag{18}$$

$$R_2 = \begin{pmatrix} e^{ik_x' d} & 0 \\ 0 & e^{-ik_x' d} \end{pmatrix} \tag{19}$$

The tunneling probability, $\Gamma_{\text{ac-field}}(E)$, is given:

$$\Gamma_{\text{ac-field}}(E) = \sum_{n=1}^{\infty} J_n^2 \left(\frac{eV_{ac}}{n\hbar\omega} \right) \cdot \left| t_{\eta\sigma} \right|^2 \tag{20}$$

Now, the conductance, $\boldsymbol{G}$, is obtained by Landauer–Buttiker formula as [37, 39–41]

$$G = \frac{e^2}{h} \int_{E_F}^{E_F + n\hbar\omega} \left(\frac{\sqrt{E^2 - (\eta\sigma\lambda_{SO} - \Delta_z)^2}}{\hbar v_F} \right) . W . dE \int_{-\pi/2}^{\pi/2} \Gamma_{\text{ac-field}}(E)$$

$$\cdot \left(-\frac{\partial f_{\text{FD}}}{\partial E} \right) . \cos\theta . d\theta \tag{21}$$

where W represents width of junction in y-direction and $\left(-\frac{\partial f_{\text{FD}}}{\partial E} \right)$ represents first derivative of Fermi–Dirac distribution function which are calculated as follows:

$$\left(-\frac{\partial f_{\text{FD}}}{\partial E} \right) = (4k_B T)^{-1} \cos h^{-2} \left(\frac{E - E_F' + n\hbar\omega}{2k_B T} \right) \tag{22}$$

in which T represents the temperature and k_B represents Boltzmann constant. In the silicene, the conductance, G, is classified into two cases [40–43].

1. Valley-resolved conductance, $G_{K(K')}$, is

$$G_{K(K')} = \frac{G_{K(K')\uparrow\uparrow} + G_{K(K')\uparrow\downarrow}}{2} \tag{23}$$

The indications $\uparrow\uparrow$ and $\uparrow\downarrow$ denote parallel and antiparallel spin alignments.

2. Spin-resolved conductance, $G_{\uparrow\uparrow\,(\uparrow\downarrow)}$, is

$$G_{\uparrow\uparrow(\uparrow\downarrow)} = \frac{G_{K\uparrow\uparrow(\uparrow\downarrow)} + G_{K'\uparrow\uparrow(\uparrow\downarrow)}}{2} \tag{24}$$

The conductance of the Dirac fermion for both cases of spin and valley, respectively, are

$$G_{D1} = G_{\uparrow\uparrow} + G_{\uparrow\downarrow} \tag{25}$$

and

$$G_{D2} = G_K + G_{K'} \tag{26}$$

The spin polarization, SP, is

$$\mathrm{SP} = \frac{\left(G_{K\uparrow\uparrow} + G_{K'\uparrow\uparrow}\right) - \left(G_{K\uparrow\downarrow} + G_{K'\uparrow\downarrow}\right)}{G_{D1}} \tag{27}$$

Also, the valley polarization, VP, is

$$\mathrm{VP} = \frac{\left(G_{K\uparrow\uparrow} + G_{K\uparrow\downarrow}\right) - \left(G_{K'\uparrow\uparrow} + G_{K'\uparrow\downarrow}\right)}{G_{D2}} \tag{28}$$

3 Results and Discussions

The present silicene-based nanostructure model parameters were calculated numerically as follows: the valley-resolved conductance, G_K and $G_{K'}$ (Eq. 23); spin-resolved conductance, $G_{\uparrow\uparrow}$ and $G_{\uparrow\downarrow}$ (Eq. 24); spin polarization, SP (Eq. 27); and valley polarization, VP (Eq. 28). The computations are made by taking $N = 1$, that is, this nanostructure is normal silicene lead/ferromagnetic silicene/normal silicene/ferromagnetic silicene/normal silicene lead. The values of the following parameters are $V_{\mathrm{ac}} = 0.25$ V [37–40], $V_F \approx 5.5 \times 10^5$ m/s, $\lambda_{\mathrm{SO}} = 3.9$ meV, and $g = 2$ [16, 24–36]. The values of barrier height, V_b, was taken as 50 meV and $d = 30$ nm, $W = 20$ nm, and $\Delta_z = 2$ meV [16, 28], since the parameter n' is tuned by proximity effect of the single-layer ferromagnetic silicene. The modulated Fermi energy, E_F' (Eq. 22), can be calculated by the carrier density of quasiparticle Dirac fermions, n', through the following equation [16, 28];

$$E_F' = \sqrt{\pi}\,\hbar V_F \sqrt{n'} \tag{29}$$

Since the parameter n' is tuned by proximity effect of the single-layer ferromagnetic silicene [16, 28, 30–32], then the parameter, n', is calculated using density function theory in order to get the optimum value of E_F'.

The IR range is expected to be the most suitable range to enhance both of the spin polarization, SP, and valley polarization, VP, of the present junction, as will be shown in the figures below.

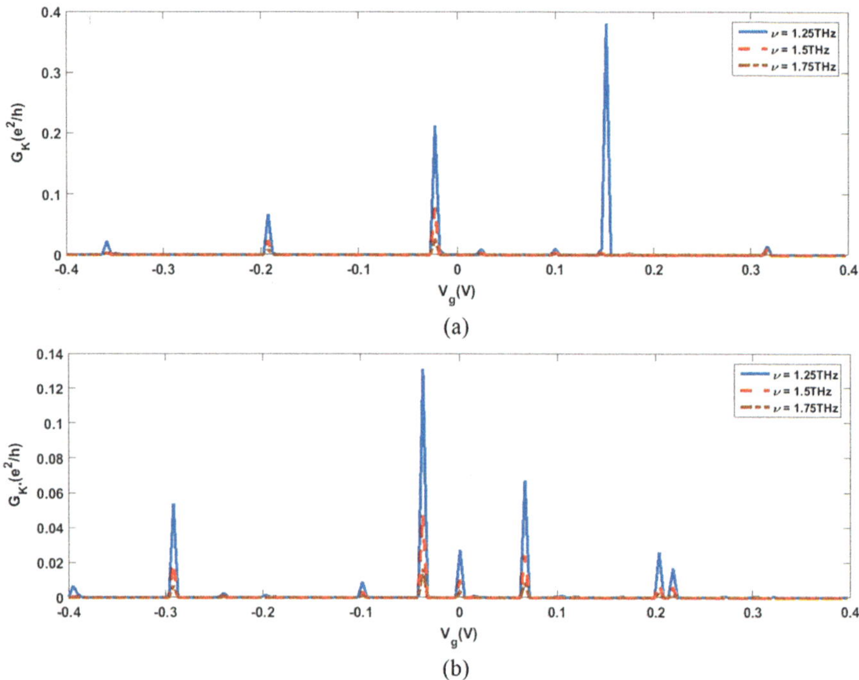

Fig. 2 Valley-resolved conductance at different frequency, ν: (**a**) for G_K and (**b**) for $G_{K'}$

Figure 2a, b explain the behavior of valley-resolved conductance versus, V_g, in case of the two valleys at various values of frequency, ν, of ac-field. It shows a selective manner at certain values of the gate voltage with maximum at 0.1519 V and equals 0.3814 $\left(\frac{e^2}{h}\right)$ for the K valley and a maximum at -0.0368 V and equals 0.1313 $\left(\frac{e^2}{h}\right)$ for the K' valley at 1.25 THz frequency. It is clear also that the increase of frequency, ν, leads to decrease of the maxima for both K and K' valleys.

Figure 3a, b explain the behavior of valley-resolved conductance versus, V_g, in the two valleys at various values of magnetic field, B. It shows a selective manner at certain values of the gate voltage with maximum at 0.1519 V and equals 0.3814 $\left(\frac{e^2}{h}\right)$ for the K valley at 0.1 T and a maximum at -0.2443 V and equals 0.1772 $\left(\frac{e^2}{h}\right)$ and at 0.7 T for the K' valley. It is clear also that the change in magnetic field, B, leads to shift in the positions of the maxima rather than values for both K and K' valleys.

Figure 4a, b explain the behavior of spin-resolved conductance versus, V_g, in the two spin alignments at various frequencies, ν, of ac-field. It shows also a selective manner at certain values of the gate voltage with maximum at 0.1519 V and equals 0.3814 $\left(\frac{e^2}{h}\right)$ for the parallel spin alignment and a maximum at 0.0009 V and equals 0.0276 $\left(\frac{e^2}{h}\right)$ for the antiparallel spin alignment at 1.25 THz frequency. It is clear also that the increase of frequency, ν, leads to decrease of the maxima for both spin alignments.

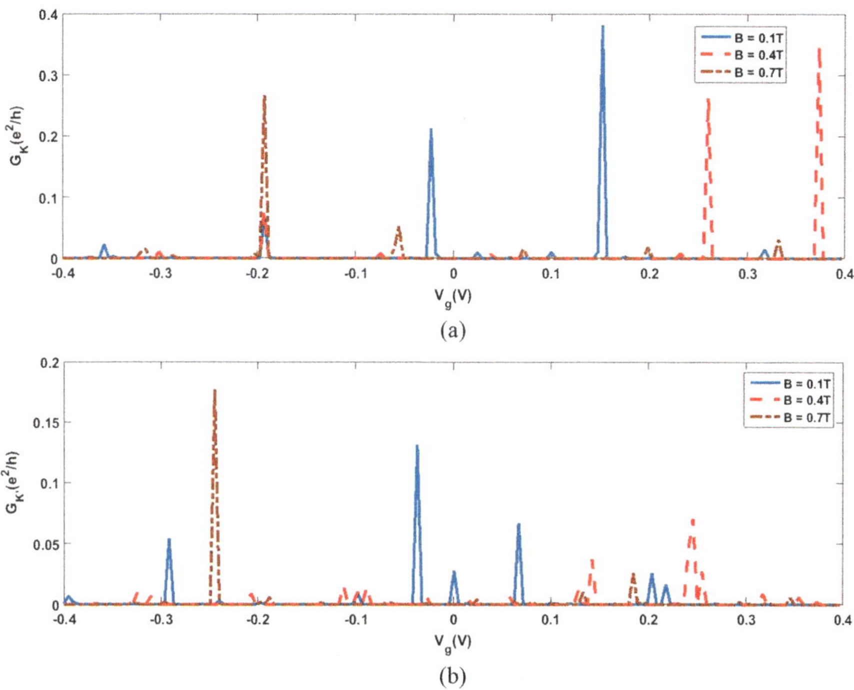

Fig. 3 Valley-resolved conductance at various magnetic field, B: (**a**) for G_K and (**b**) for $G_{K'}$

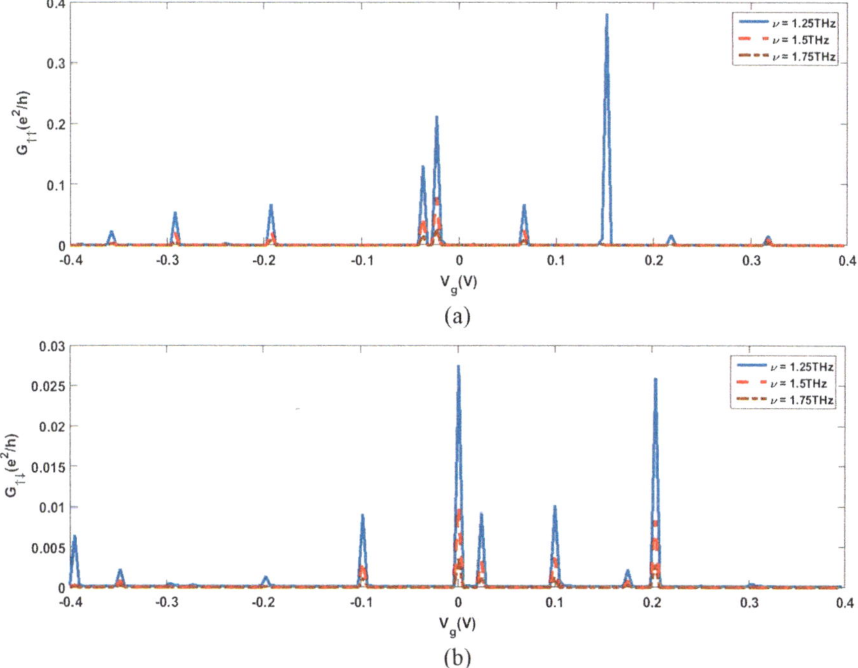

Fig. 4 Spin-resolved conductance at different frequency, ν: (**a**) for $G_{\uparrow\uparrow}$ and (**b**) for $G_{\uparrow\downarrow}$

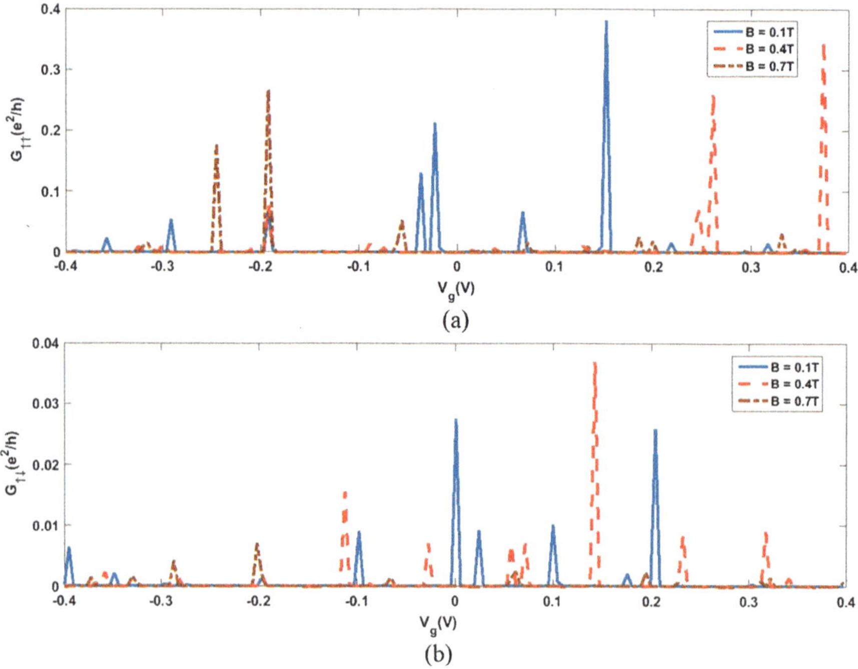

Fig. 5 Spin-resolved conductance at different magnetic field, B: (**a**) for $G_{\uparrow\uparrow}$ and (**b**) for $G_{\uparrow\downarrow}$

Figure 5a, b explain the behavior of spin-resolved conductance versus, V_g, in the two spin alignments at various values of magnetic field, B. It shows also a selective manner at certain values of the gate voltage with maximum at 0.1519 V and equals 0.3814 $(\frac{e^2}{h})$ for the parallel spin alignment at 0.1 T and a maximum at 0.1425 V and equals 0.0371 $(\frac{e^2}{h})$ for the antiparallel spin alignment at 0.4 T magnetic field. It is clear also that the change in magnetic field, B, leads to shift in the positions of the maxima rather than values for both spin alignments.

Figure 6 shows the behavior of the spin polarization with V_g at various values of the ac-field frequency, ν. The spin polarization bounces between 0% and 100% at different values of V_g. It is clear that the frequency of the ac-field does not affect the spin polarization value, although it affects the conductance value which means that the ratio does not depend on frequency.

Figure 7 shows the behavior of the spin polarization with V_g at various values of the magnetic field, B. The spin polarization bounces between 0% and 100% at different values of V_g. It is clear that the magnetic field variations lead to shift in the position of spin polarization maxima rather than values of it.

Figure 8 shows the behavior of the valley polarization with V_g at various values of the ac-field frequency, ν. The spin polarization bounces between 0% and 100% at different values of V_g. It is clear that frequency of ac-field does not affect valley polarization value, although it affects the conductance value which means that the ratio does not depend on frequency.

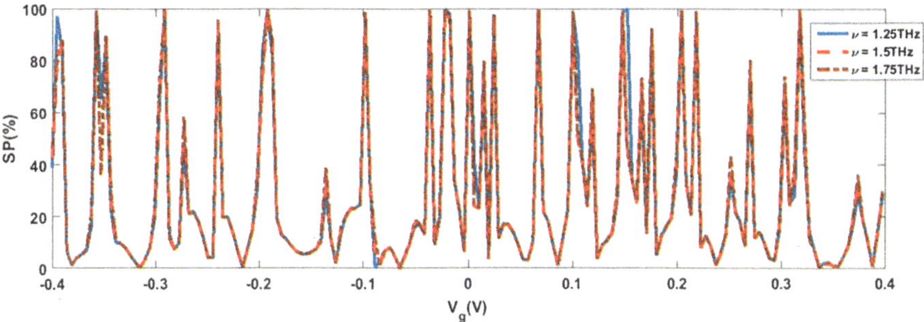

Fig. 6 Spin polarization with Vg at different frequency, ν

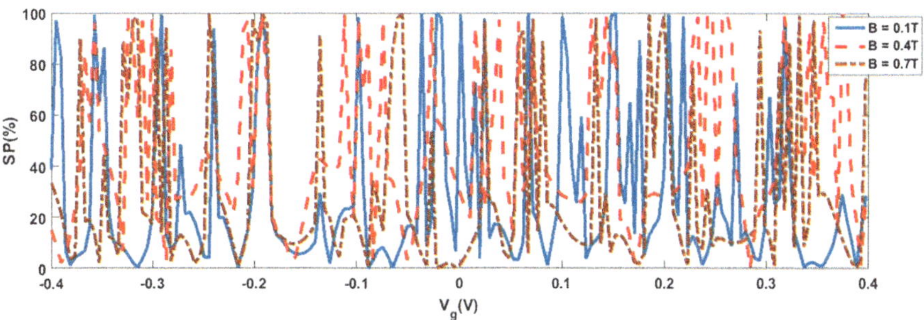

Fig. 7 Spin polarization with V_g at different magnetic field, B

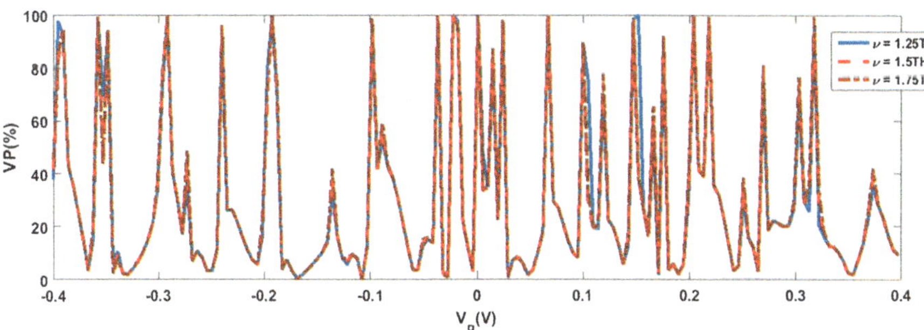

Fig. 8 Valley polarization with V_g at different frequency, ν

Figure 9 shows the behavior of the valley polarization with V_g at various values of magnetic field, B. The valley polarization bounces between 0% and 100% at different value of V_g. It is clear that the magnetic field variations lead to shift in the position of valley polarization maxima rather than values of it.

Figures 2, 3, 4, and 5 show resonant oscillatory trend with resonant peaks of different widths for both valley- and spin-resolved conductances corresponding to both K and K' and different spin alignments. This trend of oscillating behavior of the

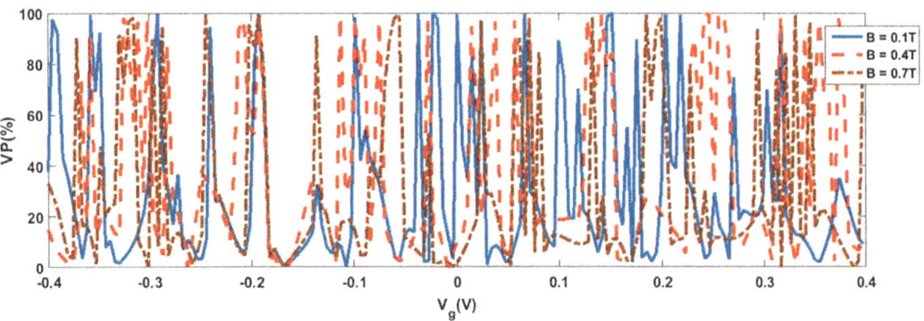

Fig. 9 Valley polarization with V_g at different magnetic field, B

valley- and spin-resolved conductances is because the valley and spin minibands interact with induced photon energy and split up according to the spin and valley alignments when Dirac fermions enter into the ferromagnetic silicene region. These effects take place alongside as ferromagnetic region potential modulates the Fermi energy with applied magnetic field and energy of induced photons [37–40]. Also, the peak height and its widths of such oscillations might be controlled by electric field and magnetic proximity effect of ferromagnetic insulator EuO [16, 44–47]. It is noticed that magnitudes of valley and spin conductances for K and K' points and both spin alignments show that flipping spin is expected to deeply affect manipulating and encoding digital quantum information processing and valleytronics applications. The proposed structure might serve as spin filters, and consequently silicene superlattice is a good candidate as nanomaterial in spin filtering [28–36]. Also results explain that values of both valley polarization, VP, and spin polarization, SP (Figs. 6, 7, 8, and 9) are enhanced due to the parameters studied in the present chapter [16, 28–36], and a perfect valley polarization could be obtained because of its sensitivity to electric field even if exchange field is greater than staggered electric field. The present results for valley and spin dependent of the present investigated ferromagnetic silicene superlattice-based nanostructure are found in agreement with those in the literature [16, 28–36].

4   Conclusion

Spin- and valley-dependent transport behavior of ferromagnetic silicene superlattice nanostructure is explored under magnetic field effect and induced ac-fields with range of frequencies. The valley- and spin-resolved conductances are deduced using transfer matrix method and Landuaer–Butticker formula. Results show oscillatory resonant trend with peaks of certain widths for both valley- and spin-resolved conductances verifying mechanism of spin filtering and valleytronics. An oscillatory characteristic resulted for both valley and spin polarizations as varying the gate voltage.

This work is very significant in various nanodevices for spin logic applications based on ferromagnetic silicene superlattice, magnetic sensors, and quantum information. The explored silicene superlattice nanostructure for spintronic and valleytronics nanodevices could be experimentally realized in the field of nanotechnology.

References

1. Ziese, M., & Thornton, M. (Eds.). (2001). *Spin electronics*. Berlin: Springer.
2. Maekawa, S., & Shinjo, T. (2002). *Spin dependent transport in magnetic nanostructures*. London: Taylor and Francis.
3. Zuti'c, I., Fabian, S., & Das Sarma, J. (2004). Spintronics: Fundamentals and applications. *Reviews of Modern Physics, 76*, 323–410.
4. Chappert, C., Fert, A., & van Dau, F. (2007). The emergence of spin electronics in data storage. *Nature Materials, 6*, 813.
5. Fert, A. (2008). The present and the future of spintronics. *Thin Solid Films, 517*, 2–5.
6. Xu, Y. D., Awschalom, J., & Nitta, D. (2015). *Handbook of Spintronics*. Berlin, Heidelberg: Springer.
7. Endoh, T. (2015). Nonvolatile logic and memory devices based on Spintronics. In: *Proceeding of IEEE International Symposium Circ. Syst.*, pp. 13–16.
8. Sverdlov, V., Weinbub, J., & Selberherr, S. (2017). Spintronics as a non-volatile complement to modern microelectronics. *Journal of Microelectronics, Electronic Components and Materials, 47*(4), 195–210.
9. Joshi, V. (2016). Spintronics: A contemporary review of emerging electronics devices. *Engineering Science and Technology, an International Journal, 19*, 1503–1513.
10. Gautam, A. (2012). Spintronics-A new hope for the digital world. *International Journal of Scientific and Research Publications, 2*(8), 1–5.
11. Novoselov, K., Geim, A., Morozov, S., Jiang, D., Zhang, Y., Dubonos, S., Grigorieva, I., & Firsov, A. (2004). Electric field effect in atomically thin carbon films. *Science, 306*, 666–669.
12. Li, L., Yu, Y., Ye, G., Ge, Q., Ou, X., Wu, H., Feng, D., Chen, X., & Zhang, Y. (2014). Black phosphorus field-effect transistors. *Nature Nanotechnology, 9*, 372.
13. Xiao, D., Liu, G., Feng, W., Xu, X., & Yao, W. (2012). Coupled spin and valley physics in monolayers of MoS2 and other group-VI dichalcogenides. *Physical Review Letters, 108*, 196802.
14. Fuhrer, M., & Hone, J. (2013). Measurement of mobility in dual-gated MoS_2 transistors. *Nature Nanotechnology, 8*, 146–147.
15. Kara, A., Enriquez, H., Seitsonen, A., Lew Yan Voon, L., Vizzini, S., Aufray, B., & Oughaddoub, H. (2012). A review on silicene—New candidate for electronics. *Surface Science Reports*, 671–618.
16. Zhao, J., Liu, H., Yu, Z., Quhe, R., Zhou, S., Wang, Y., Liu, C., Zhong, H., Han, N., Lu, J., Yao, Y., & Wu, K. (2016). Rise of silicene: A competitive 2D material. *Progress in Materials Science, 83*, 24–151.
17. Chen, L., Li, H., Feng, B., Ding, Z., Qiu, J., Cheng, P., Wu, K., & Meng, S. (2013). Spontaneous symmetry breaking and dynamic phase transition in monolayer Silicene. *Physical Review Letters, 110*, 085504.
18. Liu, F., Liu, C., Wu, K., Yang, F., & Yao, Y. (2013). D + id' chiral superconductivity in bilayer silicene. *Physical Review Letters, 111*(6), 066804.
19. Liu, C., Feng, W., & Yao, Y. (2011). Quantum spin hall effect in silicene and two-dimensional germanium. *Physical Review Letters, 107*, 076802.

20. Drummond, N., Zólyomi, V., & Fal'ko, V. (2012). Electrically tunable band gap in silicene. *Physical Review B, 85,* 075423.
21. Ezawa, M. (2012). A topological insulator and helical zero mode in silicene under an inhomogeneous electric field. *New Journal of Physics, 14,* 033003.
22. Ezawa, M. (2012). Valley-polarized metals and quantum anomalous hall effect in silicene. *Physical Review Letters, 109,* 055502.
23. Spencer, M., & Morishita, T. (Eds.). (2016). *Silicene structure, properties and applications.* Cham, Switzerland: Springer International Publishing.
24. Rohling, N., Russ, M., & Burkard, G. (2014). Hybrid spin and valley quantum computing with singlet-triplet qubits. *Physical Review Letters, 113,* 176801.
25. Tahir, M., & Schwingenschlögl, U. (2013). Valley polarized quantum hall effect and topological insulator phase transitions in silicene. *Scientific Reports, 3,* 1075.
26. Chen, L., Liu, C., Feng, B., He, X., Cheng, P., Ding, Z., Meng, S., Yao, Y., & Wu, K. (2012). Evidence for Dirac fermions in a honeycomb lattice based on silicon. *Physical Review Letters, 109,* 056804.
27. Missault, N., Vasilopoulos, P., Vargiamidis, V., Peeters, F., & Van Duppen, B. (2015). Spin and valley dependent transport through arrays of ferromagnetic silicene junctions. *Physical Review B, 92,* 195423.
28. Yang-Yang, W., Ru-Ge, Q., Da-Peng, Y., & Jin, L. (2015). Silicene Spintronics—A concise review. *Chinese Physics B, 24*(8), 087201.
29. Missault, N., Vasilopoulos, P., Peeters, F., & Van Duppen, B. (2016). Spin- and valley-dependent miniband structure and transport in silicene superlattices. *Physical Review B, 93,* 125425–125426.
30. Zhang, Y., Sun, J., & Guo, Y. (2018). Spin-valley decoupling in magnetic silicene superlattices. *Journal of Physics D: Applied Physics, 51,* 045303.
31. Rashidian, Z., Lorestaniweissa, Z., Hajatic, Y., Rezaeipourb, S., & Rashedi, G. (2017). Valley polarized current and Fano factor in a ferromagnetic/normal/ferromagnetic silicene superlattice junction. *Journal of Magnetism and Magnetic Materials, 442,* 15–24.
32. Rashidian, Z., Hajati, Y., Rezaeipour, S., & Baher, S. (2017). Controllable spin and valley polarized current through a superlattice of normal/ferromagnetic/normal silicene junction. *Physica E Low Dimensional Systems and Nanostructures, 86,* 111–116.
33. Li, W., Lua, W., Li, Y., & Han, H. (2017). Defect enhanced spin and valley polarizations in silicene superlattices. *Physica E Low Dimensional Systems and Nanostructures, 88,* 284–288.
34. Lorestaniweiss, Z., & Rashidian, Z. (2017). Fano factor for Dirac electrons in a superlattice of normal/ferromagnetic/normal silicene junction. *Superlattices and Microstructures, 106,* 197–205.
35. Lu, W., Li, Y., & Tian, H. (2018). Spin- and valley-dependent electronic structure in Silicene under periodic potentials. *Nanoscale Research Letters, 13,* 84.
36. Niu, Z., Zhang, Y., & Dong, S. (2015). Enhanced valley-resolved thermoelectric transport in a magnetic silicene superlattice. *New Journal of Physics, 17,* 073026.
37. Asham, M., Zien, W., & Phillips, A. (2012). Photo-induced spin dynamics in Nanoelectronic devices. *Chinese Physics Letters, 29*(10), 108502.
38. Platero, G., & Aguado, R. (2004). Photon assisted transport in semiconductor nanostructures. *Physics Reports, 395,* 1–157.
39. Abdelrazek, A., Elbanna, M., & Phillips, A. (2016). Photon-spin coherent manipulation of piezotronic nanodevice. *Micro & Nano Letters, 11*(12), 876–880.
40. Ahmed, I., Asham, M., & Phillips, A. (2018). Spin-valleytronics of silicene based nanodevices (SBNs). *Journal of Magnetism and Magnetic Materials, 456,* 199–203.
41. Buttiker, M. (1986). Four-terminal phase-coherent conductance. *Physical Review Letters, 57,* 1761.
42. Datta, S. (1995). *Electronic transport in Mesoscopic systems.* New York: Cambridge University Press.

43. Wu, X., & Meng, H. (2015). Gate–tunable valley spin filtering in Silicene with magnetic barrier. *Journal of Applied Physics, 117,* 203903.
44. Yokoyama, T. (2013). Controllable valley and spin transport in ferromagnetic silicene junctions. *Physical Review B, 87,* 241409.
45. Yokoyama, T. (2014). Spin and valley transports in junctions of Dirac fermions. *New Journal of Physics, 16,* 085005.
46. Ahmed, I., Asham, M., & Phillips, A. (2017). Coherent spin-valley polarization characteristics of silicene field effect transistor. *Journal of Multidisciplinary Engineering Science and Technology, 4*(2), 6701–6708.
47. Abdelrazek, A., & Phillips, A. (2019). Spin coherent transport in graphene superlattice based nanostructure. *Journal of Multidisciplinary Engineering Science and Technology, 6*(6), 1190–1196.

3D Analytical Modeling of Potential, Drain Current, and Threshold Characteristics for Long-Channel Square Gate-All-Around (SGAA) MOSFETs

Hamdy Abdelhamid, Azza M. Anis, Mohamed E. Aboulwafa, and Mohamed I. Eladawy

Abstract Gate-all-around (GAA)-based field effect transistors (FETs) are considered to be one of the dominant structures that overcome the performance degradation problems that face complementary metal-oxide semiconductor (CMOS) technology in the nanometer scale. This chapter presents a three-dimensional (3D) analytical model for electrostatic potential in the channel of the lightly doped n-channel square GAA MOSFETs. The model is based on the solution of the 3D Poisson's equation with mobile carriers. Based on the developed potential model and the current continuity equation, models for the drain current, the transconductance, and the output conductance are presented. Additionally, the threshold voltage and the short-channel characteristics such as threshold voltage roll-off (roll-off), drain-induced barrier lowering (DIBL), and subthreshold swing (SS) are also analyzed at different biasing values and device parameters. The results of the proposed models are compared with those obtained by COMSOL 3D simulations, and the results show reasonable agreement.

Keywords GAA MOSFETs · Modeling · 3D electrostatic potential · Drain current · Short-channel effects

H. Abdelhamid
Center of Nano-Electronics and Devices (CND), Zewail City of Science and Technology, 6th October City, Egypt

Faculty of Engineering, Electrical Engineering Department, Ajman University, Ajman, United Arab Emirates

A. M. Anis (✉) · M. E. Aboulwafa · M. I. Eladawy
Faculty of Engineering, Electronics, Communications, and Computers Engineering Department, Helwan University, Helwan, Egypt
e-mail: azza_anis@h-eng.helwan.edu.eg

© Springer Nature Switzerland AG 2020
M. H. Farouk, M. A. Hassanein (eds.), *Recent Advances in Engineering Mathematics and Physics*, https://doi.org/10.1007/978-3-030-39847-7_9

1 Introduction

According to the predictions of the International Roadmap for Devices and Systems (IRDS), the gate-all-around (GAA) field effect transistor (FET) is a promising structure for future technology nodes beyond FinFETs [1]. GAA FETs offer excellent electrostatic control of the channel through the fully surrounding gate that recommends them to be an alternative of FinFETs over the next generations [2–5]. GAA FETs have high immunity to short-channel effects, ideal subthreshold slope, low leakage current, and high on/off current ratio over other multigate devices [6–9].

Several numerical simulations have been carried out in quantum level, such as [10–12] for square, [13–15] for circular, [16, 17] for hexagonal, and [18] for elliptical nanowire FETs for digital applications. In the level of drift-diffusion transport, two-dimensional (2D) surface potential, threshold voltage, drain current, transconductance, charges, and capacitances models have been proposed in [19–23] for cylindrical GAA FETs.

The solution of the nonlinear 3D Poisson's equation is sometimes difficult when incorporating mobile carriers [24, 25]. Therefore, few analytical models have been suggested for long-channel square/rectangular GAA FETs. In [26, 27], potential, subthreshold current, and subthreshold swing have been developed based on the fixed-charges term in Poisson's equation. Another model has been reported in [28], but the charge carrier term has been neglected in the threshold voltage calculations.

In this chapter, we develop a complete solution for the nonlinear 3D Poisson's equation in lightly doped long-channel square GAA metal-oxide semiconductor (MOS) FETs. Based on that potential model and the current continuity equation, analytical modeling of drain current, transconductance, and drain conductance are presented. Additionally, the subthreshold parameters have been developed.

These models are based on the drift-diffusion transport mechanism and valid in subthreshold, linear, and saturation operation modes at different device parameters. The other operating conditions (ballistic transport and the breakdown modes) and physical phenomena such as tunneling and quantum confinement effects are not included in the analysis.

The proposed models can be used in SPICE-based simulators to implement square/rectangular GAA circuits for analog/radio-frequency applications. These models can predict the physical behavior of the device and the effect of the device parameters on the final design.

This chapter is organized as follows: in Sect. 2, 3D potential model is developed for the lightly doped n-type SGAA MOSFETs including mobile charges effect. Section 3 presents models for charge-based drift-diffusion drain-to-source current, transconductance, and output conductance. Threshold voltage and short-channel effects such as threshold voltage roll-off, drain-induced barrier lowering, and subthreshold swing are analyzed in Sect. 4. The results of the proposed models are compared with the 3D numerical simulations in Sect. 5. Section 6 outlines the conclusions of this work.

2 Channel Potential Modeling

The 3D Poisson's equation that describes the channel potential $\varphi_c(x, y, z)$ of the lightly doped n-channel SGAA MOSFET given in Fig. 1, including the mobile-charges, reads

$$\frac{\partial^2 \varphi_c(x, y, z)}{\partial x^2} + \frac{\partial^2 \varphi_c(x, y, z)}{\partial y^2} + \frac{\partial^2 \varphi_c(x, y, z)}{\partial z^2} = \frac{q n_i^2}{\varepsilon_c N_c} e^{(\varphi_c(x,y,z) - V)/\varphi_t} \tag{1}$$

where V is the electron quasi-Fermi potential, φ_t is the thermal-potential, n_i is the intrinsic concentration, and ε_c and N_c are the channel permittivity and the doping concentration, respectively.

The boundary conditions for $\varphi_c(x, y, z)$ are given as follows:

$$\frac{\pm \varepsilon_{ox}}{\varepsilon_c T_{ox}} \left(V_{gs} - \varphi_{ms} - \varphi_c(x, y, z)\big|_{x=\mp\frac{W}{2}} \right) = \frac{\partial \varphi_c(x, y, z)}{\partial x}\bigg|_{x=\mp\frac{W}{2}} \tag{2}$$

$$\frac{\pm \varepsilon_{ox}}{\varepsilon_c T_{ox}} \left(V_{gs} - \varphi_{ms} - \varphi_c(x, y, z)\big|_{y=\mp\frac{H}{2}} \right) = \frac{\partial \varphi_c(x, y, z)}{\partial y}\bigg|_{y=\mp\frac{H}{2}} \tag{3}$$

$$\varphi_c(x, y, z)\big|_{z=0} = \varphi_{bi} \tag{4}$$

$$\varphi_c(x, y, z)\big|_{z=L} = \varphi_{bi} + V_{ds} \tag{5}$$

V_{gs} and V_{ds} are the gate-to-source and the drain-to-source applied voltages; φ_{bi} is the built-in potential between source/drain and channel; φ_{ms} is the difference between gate and channel work functions; W, L, and H are width, length, and height of channel region; and ε_{ox} and T_{ox} are the permittivity and the thickness of oxide layer.

The channel potential $\varphi_c(x, y, z)$ can be split into two components as shown in Fig. 1:

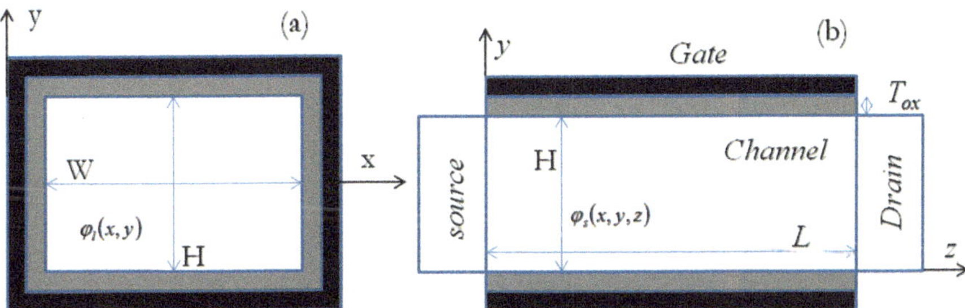

Fig. 1 Cross-section of SGAA MOSFET: (**a**) x–y plane and (**b**) y–z plane, where T_{ox} is oxide thickness and L, H, W, are length, height, and width of channel region, respectively

$$\varphi_c(x, y, z) = \varphi_l(x, y) + \varphi_s(x, y, z) \tag{6}$$

$\varphi_l(x, y)$ is the long-channel potential in the x-y plane and assumed to be [29–32]:

$$\varphi_l(x, y) = \varphi_0(x) + \varphi_1(x)y + \varphi_2(x)y^2 \tag{7}$$

Expressions for $\varphi_1(x)$ and $\varphi_2(x)$ can be calculated from

$$\frac{\pm \varepsilon_{ox}}{\varepsilon_c T_{ox}} \left(V_{gs} - \varphi_{ms} - \varphi_l(x, y)|_{y=\mp\frac{H}{2}} \right) = \left.\frac{\partial \varphi_l(x, y)}{\partial y}\right|_{y=\mp\frac{H}{2}} \tag{8}$$

Gives

$$\varphi_1(x) = 0 \tag{9}$$

$$\varphi_2(x) = \frac{V_{gs} - \varphi_{ms} - \varphi_0(x)}{1 - \frac{4\varepsilon_c T_{ox}}{\varepsilon_{ox} H}} \tag{10}$$

$\varphi_0(x)$ can be evaluated by double integration of Poisson equation along x-direction:

$$\frac{d^2 \varphi_0(x)}{dx^2} = \frac{q n_i^2}{\varepsilon_c N_c} e^{(\varphi_0(x)-V)/\varphi_t} \tag{11}$$

Subjected to the boundary conditions

$$\frac{\pm \varepsilon_{ox}}{\varepsilon_c T_{ox}} \left(V_{gs} - \varphi_{ms} - \varphi_0(x)|_{x=\mp\frac{W}{2}} \right) = \left.\frac{d\varphi_0(x)}{dx}\right|_{x=\mp\frac{W}{2}} \tag{12}$$

$$\left.\frac{d\varphi_0(x)}{dx}\right|_{x=0} = 0 \tag{13}$$

After some mathematical manipulations [33], we obtain

$$\varphi_0(x) = \varphi_t \ln \left(\frac{8 N_c \varepsilon_c \varphi_t}{q n_i^2 W^2} \omega^2 \sec^2(\omega x) \right) + V \tag{14}$$

ω is calculated from Eq. (12) by

$$\frac{V_{gs} - V_0 - V}{2\varphi_t} - \frac{2\varepsilon_c T_{ox}}{\varepsilon_{ox} W} \omega \tan(\omega) = \ln \left(\frac{\omega}{\cos(\omega)} \right) \tag{15}$$

$$V_0 = \varphi_{ms} + \varphi_t \ln \left(\frac{8N_c \varepsilon_c \varphi_t}{q n_i^2 W^2} \right) \tag{16}$$

The short-channel potential component $\varphi_s(x, y, z)$ is obtained by solving

$$\nabla^2 \varphi_s(x, y, z) = 0 \tag{17}$$

The solution for $\varphi_s(x, y, z)$ can take the form

$$\varphi_s(x, y, z) = \cos\left(\beta_x x\right)\left[Ae^{\beta_z(z-L)} + Be^{-\beta_z z}\right]\left[C\cos\left(\beta_y y\right) + D\sin\left(\beta_y y\right)\right] \tag{18}$$

The constants A, B, C, and D can be determined using the boundary conditions given by Eqs. (4) and (5). Finally, the channel potential $\varphi_c(x, y, z)$ can be written as

$$\varphi_c(x, y, z) = \varphi_0(x) + \varphi_2(x)y^2 + \varphi \cdot \varphi_3(x, y, z) + \varphi_4(x, y, z) \tag{19}$$

$$\varphi_3(x, y, z) = -\cos\left(\beta_x x\right)\cos\left(\beta_y y\right)\cos\left(0.5\beta_x\right)\frac{D_0(z)}{K_1} \tag{20}$$

$$\varphi_4(x, y, z) = \cos\left(\beta_x x\right)\cos\left(\beta_y y\right)\left[\varphi_{bi}D_0(z) + V_{ds}D_1(z)\right]\frac{K_0}{K_1} \tag{21}$$

$$\varphi = 4K_2\varphi_0(0.5) + 8K_2K_3\varphi_2(0.5) - \varphi_t \tag{22}$$

$$D_0(z) = \frac{\sinh\left(\beta_z z\right) - \sinh\left(\beta_z(z - L)\right)}{\sinh\left(\beta_z L\right)} \tag{23}$$

$$D_1(z) = \frac{\sinh\left(\beta_z z\right)}{\sinh\left(\beta_z L\right)} \tag{24}$$

Eigenvalues $(\beta_x, \beta_y, \beta_z)$ and the intermediate parameters $(K_0 : K_3)$ are given in the Appendix.

3   Drain Current Modeling

Using the drift-diffusion transport mechanism, the drain-to-source current for long-channel SGAA MOSFETs is [23, 34, 35]

$$I_{ds} = \frac{2(H + W)\mu}{L} \int_0^{V_{ds}} Q dV \tag{25}$$

μ is the electron mobility.

From Gauss' law, the inversion charge density (Q) in the body of the SGAA MOSFET is represented by

$$\frac{\varepsilon_{ox}}{T_{ox}} \int_{y=-\frac{H}{2}}^{y=\frac{H}{2}} \left(V_{gs} - \varphi_{ms} - \varphi_l(x,y)|_{x=\frac{W}{2}}\right) dy = Q = \varepsilon_c \int_{y=-\frac{H}{2}}^{y=\frac{H}{2}} E_x dy \qquad (26)$$

$$E_x = \left.\frac{\partial \varphi_l(x,y)}{\partial x}\right|_{x=\frac{W}{2}} \qquad (27)$$

Using the right-hand side in Eq. (26), Q can be simplified into

$$Q = \varepsilon_c H (1 - \Delta) \left.\frac{d\varphi_0(x)}{dx}\right|_{x=\frac{W}{2}} \qquad (28)$$

$$\Delta = \left[\frac{3}{4} - \frac{3\varepsilon_c T_{ox}}{\varepsilon_{ox} H}\right]^{-1} \qquad (29)$$

Differentiating Eq. (14) and using the obtained results in Eq. (28) yield

$$Q = Q_0 \omega \tan(\omega) \qquad (30)$$

From the left-hand side in Eq. (26), we obtain

$$Q = C'_{ox} H \left[V_{gs} - \varphi_t \ln\left(\frac{Q}{Q_0} + \left(\frac{Q}{Q_0}\right)^2\right) - V - V_0\right] \qquad (31)$$

where $Q_0 = 4\varphi_t \varepsilon_c H (1 - \Delta)/W$ and $C'_{ox} = \varepsilon_{ox}(1 - \Delta)/T_{ox}$.

Differentiate Eq. (31) to get dV, then substitute into Eq. (25). Integrating the obtained result from Q_s up to Q_d results in the charge-based drain current, I_{ds}, as

$$I_{ds} = \frac{2(H + W)\mu}{L} \left[2\varphi_t(Q_s - Q_d) + \frac{Q_s^2 - Q_d^2}{2C'_{ox}H} + \varphi_t Q_0 \ln\left(\frac{Q_d + Q_0}{Q_s + Q_0}\right)\right] \qquad (32)$$

Here $Q_s = Q(V = 0)$ and $Q_d = Q(V = V_{ds})$.

Differentiating I_{ds} formula given by Eq. (32) with respect to V_{gs} and V_{ds} results in the transconductance (g_m) and the drain-to-source conductance (g_{ds}), respectively [36, 37]:

$$g_m = \left.\frac{\partial I_{ds}}{\partial V_{gs}}\right|_{V_{ds}} = \frac{2(H + W)\mu}{L}(Q_s - Q_d) \qquad (33)$$

$$g_{ds} = \left. \frac{\partial I_{ds}}{\partial V_{ds}} \right|_{V_{gs}} = \frac{2(H + W)\mu}{L} Q_d \tag{34}$$

4 Threshold Voltage Modeling

Threshold voltage has been defined as the gate voltage at which the inversion charge density at a point called "virtual cathode" equals a critical value (Q_c) sufficient for the turn-on condition of the device [25, 38, 39].

The virtual cathode point along the channel (z_{min}) can be obtained by differentiating the channel potential given by Eq. (19) with respect to z and equating the results to zero, i.e., $\partial \varphi_c(x, y, z)/\partial z|_{z=z_{min}} = 0$, yields

$$z_{min} = \frac{L}{2} - \frac{1}{2\beta_z} \ln\left(\frac{K_4}{K_5}\right) \tag{35}$$

$$K_4 = \frac{K_0 \left[\varphi_{bi}\left(1 - e^{-\beta_z L}\right) + V_{ds} \right] - \varphi \cos\left(0.5\beta_z\right)\left(1 - e^{-\beta_z L}\right)}{K_1 \left(1 - e^{-2\beta_z L}\right)} \tag{36}$$

$$K_5 = \frac{K_0}{K_1} \left[\varphi_{bi} - \frac{\left[\varphi_{bi}\left(1 - e^{-\beta_z L}\right) + V_{ds} \right]}{2 \sinh\left(\beta_z L\right)} \right] - \frac{\varphi \cos\left(0.5\beta_z\right)}{K_1} \left[1 - \frac{\left(1 - e^{-\beta_z L}\right)}{2 \sinh\left(\beta_z L\right)} \right] \tag{37}$$

The total charge density at virtual cathode, Q_{vc}, will be represented by [40–42]

$$Q_{vc} = \frac{qn_i^2}{N_c} \int_{y=\frac{-H}{2}}^{y=\frac{H}{2}} \int_{x=\frac{-W}{2}}^{x=\frac{W}{2}} e^{\frac{\varphi_{vc}\left(x_c, y_c, z_{min}\right)}{\varphi_t}} dxdy \tag{38}$$

By replacing z by z_{min} in the potential model, Eq. (19), the potential at the virtual cathode point, $\varphi_{vc}(x_c, y_c, z_{min})$, will become

$$\varphi_{vc}\left(x_c, y_c, z_{min}\right) = \varphi_l\left(x_c, y_c\right) + \varphi \cdot \varphi_3\left(x_c, y_c, z_{min}\right) + \varphi_4\left(x_c, y_c, z_{min}\right) \tag{39}$$

where x_c and y_c are the position of the conduction path in the x and y directions.

Substituting Eq. (39) into Eq. (38), Q_{vc} will be written as

$$Q_{vc} = \rho \cdot e^{\left(\varphi_l\left(x_c, y_c\right) + \varphi \cdot \varphi_3\left(x_c, y_c, z_{min}\right) + \varphi_4\left(x_c, y_c, z_{min}\right)\right)/\varphi_t} \tag{40}$$

with $\rho = qn_i^2 WH/N_c$.

By equating Eq. (40) to Q_c, the threshold voltage is written as

$$V_{th} = \frac{\varphi_t \ln (Q_c/\rho) + \varphi_t \varphi_3(x_c, y_c, z_{min}) - \varphi_4(x_c, y_c, z_{min})}{1 + 4K_2 \varphi_3(x_c, y_c, z_{min})} \tag{41}$$

By subtracting the long-channel threshold voltage, $V_{th,\,long} = \varphi_t \ln (Q_c/\rho)$, from the obtained V_{th} model, Eq. (41), the threshold voltage roll-off (Roll − Off) can be calculated by

$$(\text{Roll} - \text{Off}) = \frac{\varphi_t \varphi_3(x_c, y_c, z_{min})[1 - 4K_2 \ln (Q_c/\rho)] - \varphi_4(x_c, y_c, z_{min})}{1 + 4K_2 \varphi_3(x_c, y_c, z_{min})} \tag{42}$$

The drain-induced barrier lowering (DIBL) of the proposed SGAA MOSFET can be determined from the difference between the threshold voltage ($V_{th,\,high}$) at high drain-to-source voltage ($V_{ds,\,high}$) and the threshold voltage ($V_{th,\,low}$) at low drain-to-source voltage ($V_{ds,\,low}$) as [43–45]

$$\text{DIBL} = \frac{V_{th,high} - V_{th,low}}{V_{ds,low} - V_{ds,high}} \tag{43}$$

The subthreshold swing (SS) is defined as the change in the gate voltage required to increase the output current by one decade and can be expressed by [46–49]

$$SS = \varphi_t \ln (10) \left[\frac{\partial \varphi_{vc}(x_c, y_c, z_{min})}{\partial V_{gs}} \right]^{-1} \tag{44}$$

By substituting Eq. (39) into Eq. (44), SS can be written as

$$SS = \frac{\varphi_t \ln (10)}{1 + 4K_2 \varphi_3(x_c, y_c, z_{min})} \tag{45}$$

5 Results

In our analysis, we assumed silicon (Si) as a channel material, and the gate insulator is silicon dioxide (SiO_2). The parameters of the considered SGAA MOSFET are listed in Table 1. The structure of SGAA MOSFET is modeled using the 3D finite element simulations performed in COMSOL Multiphysics [50]. The 3D Poisson's equation, diffusion, and convection for electrons have been simulated self-consistently using the "general-form partial differential equation" application mode.

Table 1 Parameters of SGAA MOSFET

Parameters	Symbols	Values
Channel length	L	50 nm
Channel width	W	10 nm
Channel height	H	10 nm
Oxide thickness	T_{ox}	1.5 nm
Relative permittivity of silicon	ε_c	1.045×10^{-10} F/m
Relative permittivity of oxide	$\varepsilon_{\mathrm{ox}}$	3.453×10^{-11} F/m
Intrinsic concentration of silicon	n_i	1.5×10^{10} cm^{-3}
Channel doping concentration	N_c	10^{14} cm^{-3}
Source/drain doping concentration	N_d	10^{18} cm^{-3}
Gate-to-source voltage	V_{gs}	0.2 V : 1 V
Drain-to-source voltage	V_{ds}	0.1 V : 1 V
Built-in potential	φ_{bi}	0.7 V
Thermal potential	φ_t	26 mV
Metal work function of gate	φ_m	4.8 eV
Electron affinity of silicon	χ	4.05 eV
Energy gap of silicon	E_g	1.12 eV

The 3D numerical simulations indicate that the inversion charge density at threshold condition, $Q_c \approx 4.5 \times 10^{10}$ WHm^{-1}, and the location of the conduction path at $x_c = W/4$ and $y_c = H/4$.

Figure 2 shows the results of the potential model derived in Eq. (19) along channel length at $x = W/2$, $y = H/2$ and different values of V_{ds} and V_{gs}. The analytical results of the potential model agree well with the 3D numerical results obtained from COMSOL without using any fitting parameters.

The transfer ($I_{\mathrm{ds}} - V_{\mathrm{gs}}$) and the output ($I_{\mathrm{ds}} - V_{\mathrm{ds}}$) characteristics of SGAA MOSFET calculated using the model given by Eq. (32) are plotted in Figs. 3 and 4, respectively. The drain current is valid in subthreshold, linear, and saturation regions of device operation for different values of applied voltages. The results of the developed current model are compared with the 3D numerical simulation results and show reasonable agreement.

Figures 5 and 6 display the transconductance (g_m) and the drain conductance (g_{ds}) using Eq. (33) and Eq. (34) as a function of gate voltage (V_{gs}) and drain voltage (V_{ds}), respectively. The proposed models are verified against 3D COMSOL simulations. The figures show an excellent matching between the developed models and the numerical simulation results.

Figure 7 presents the variation of threshold voltage model derived in Eq. (41) with channel length for different values of V_{ds}. As shown in Fig. 7, the threshold voltage of the proposed SGAA MOSFET is constant for channel length greater than 40 nm, and variation is observed for devices of smaller lengths. The proposed model is highly consistent with the 3D numerical results.

The threshold voltage roll-off (Roll − Off), the drain-induced barrier lowering (DIBL), and the subthreshold swing (SS) are indicated versus the length of the

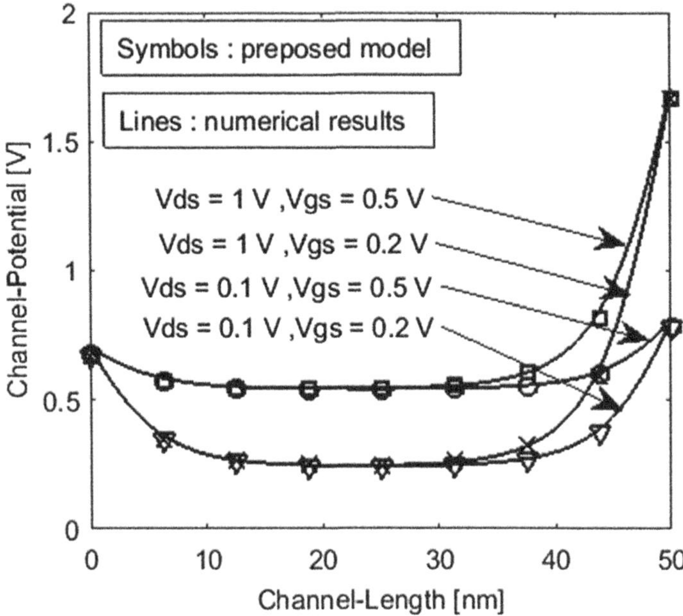

Fig. 2 Channel potential of SGAA MOSFET as a function of channel length with applied drain-to-source voltage (V_{ds}) of 0.1 and 1 V and gate-to-source voltage (V_{gs}) of 0.2 and 0.5 V

channel of the proposed SGAA MOSFET in Figs. 8, 9, and 10. As seen, the calculated models of the short-channel effects agree very well with COMSOL results.

6   Conclusion

An analytical model for electrostatic potential along the channel of the lightly doped square gate-all-around MOSFETs is developed from the solution of the 3D Poisson's equation with mobile carriers. Based on the obtained potential model and the drift-diffusion mechanism, the transfer and the output characteristics are presented. The transconductance and the drain conductance are also modeled. Threshold voltage and short-channel effects such as threshold voltage roll-off, drain-induced barrier lowering, and subthreshold swing have been found at different biasing values and device parameters. An excellent agreement with 3D numerical simulations has been observed.

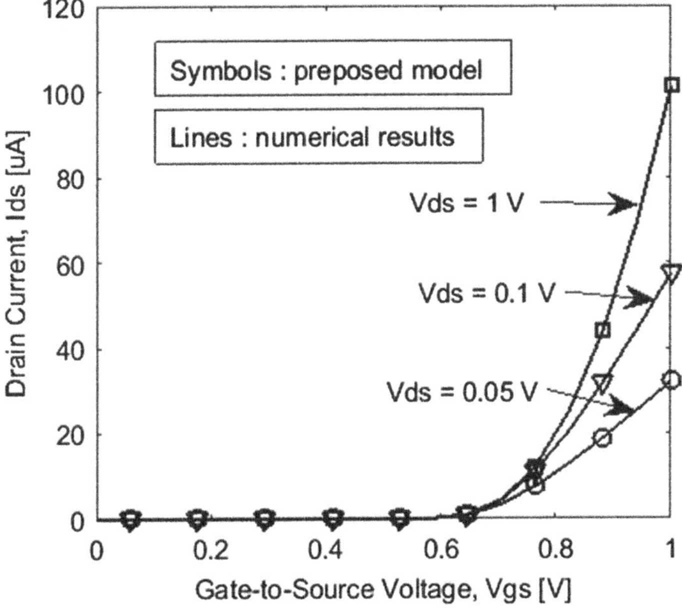

Fig. 3 Drain current (I_{ds}) of SGAA MOSFET as a function of the gate-to-source voltage (V_{gs}) for different values of drain-to-source voltage (V_{ds})

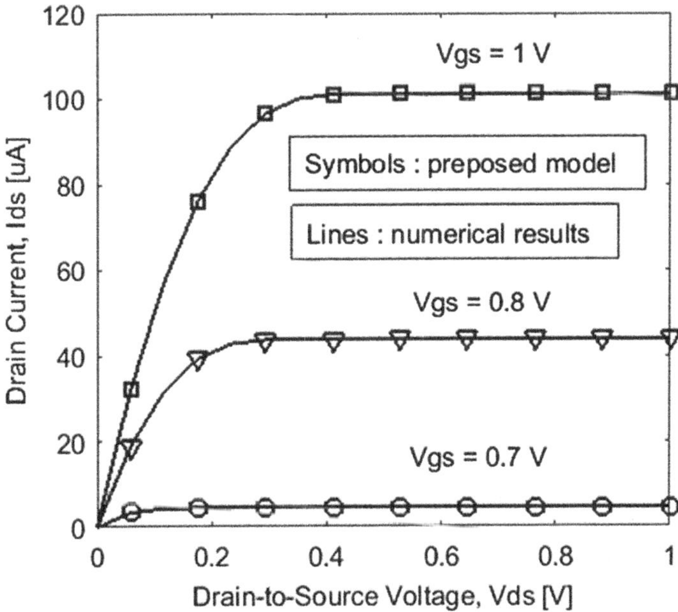

Fig. 4 Drain current (I_{ds}) of SGAA MOSFET as a function of drain-to-source voltage (V_{ds}) for different values of gate-to-source voltage (V_{gs})

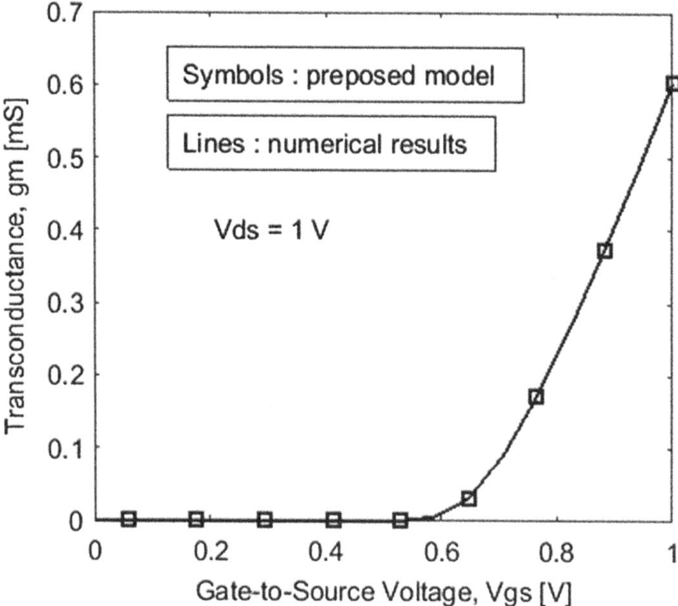

Fig. 5 Transconductance (g_m) as a function of gate-to-source voltage (V_{gs})

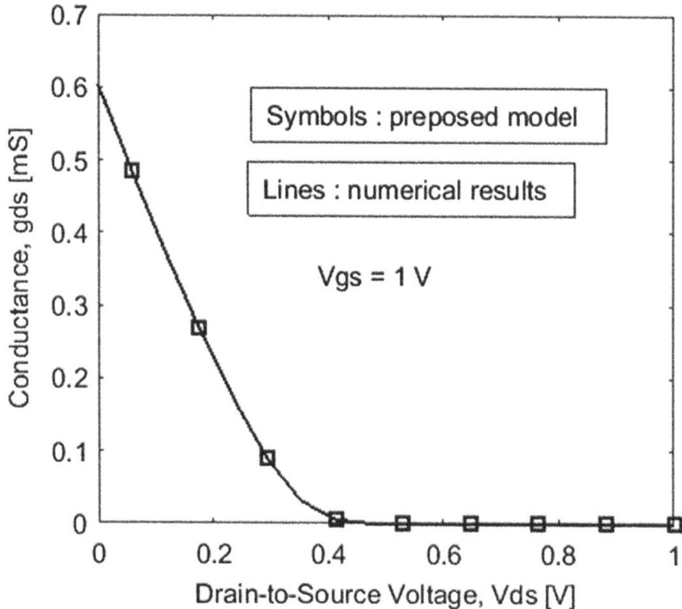

Fig. 6 Drain conductance (g_{ds}) as a function of drain-to-source voltage (V_{ds})

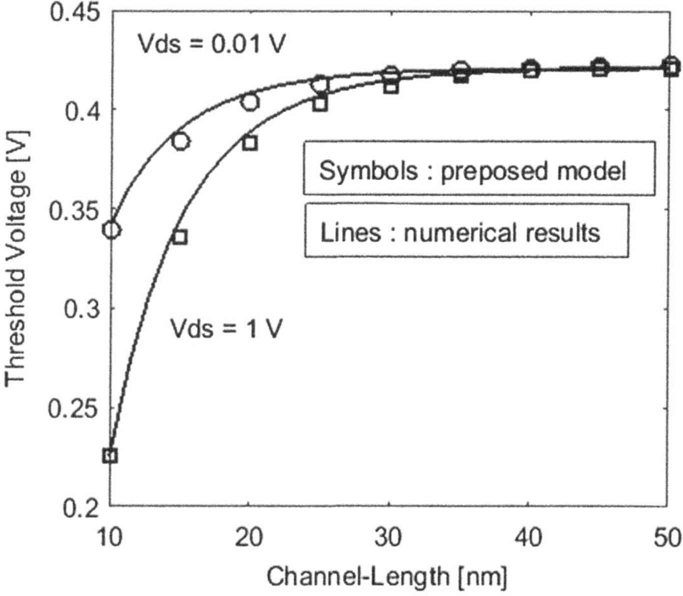

Fig. 7 Threshold voltage (V_{th}) as a function of channel length at different drain-to-source voltage (V_{ds})

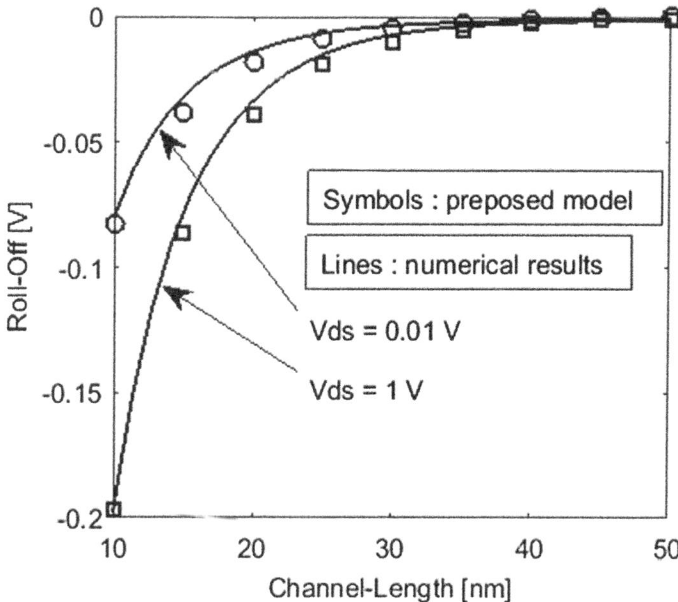

Fig. 8 Roll off as a function of channel length at different drain-to-source voltage (V_{ds})

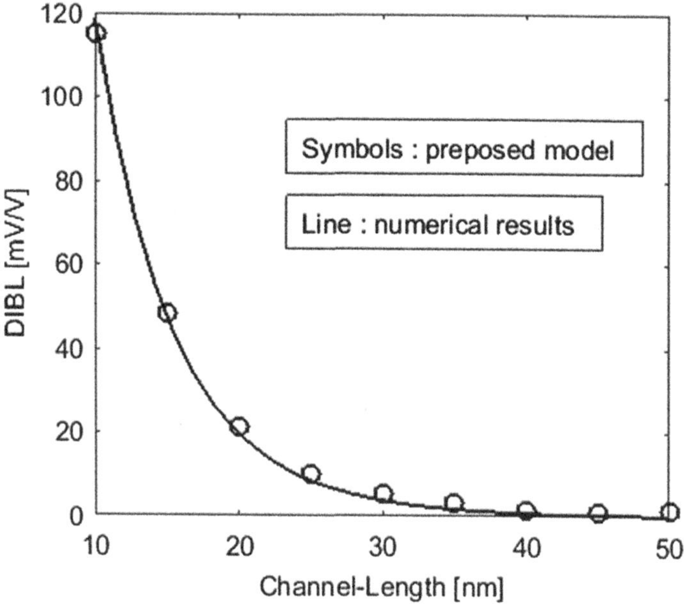

Fig. 9 DIBL variation versus channel length

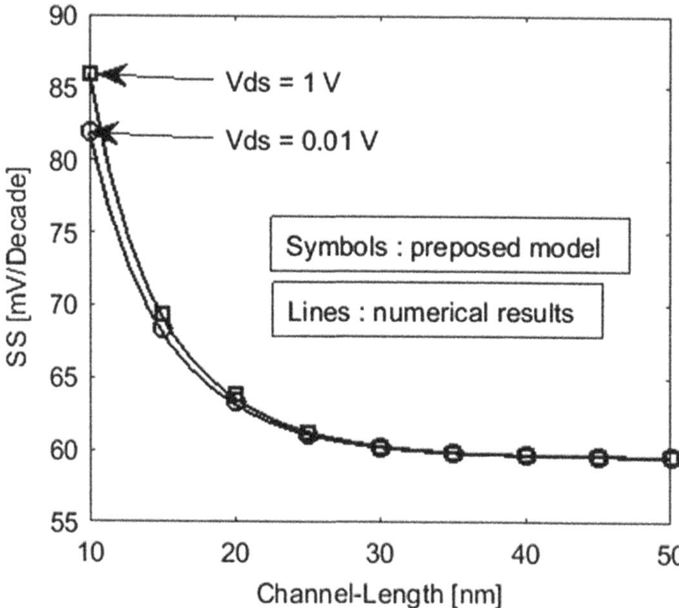

Fig. 10 Subthreshold swing (SS) as a function of channel length for different values of drain-to-source voltage (V_{ds})

Appendix

The eigenvalues, β_x, β_y, and β_z, can be found from

$$\frac{\varepsilon_{ox}W}{2\varepsilon_c T_{ox}} + \beta_x \tan(\beta_x) = 0 \tag{46}$$

$$\frac{\varepsilon_{ox}H}{2\varepsilon_c T_{ox}} + \beta_y \tan(\beta_y) = 0 \tag{47}$$

$$\beta_z = \sqrt{\left(\frac{2\beta_x}{W}\right)^2 + \left(\frac{2\beta_y}{H}\right)^2} \tag{48}$$

Expressions on the text can be calculated by

$$K_0 = \frac{4\sin(\beta_x)\sin(\beta_y)}{\beta_x \beta_y} \tag{49}$$

$$K_1 = \frac{\sin(2\beta_y) + 2\beta_y}{2\beta_y} \tag{50}$$

$$K_2 = \frac{\sin(\beta_y)}{\beta_y} \tag{51}$$

$$K_3 = \frac{1}{2} - \frac{1}{\beta_y^2} + \frac{\cos(\beta_y)}{\beta_y \sin(\beta_y)} \tag{52}$$

$$\varphi_t = KT/q \tag{53}$$

$$\varphi_{bi} = \varphi_t \ln\left(\frac{N_c N_d}{n_i^2}\right) \tag{54}$$

$$\varphi_{ms} = \varphi_m - \left(\chi + \frac{E_g}{2} + \varphi_t \ln\left(\frac{N_c}{n_i}\right)\right) \tag{55}$$

K is the Boltzmann constant, T is the temperature, q is the electron charge, N_d is the donor concentration in the source/drain region, φ_m is the gate work function, and χ and E_g are the electron-affinity and the energy gap of the channel material, respectively.

References

1. IRDS™. (2017). Edition. Retrieved May 23, 2019, from http://irds.ieee.org/editions/2017.
2. Kim, S.-D., Guillorn, M., Lauer, I., Oldiges, P., Hook, T., & Na, M.-H. (2015). Performance trade-offs in FinFET and gate-all-around device architectures for 7 nm-node and beyond. In: *2015 IEEE SOI-3D-Subthreshold Microelectronics Technology Unified Conference (S3S), Rohnert Park, CA, USA.* pp. 1–3.
3. Collaert, N. (2016). *Device architectures for the 5nm technology node and beyond.* Taiwan: Semicon.
4. Asenov, A., Wang, Y., Cheng, B., Wang, X., Asenov, P., Al-Ameri, T., & Georgiev, V.P. (2016). Nanowire transistor solutions for 5nm and beyond. In: *17th International Symposium on Quality Electronic Design (ISQED), Santa Clara, CA, USA.* pp. 269–274.
5. Nagy, D., Indalecio, G., García-Loureiro, A. J., Elmessary, M. A., Kalna, K., & Seoane, N. (2018). FinFET versus gate-all-around nanowire FET: Performance, scaling, and variability. *Journal of the Electron Devices Society, 6,* 332–340.
6. Jena, B., Pradhan, K. P., Sahu, P. K., Dash, S., Mishra, G. P., & Mohapatra, S. K. (2015). Investigation on cylindrical gate all around (GAA) to nanowire MOSFET for circuit application. *Electronics and Energetics, 28*(4), 637–643.
7. Panda, S.R., Sharma, R., Pradhan, K.P., & Sahu, P.K. (2016). Junctionless GAA nanowire transistor: Towards circuit application. In: 3rd International Conference on Emerging Electronics (ICEE), Mumbai, India. pp. 1–4.
8. Doornbos, G., Holland, M., Vellianitis, G., Dal, M. J. H. V., Duriez, B., Oxland, R., Afzalian, A., Chen, T.-K., Hsieh, G., Passlack, M., & Yeo, Y.-C. (2016). High performance InAs gate-all-around nanowire MOSFETs on 300 mm Si substrates. *Journal of the Electron Devices Society, 4*(5), 253–259.
9. Sahay, S., & Kumar, M. J. (2016). A novel gate-stack-engineered nanowire FET for scaling to the sub-10-nm regime. *IEEE Transactions on Electron Devices, 63*(12), 5055–5059.
10. Shin, M., Lee, S., & Klimeck, G. (2010). Computational study on the performance of Si nanowire PMOSFETs based on the k.p method. *IEEE Transactions on Electron Devices, 57* (9), 2274–2283.
11. Bayani, A. H., Voves, J., & Dideban, D. (2018). Effective mass approximation versus full atomistic model to calculate the output characteristics of a gate-all-around germanium nanowire field effect transistor (GAA-GeNW-FET). *Superlattices and Microstructures, 113,* 769–776.
12. Sadi, T., Medina-Bailon, C., Nedjalkov, M., Lee, J., Badami, O., Berrada, S., Carrillo-Nuñez, H., Georgiev, V., Selberherr, S., & Asenov, A. (2019). Simulation of the impact of ionized impurity scattering on the total mobility in Si nanowire transistors. *Materials, 12*(1), 124–135.
13. Padmanaban, B., Ramesh, R., Nirmal, D., & Sathiyamoorthy, S. (2015). Numerical modeling of triple material gate stack gate all-around (TMGSGAA) MOSFET considering quantum mechanical effects. *Superlattices and Microstructures, 82,* 40–54.
14. Bayani, A. H., Dideban, D., Voves, J., & Moezi, N. (2017). Investigation of sub-10nm cylindrical surrounding gate germanium nanowire field effect transistor with different cross-section areas. *Superlattices and Microstructures, 105,* 110–116.
15. Abadi, R. M. I., & Saremi, M. (2018). A resonant tunneling nanowire field effect transistor with physical contractions: A negative differential resistance device for low power very large scale integration applications. *Journal of Electronic Materials, 47*(2), 1091–1098.
16. Degtyarev, V. E., Khazanova, S. V., & Demarina, N. V. (2017). Features of electron gas in InAs nanowires imposed by interplay between nanowire geometry, doping and surface states. *Scientific Reports, 7*(1), 3411.
17. Wojcik, P., Bertoni, A., & Goldoni, G. (2018). Tuning Rashba spin-orbit coupling in homogeneous semiconductor nanowires. *Physical Review B, 97,* 165401.
18. Holovatsky, V. A., & Gutsul, V. I. (2007). Electron energy spectrum and wave functions in complicated elliptic quantum wires. *Journal of Optoelectronics and Advanced Materials, 9*(5), 1437–1441.

19. Pradhan, K. P., Kumar, M. R., Mohapatra, S. K., & Sahu, P. K. (2015). Analytical modeling of threshold voltage for cylindrical gate all around (CGAA) MOSFET using center potential. *Ain Shams Engineering Journal, 6*(4), 1171–1177.
20. Verma, J. H. K., Pratap, Y., Haldar, S., Gupta, R. S., & Gupta, M. (2015). Capacitance modeling of gate material engineered cylindrical/surrounded gate MOSFETs for sensor applications. *Superlattices and Microstructures, 88*, 271–280.
21. Kumar, S., Kumari, A., & Das, M. K. (2016). Modeling gate-all-around Si/SiGe MOSFETs and circuits for digital applications. *Journal of Computational Electronics, 16*(1), 47–60.
22. Pratap, Y., Kumar, M., Kabra, S., Haldar, S., Gupta, R. S., & Gupta, M. (2017). Analytical modeling of gate-all-around junctionless transistor based biosensor for detection of neutral biomolecule species. *Journal of Computational Electronics, 17*(1), 1–9.
23. Gaidhane, A. D., Pahwa, G., Verma, A., & Chauhan, Y. S. (2018). Compact modeling of drain current, charges, and capacitances in long-channel gate-all-around negative capacitance MFIS transistor. *IEEE Transactions on Electron Devices, 65*(5), 2024–2032.
24. Sharma, D., & Vishvakarma, S. K. (2012). Analytical modeling for 3D potential distribution of rectangular gate (RecG) gate-all-around (GAA) MOSFET in subthreshold and strong inversion regions. *Microelectronics Journal, 43*(6), 358–363.
25. Samoju, V. R., & Tiwari, P. K. (2016). Threshold voltage modeling for dual-metal quadruple-gate (DMQG) MOSFETs. *International Journal of Numerical Modelling, 29*, 695–706.
26. He, L., Chiang, T.-K., Liou, J. J., Zheng, W., & Liu, Z. (2014). A new analytical subthreshold potential/current model for quadruple-gate junctionless MOSFETs. *IEEE Transactions on Electron Devices, 61*(6), 1972–1978.
27. Samoju, V. R., Dubey, S., & Tiwari, P. K. (2015). Quasi-3D subthreshold current and subthreshold swing models of dual-metal quadruple-gate (DMQG) MOSFETs. *Journal of Computational Electronics, 14*(2), 582–592.
28. Pandian, M. K., & Balamurugan, N. B. (2014). Analytical threshold voltage modeling of surrounding gate silicon nanowire transistors with different geometries. *Journal of Electrical Engineering and Technology, 9*(6), 2079–2088.
29. Colinge, J.-P. (2007). *FinFETs and other multi-gate transistors*. New York, NY: Springer.
30. Saha, P., Sarkhel, S., & Sarkar, S. K. (2017). Compact 2D threshold voltage modeling and performance analysis of ternary metal alloy work-function-engineered double-gate MOSFET. *Journal of Computational Electronics, 16*(3), 648–657.
31. Kumar, S., Kumari, V., Singh, S., Saxena, M., & Gupta, M. (2018). Sub-threshold drain current model of double gate ringFET (DG-RingFET) architecture: An analog and linearity performance investigation for RFIC design. *IETE Technical Review, 35*(2), 169–179.
32. Banerjee, P., & Sarkar, S. K. (2019). Modeling and analysis of a front high-k gate stack dual-material tri-gate schottky barrier silicon-on-insulator MOSFET with a dual-material bottom gate. *SILICON, 11*(1), 513–519.
33. Gradshteyn, I. S., & Ryzhik, I. M. (2014). *Tables of integrals, series and products* (8th ed.). New York: Academic Press.
34. Balestra, F. (2010). *Nanoscale CMOS: Innovative materials, modeling and characterization*. London, UK/Hoboken, NJ: ISTE/Wiley.
35. Duarte, J. P., Choi, S.-J., Moon, D.-I., Ahn, J.-H., Kim, J.-Y., Kim, S., & Choi, Y.-K. (2013). A universal core model for multiple-gate field-effect transistors. Part II: Drain current model. *IEEE Transactions on Electron Devices, 60*(2), 848–855.
36. Ghosh, P., Haldar, S., Gupta, R. S., & Gupta, M. (2012). An analytical drain current model for dual material engineered cylindrical/surrounded gate MOSFET. *Microelectronics Journal, 43*(1), 17–24.
37. Kaur, N., Rattan, M., & Gill, S. S. (2019). Design and optimization of novel shaped FinFET. *Arabian Journal for Science and Engineering, 44*(4), 3101–3116.
38. Samy, O., Abdelhamid, H., Ismail, Y., & Zekry, A. (2016). A 2D compact model for lightly doped DG MOSFETs (P-DGFETs) including negative bias temperature instability (NBTI) and short channel effects (SCEs). *Microelectronics and Reliability, 67*, 82–88.

39. Saha, R., Bhowmick, B., & Baishya, S. (2019). Analytical threshold voltage and subthreshold swing model for TMG FinFET. *International Journal of Electronics, 106*(4), 553–566.
40. Abdelhamid, H., Iniguez, B., & Guitart, J. R. (2007). Analytical model of the threshold voltage and subthreshold swing of undoped cylindrical gate-all-around-based MOSFETs. *IEEE Transactions on Electron Devices, 54*(3), 572–579.
41. Tsividis, Y., & McAndrew, C. (2011). *Operation and modeling of the MOS transistor* (3rd ed.). New York, NY: Oxford University Press.
42. Nandi, A., Pandey, N., & Dasgupta, S. (2017). Analytical modeling of DG-MOSFET in subthreshold regime by green's function approach. *IEEE Transactions on Electron Devices, 64*(8), 3056–3062.
43. Hu, G., Xiang, P., Ding, Z., Liu, R., Wang, L., & Tang, T.-A. (2014). Analytical models for electric potential, threshold voltage, and subthreshold swing of junctionless surrounding-gate transistors. *IEEE Transactions on Electron Devices, 61*(3), 688–695.
44. Hu, G., Hu, S., Feng, J., Liu, R., Wang, L., & Zheng, L. (2016). Analytical models for channel potential, threshold voltage, and subthreshold swing of junctionless triple-gate FinFETs. *Microelectronics Journal, 50*, 60–65.
45. Passi, V., & Raskin, J. P. (2017). Review on analog/radio frequency performance of advanced silicon MOSFETs. *Semiconductor Science and Technology, 32*(12), 123004.
46. Jiang, C., Liang, R., Wang, J., & Xu, J. (2015). A two-dimensional analytical model for short channel junctionless double-gate MOSFETs. *AIP Advances, 5*(5), 057122.
47. Banerjee, P., & Sarkar, S. K. (2017). 3-D analytical modeling of high-k gate stack dual-material tri-gate strained silicon-on-nothing MOSFET with dual-material bottom gate for suppressing short channel effects. *Journal of Computational Electronics, 16*(3), 631–639.
48. Banerjee, P., Kumari, T., & Sarkar, S. K. (2018). 2-D modeling and analysis of short-channel behavior of a front high-K gate stack triple-material gate SB SON MOSFET. *Applied Physics A: Materials Science & Processing, 124*(1–8).
49. Maduagwu, U. A., & Srivastava, V. M. (2019). Analytical performance of the threshold voltage and subthreshold swing of CSDG MOSFET. *Journal of Low Power Electronics, 9*(1), 1–20.
50. COMSOL Multiphysics. (2019). Retrieved May 27, 2019, from https://www.comsol.com.

Optimal Parameter Estimation of Solid Oxide Fuel Cell Model Using Coyote Optimization Algorithm

Amlak Abaza, Ragab A. El Sehiemy,
and Ahmed Saeed Abdelrazek Bayoumi

Abstract A coyote optimization algorithm (COA) is used to estimate accurate model parameters of SOFC stack. The COA is a new bioinspired optimization algorithm dependent on the behavior of a population of coyotes. In COA, the coyotes adapt in intelligent manner their social behavior, interact, and exchange experiences among them to reach the objective. COA is rapid, smooth, and steady in convergence process. In addition, COA differs from other algorithms, those that require more efforts to adjust the control variable of the algorithm. In this chapter, the COA results are compared to parameters estimated from a modern applied technique ranking teaching-learning optimizer RTLBO. The proposed COA leads to more accurate parameters with good convergences for different operating conditions of SOFC.

Keywords Coyote optimization algorithm · Fuel cell · SOFC parameter estimation · SOFC operation conditions

1 Introduction

With the utilization of electron spin and electron charge in combination with photons, the growing need for energy made the fuel cell idea rise again to the top of research issues. The fuel cell simply is an electrochemical reactor which consists of anode, cathode, and electrolyte which generates electricity as long as it is supplied with fuel. Different types of fuel cells are used and divided according to the fuel and electrolyte used. The most promising and attractive one is the solid oxide fuel cell

A. Abaza · R. A. El Sehiemy
Faculty of Engineering, Electrical Engineering Department, Kafr-Elsheikh University, Kafr-Elsheikh, Egypt

A. S. A. Bayoumi (✉)
Faculty of Engineering, Department of Engineering Physics and Mathematics, Kafr-Elsheikh University, Kafr-Elsheikh, Egypt
e-mail: ahmed.bayoumi@eng.kfs.edu.eg

M. H. Farouk, M. A. Hassanein (eds.), *Recent Advances in Engineering Mathematics and Physics*, https://doi.org/10.1007/978-3-030-39847-7_10

(SOFC) [1–4]. The merits of SOFCs like fuel flexibility, low emission, and high electrical efficiency make them promising candidates in many fields of application especially the cogeneration of heat and electricity [5–9]. The performance improvement of SOFC directs the research to develop accurate models for it and use various materials. For any material used or modeled, the more accurate the model parameters, the more efficient the performance improvement as the voltage of the SOFC is closely dependent on the parameters which model the internal processes [10, 11].

The optimization of model parameters of SOFC is a very important issue because it is a multivariable coupled system.

The meta-heuristic optimization techniques are very important, promising, and powerful to obtain accurate model parameters of SOFC because of their ease of implementation, reproducibility, simplicity, and robustness.

The genetic algorithm (GA) was improved (IGA) by Yang et al. [12] and applied to estimate parameters for tubular SOFC stack consisting of fine adjustment, speed cycle, and renascence. Jiang et al. [13] develop the particle swarm optimization (PSO) to a breed PSO for parameter estimation of a steam reformer model. An optimization algorithm based on biogeography with mutation strategies was proposed by Niu et al. [14] for proton exchange membrane fuel cells (PEMFC). Combined backtracking search algorithm with Burger's chaotic map for PEMFC was proposed by Askarzadeh and Coelho [15]. A developed adaptive differential evolution algorithm for SOFC, in which three strategies, i.e., crossover rate repairing technique, ranking-based vector selection, and parameter adaptation, are combined to improve performance of the algorithm is presented by Gong et al. [16]. A different strategy of cooperative coevolution decomposes objective function of parameter identification of SOFCs into multi-relative subfunctions first, then utilized a hybrid learning-based barebone PSO to solve each subfunction was employed by Jiang et al. [17]. A cuckoo search algorithm for area-specific resistance was applied by Ding et al. [18] for direct methanol fuel cells. A salp swarm optimizer has been utilized by El-Fergany [19] for polymer exchange membrane fuel cells. Both steady-state and dynamic models' parameters of SOFCs have been identified by El-Hay et al. [20] by satin bower bird optimizer. Ranking teaching-learning-based algorithm was implemented to estimate the SOFC's model parameters by Guojiang Xiong et al. [21]. Recently great attention to the meta-heuristic optimization methods has been gained beside its application in SOFCs in solving parameter identification problems of other fields.

It is clear from the aforementioned that there is always a development in field of optimization algorithms. This encourages a lot of researchers to apply and achieve the raised merits to their own problems in many engineering fields. A new bioinspired algorithm called coyote optimization algorithm (COA) is one of these algorithms which was developed by Juliano Pierezan and Leandro dos Santos Coelho (2018) [22] based on the social behavior of population. The act of coyote's population is the base of COA. The swarm intelligence and evolutionary heuristic represent the coyote's social organization and behavior. In addition of prey exchange, coyotes also exchange experiences to adapt environmental conditions.

The salient features of this effort summarized as

- COA is proposed as an effective candidate technique to optimize model parameters of SOFC.
- COA is used to explore 5-kW SOFC stack at different operation conditions, and results are compared to a modern technique RTLBO.
- The transcendence of COA of RTLBO is clarified through comparing with RTLBO which already shows its superiority [21].

The rest of this effort is represented as follows: Sect. 2 provides a theoretical model describing SOFC; in Sect. 3, COA is proposed in detail; simulation results and discussions are reported in Sect. 4; at last, conclusion is presented at Sect. 5.

2 Theoretical Model

The fuel cell is an electrochemical energy conversion device. It continuously converts chemical energy to electrical energy and heat. The fuel cell burns the chemical energy supplied from a fuel and oxidant producing water and electrical form of energy. The fuel cell differs from the battery as it does not store the energy, but it operates in a continuous mode with supplying the fuel. On the other hand, fuel cell is similar to an engine in the energy conversion concept, but in fuel cell, there is no output greenhouse gases emission. Thus, fuel cell is considered as a clean and efficient source of energy [23].

SOFC uses hydrogen as a fuel and oxygen as an oxidant. It is a very high-temperature type that operates around 1000 °C. The SOFC consists of three basic elements, a fuel electrode (anode), an oxide ion conducting electrolyte, and an oxidant electrode (cathode). The reactions of SOFC are given as follows [23–25]:

$$\text{Anode}: \quad H_2 + O^{2-} \rightarrow H_2O + 2e^- \tag{1}$$

$$\text{Cathode}: \quad \tfrac{1}{2}O_2 + 2e^- \rightarrow O^{2-} \tag{2}$$

$$\text{Overall reaction}: \quad H_2 + \tfrac{1}{2}O_2 \rightarrow H_2O \tag{3}$$

At the cathode site, a reduction of oxygen molecules occurs resulting in negative oxygen ions. The negative ions transport through the ionic conduction electrolyte, while the electrolyte prevents conducting the electrons. At anode site, the hydrogen reacts with the passed negative ions producing water and electrons which conduct through the connected electrical loads between anode and cathode. Figure 1 shows a schematic of SOFC [26].

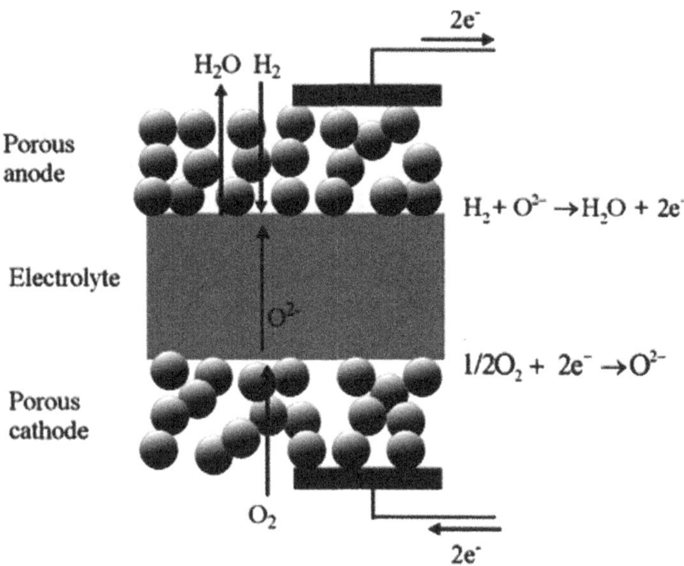

Fig. 1 Schematic of solid oxide fuel cell [26]

2.1 Output Voltage of SOFC

The output voltage of a single cell of SOFC, V_{cell}, can be calculated using the thermodynamic potential, E_{Nernst}, that resulted from the chemical reaction at no load and the voltage losses resulting during conversion process [27–30]:

$$V_{cell} = E_{Nernst} - (\text{All losses in a fuel cell}) \tag{4}$$

The thermodynamic potential E_{Nernst} is given from Nernst equation for hydrogen/oxygen fuel cell at actual operating conditions, temperature (T in Kelvin), and partial pressure of hydrogen, oxygen, and water ($P^*_{H_2}$, $P^*_{O_2}$ and $P^*_{H_2O}$, respectively), as follows [21, 30–33]:

$$E_{Nernst} = E_0 + \frac{RT}{2F} \ln \left(\frac{P^*_{H_2} \times \sqrt{P^*_{O_2}}}{P^*_{H_2O}} \right) \tag{5}$$

where E_0 is the reversible standard potential, R represents universal gas constant (8.3145 J/mol/K), and F represents Faraday constant (96485 C/mol).

The losses associated with SOFC are three types: activation loss, ohmic resistance loss, and concentration over potential. These losses depend on the operating and loading conditions. The activation loss occurs at the beginning of the chemical reaction. It is caused due to the energy barrier that is required to be overcome before the reaction starts. Activation voltage loss ($V_{\text{activation}}$) is given by the Butler–Volmer equation as follows:

$$V_{\text{activation}} = A \sin h^{-1}\left(\frac{J_{\text{load}}}{2J_{o,a}}\right) + A \sin h^{-1}\left(\frac{J_{\text{load}}}{2J_{o,c}}\right) \tag{6}$$

where, J_{load}, $J_{o,\,a}$, and $J_{o,\,c}$ are load current density, anode exchange current density, and cathode exchange current density, respectively, in mA/cm^2. A represents slope of Tafel line.

The ohmic voltage loss is the ohmic resistance drop due to the resistance of ionic flow through electrolyte and the resistance to flow of electrons in electrodes of the cell. It can be calculated as follows:

$$V_{\text{ohmic}} = J_{\text{load}}R_{\text{ohm}} \tag{7}$$

where R_{ohm} is the ionic resistance in k Ω cm^2.

The concentration loss, $V_{\text{concentraion}}$, is due to concentration gradients during the reaction. It takes place when the operating current density approach limits current (J_{max}):

$$V_{\text{concentraion}} = -b \times \ln\left((J_{\text{max}} - J_{\text{load}})/J_{\text{max}}\right) \tag{8}$$

where b is unknown parametric coefficient, depending on the fuel cell operating conditions.

The output voltage of SOFC is determined by Eq. (9). The resulting stack voltage of n_{cell}, connected in series, will be given by Eq. (9):

$$V_{\text{cell}} = E_{\text{Nernst}} - (V_{\text{activation}} + V_{\text{ohmic}} + V_{\text{concentraion}}) \tag{9}$$

$$V_{\text{stack}} = n_{\text{cell}} \times V_{\text{cells}} \tag{10}$$

3 Parameter Optimization of SOFC with COA

3.1 *Coyote Optimization Algorithm (COA)*

COA is a population-based metaheuristic technique. COA is naturally inspired on social behavior of coyotes. The COA can be considered evolutionary heuristic in addition to swarm intelligence. According to the proposed objective function (minimization or maximization), coyotes adapt their social structure and exchange experiences among them to realize it. Coyotes' population divide themselves into packs: $N_P \in N^*$ packs. Each pack consists of a number of coyotes, $N_C \in N^*$ coyotes. The number of N_C is limited to 14 coyotes inside the pack [22].

All packs adapt their social conditions and interact exchanging the culture and experiences among them. The solution of any optimization problem depends on social conditions of each coyote, **soc'**, which represents decision variables X? of the optimization problem [29].

The space search of COA is d-dimension. At tth instant of time, social condition ' soc', of cth coyote that belongs to the Pth pack ($\mathrm{soc}_C^{P,t}$) can be written as

$$\mathrm{soc}_C^{P,t} = \overline{X} = (x_1, x_2, x_{31}, \dots, x_d) \tag{11}$$

COA solves any global optimization problem according to the following steps:

Step 1: Initialization of Global Population of Coyotes

The random initialization of social conditions is achieved for each coyote cth of Pth at instant tth and jth dimension as follows:

$$\mathrm{soc}_{C,j}^{P,t} = LB_j + r_j \times \left(UB_j - LB_j\right) \tag{12}$$

where LB_j and UB_j are the lower and upper bounds of the control variable jth of the search space, d. $'r_j'$ is a real random number which lies in the range of [0,1], generated by uniform probability.

Step 2: Evaluation of Initial Solution and the Global Best

The evaluation of the initial adaptation of coyotes; conditions is determined according to Eq. (13):

$$\mathrm{fit}_C^{P,t} = f\left(\mathrm{soc}_C^{P,t}\right) \tag{13}$$

The optimal social condition of global population is determined as, "alpha." It is only one for the global population. It is given as

$$\mathrm{alpha}^{P,t} = \left\{\mathrm{soc}_C^{P,t} \mid \arg_{C=\{1,2,\dots,N_C\}} \min f\left(\mathrm{soc}_C^{P,t}\right)\right\} \tag{14}$$

Step 3: Determination of the Cultural Tendency of Each Pack

Intelligent behavior of the coyotes enables them to share the social conditions and link all information from the global population. The t median social conditions of all coyotes from a defined pack is computed which represent the cultural tendency of the pack, $\mathrm{cult}_j^{P,t}$ at tth instant:

$$\mathrm{cult}_j^{P,t} = \begin{cases} O_{\frac{(N_C+1)}{2},j}^{P,t}, & N_C \text{ is odd} \\[2mm] \dfrac{O_{\frac{N_C}{2},j}^{P,t} + O_{\frac{(N_C+1)}{2},j}^{P,t}}{2} & \text{otherwise} \end{cases} \tag{15}$$

where $O^{P,t}$ is the ranked decision variables (i.e., social conditions) of all coyotes inside the pth pack at tth instant for every j in the space of decision variables, d.

Step 4: Updating the Social Conditions of Coyotes

The intelligence of coyotes, interaction, and the good organization help them to update their social conditions. Coyotes exchange the global best 'alpha$^{P,\,t}$' and use cultural tendency of the pack, cult$^{P,\,t}$, to update their conditions, according to the following equations:

$$\delta_1 = \mathrm{alpha}^{P,t} - \mathrm{soc}^{P,t}_{\mathrm{Cr1}} \tag{16}$$

$$\delta_2 = \mathrm{cult}^{P,t} - \mathrm{soc}^{P,t}_{\mathrm{Cr2}} \tag{17}$$

where the factor δ_1 represents the influence of the difference from a random coyote (Cr1) inside the pack to alpha coyote in the same pack. The factor δ_2 is considered as the difference from a random coyote (Cr2) of a pack to cultural tendency of that pack.

The updating of the social conditions of the coyotes is done as follows:

$$\mathrm{new\_soc}^{P,t}_{C} = \mathrm{soc}^{P,t}_{C} + r_1\delta_1 + r_2\delta_2 \tag{18}$$

where r_1 and r_2 are uniformly distributed random numbers within $[0,1]$.

Step 5: Updating the Objective Function

The updating of the objective function is determined as

$$\mathrm{new\_fit}^{P,t}_{C} = f\left(\mathrm{new\_soc}^{P,t}_{C}\right) \tag{19}$$

Step 6: Making Decisions About the New Social Conditions

At the next $(t + 1)$th instant, the coyotes take a decision about the new social condition according to the updated objective function. Equation (20) explains this decision:

$$\mathrm{soc}^{P,t+1}_{C} = \begin{cases} \mathrm{new\_soc}^{P,t}_{C} & ,\mathrm{new\_fit}^{P,t}_{C} < \mathrm{fit}^{P,t}_{C} \\ \mathrm{soc}^{P,t}_{C} & \mathrm{otherwise} \end{cases} \tag{20}$$

The birth of a coyote as well as the death affect the population size. To keep the pack size static, COA tests birth and death as in [22] and computes the ages of all coyotes inside a pack (in years) as age$^{P,t}_{C} \in N$. The birth of a new coyote is represented by a combination of the social conditions of two parents inside a pack, which are chosen randomly, as follows:

$$\mathrm{Pup}^{P,t}_{j} = \begin{cases} \mathrm{soc}^{P,t}_{r1,j}, & \mathrm{rand}j < P_s \ \mathrm{or} \ j = j_1 \\ \mathrm{soc}^{P,t}_{r2,j}, & \mathrm{rand}j \geq (P_s + P_a) \ \mathrm{or} \ j = j_2 \\ R_j, & \mathrm{Otherwise} \end{cases} \tag{21}$$

where r_1 and r_2 are random coyote inside Pth pack. j_1 and j_2 represent two random dimensions of the optimization problem. P_s and P_a are the scatter and association probabilities, given by Eqs. (22) and (23). R_j is a random number which lies between the decision variable bound of the jth dimension. The value of the real random number randj lies in the range [0,1] and generated using uniform probability:

$$P_s = 1/d \tag{22}$$

$$P_a = (1 - P_s)/2 \tag{23}$$

The birth and the death of coyote are syncs to keep the population static. This is realized by applying the following steps (Algorithm #1):

1. Compute the group worse adapted to environment than pups, "w," and number of coyote in this group, "φ."
2. If φ equals 1, then go to 3, or else if φ is greater than 1, go to 4 or else the pub dies.
3. The pub survives and only coyote representing w dies.
4. The pub survives and oldest coyote in w dies. If two or more coyotes have the same age, the less adaptive coyote dies.

Figure 2 shows the flowchart of COA.

3.2 Parameter Optimization of SOFC Stack

The optimal estimation of SOFC stack parameters is solved as an optimization problem. This problem aims to optimally minimize the error between the proposed electrochemical model of SOFC stack and measurements. It is important to optimize seven parameters (E_0, A, $J_{o,\,a}$, $J_{o,\,c}$, b, $J_{\max}$, and R_{ohm}) represented in Eqs. (5)–(8), in order to determine the stack voltage of SOFC stack. It is required to minimize the mean of square error (MSE) between measured and modeled stack voltages. Coyote optimization algorithm as a global meta-heuristic optimization algorithm is proposed to achieve the objective function of the problem:

$$\mathrm{MSE} = \frac{1}{n} \sum_{k=1}^{n} \left[V_{\mathrm{measured}}(k) - V_{\mathrm{stack}}(k) \right]^2 \tag{24}$$

where n is number of measured voltages at different load current. The problem can be expressed as follows:

$$\mathrm{OF} = \min\,(\mathrm{MSE}) \tag{25}$$

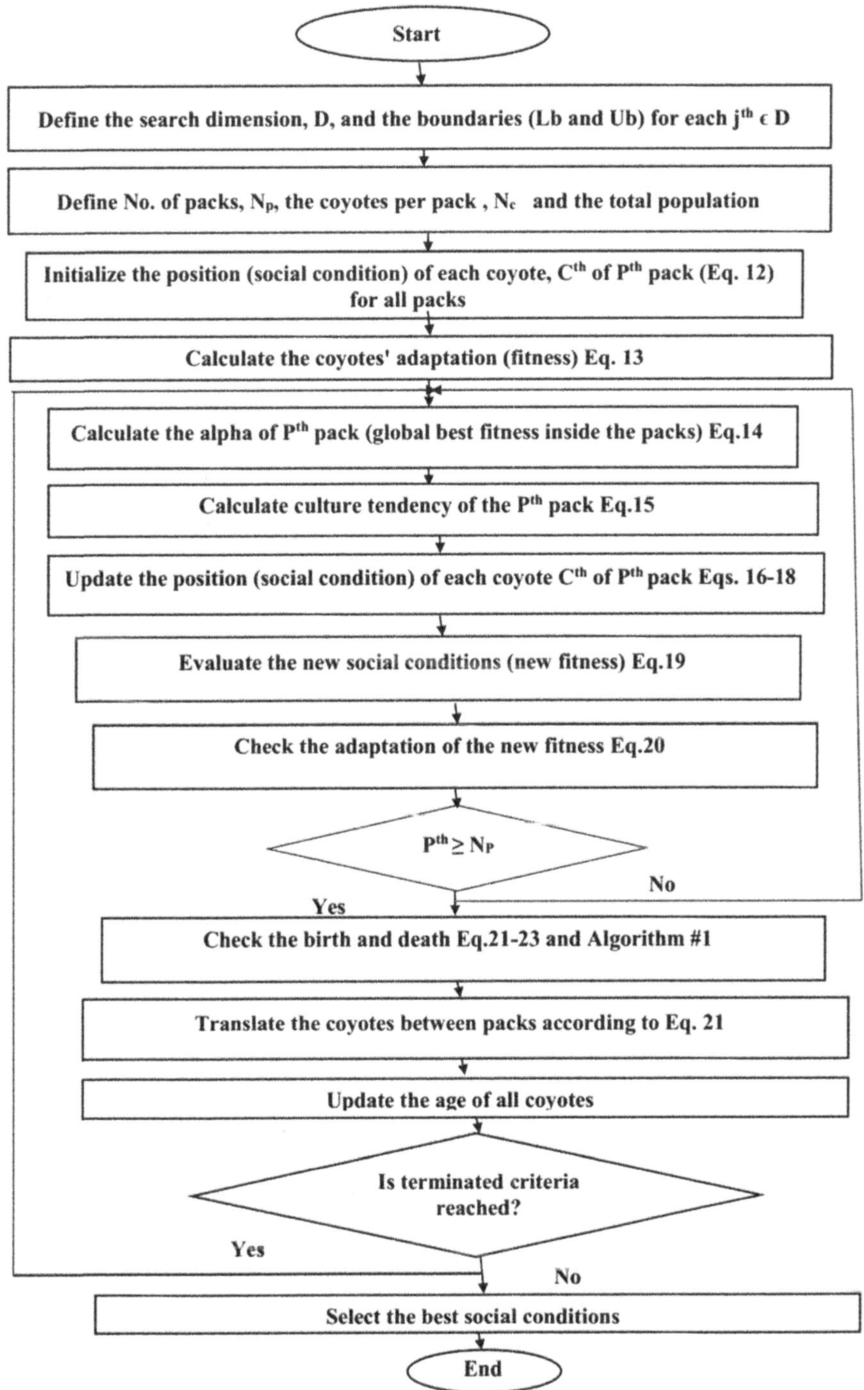

Fig. 2 Flowchart of COA

subject to :

$$E_0^{\min} \leq E_0 \leq E_0^{\max}$$

$$A^{\min} \leq A \leq A^{\max}$$

$$J_{o,i}^{\min} \leq J_{o,i} \leq J_{o,i}^{\max}, i = a, c$$

$$b^{\min} \leq b \leq b^{\max} \tag{26}$$

$$J_{\max}^{\min} \leq J_{\max} \leq J_{\max}^{\max}$$

$$R_{\text{ohm}}^{\min} \leq R_{\text{ohm}} \leq R_{\text{ohm}}^{\max}$$

4 Results and Discussions

The proposed algorithm is tested by using the data in literature [21]. The technical and operating data of the tested stack are illustrated in Table 1.

Table 2 shows the bounds of the control variables of the SOFC stack. The estimated values of the control variables (E_0, A, $J_{o, a}$, $J_{o, c}$, b, $J_{\max}$, and R_{ohm}) have been calculated by solving the optimization problem. These parameters will be used to obtain an accurate model of the SOFC stack. COA has no control parameters. In the optimization problem, it is desired to minimize difference between output voltage of the proposed model and the measured values at different values of current.

Table 3 explains the extracted optimal control variables using the proposed algorithm, COA, at different operating conditions (3 bar, 923/1073 K). The value of MSE is 9.991E–5 and 2.393E–4 at the two considered operating temperatures. Figure 3 shows a good and smooth convergence of the objective.

To confirm that the proposed COA is valid and accurate, the results obtained are compared to ranking teaching-learning optimizer (RTLBO). Table 4 explains the optimum parameters of stack model of SOFC with COA compared to the RTLBO at

Table 1 Technical data and operating conditions of SOFC stack

Technical data	
Stack parameter	Value
Number of cells in a stack, n_{cells}	96
Rated power (kW)	5
Operating conditions	
P (bar)	3
Stack temperature T (K)	923:50:1073
Partial pressure of hydrogen	0.9
Partial pressure of water	0.1
Partial pressure of oxygen	0.21
Reactants	H_2 and air

Table 2 Lower and upper bounds of SOFC stack control variables

Parameter	Bounds	
	Lower	Upper
E_0 (V)	0	1.2
A (V)	0	1
$J_{o,a}$ (mA/cm^2)	0	100
$J_{o,c}$ (mA/cm^2)	0	100
b (V)	0	1
J_{max} (mA/cm^2)	0	1000
R_{ohm} (kΩ m^2)	0	1

Table 3 Optimal decision variables obtained with COA for SOFC at 3 bar and 923 K/ 1073 K

SOFC parameters	$T = 923$ K	$T = 1073$ K
E_0 (V)	1.1197	1.1150
A (V)	0.06007	0.036940
$J_{o,a}$ (mA/cm^2)	18.00	28.762
$J_{o,c}$ (mA/cm^2)	6.258	6.7440
b (V)	0.0535	0.06502
J_{max} (mA/cm^2)	157.00	160.00
R_{ohm} (kΩ m^2)	0.00611	0.00398
MSE	9.991×10^{-5}	2.393×10^{-4}

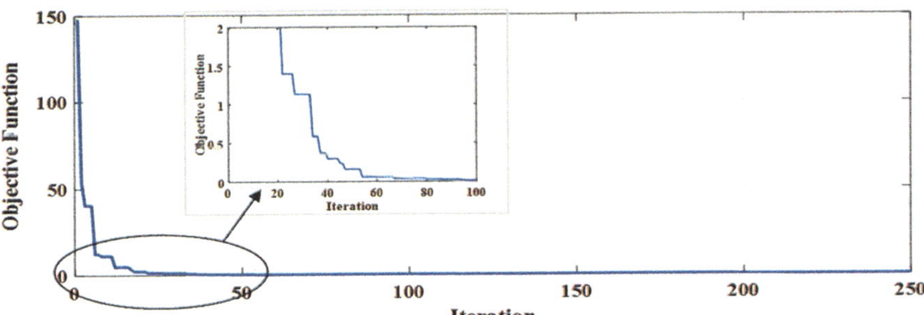

Fig. 3 Convergence of COA curve of SOFC stack

3 bar and 923/973/1023/1073 K. The value of MSE with COA is less than RTLBO over the studied conditions.

Figure 4 shows the polarization curve (V-I curve) of the 5-kW SOFC stack at 1073 K compared to the measured values. It is clear that the best agreement of model curve with measured is realized. The power of the stack model at different conditions is illustrated in Fig. 5.

At four operating conditions, two for testing the model and the other two for validating the model, a set of the polarization curves is constructed. Figure 6 explains the polarization curves (V-I curves) at 3 bar and 923/973/1023/1073 K. the effectiveness of the proposed algorithm in estimating an accurate model for the SOFC stack module is confirmed throughout the results obtained.

Table 4 Optimal decision variables obtained with RTLBO compared to COA for PEMFC at different operating conditions

SOFC parameters	$T = 923$ K		$T = 973$ K	
	COA	RTLBO [21]	COA	RTLBO [21]
E_0 (V)	1.1197	1.1196	1.1189	1.119
A (V)	0.06007	0.0596	0.0485	0.0474
$J_{o,a}$ (mA/cm^2)	18.00	17.5094	20.000	19.2418
$J_{o,c}$ (mA/cm^2)	6.258	6.2457	6. 487	6.4201
b (V)	0.0535	0.0526	0.0555	0.0539
$J_{\max}$ (mA/cm^2)	157.00	156.8305	157.9765	157.5788
R_{ohm} ($k\Omega$ m^2)	0.00611	0.0061	0.0052	0.0052
MSE	9.991×10^{-5}	1.536×10^{-4}	$\mathbf{5.6131 \times 10^{-5}}$	8.9207×10^{-4}
	$T = 1023$ K		**$T = 1073$ K**	
	COA	**RTLBO [21]**	**COA**	**RTLBO [21]**
E_0 (V)	1.1171	1.1173	1.1144	1.1148
A (V)	0.0408	0.0404	0.034976	0.0361
$J_{o,a}$ (mA/cm^2)	24	23.8021	27.0707	27.5309
$J_{o,c}$ (mA/cm^2)	6.6325	6.5017	6.4123	6.662
b (V)	0.0562	0.0558	0.0655	0.0654
$J_{\max}$ (mA/cm^2)	158.8119	158.7944	159.9075	159.9
R_{ohm} ($k\Omega$ m^2)	0.0047	0.0047	0.00402	0.004
MSE	$\mathbf{1.5373 \times 10^{-4}}$	1.4833×10^{-3}	$\mathbf{5.779 \times 10^{-4}}$	1.5876×10^{-3}

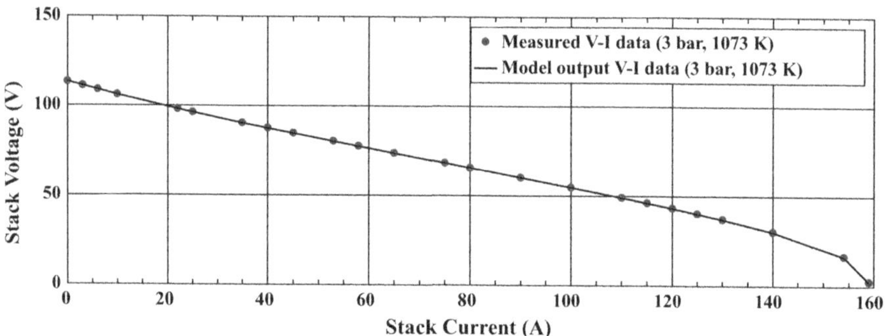

Fig. 4 Polarization (V-I) curve of SOFC 5-kW stack at 3 bar and 1073 K

5 Conclusion

A naturally inspired coyote optimization algorithm (COA) is proposed for estimating accurate model parameters of SOFC stack and compared to results from a modern applied technique (RTLBO). The proposed COA estimated parameter of the electrochemical model considered to minimize mean square error (MSE) between measured and the modeled stack voltages. In COA, the coyotes adapt in intelligent

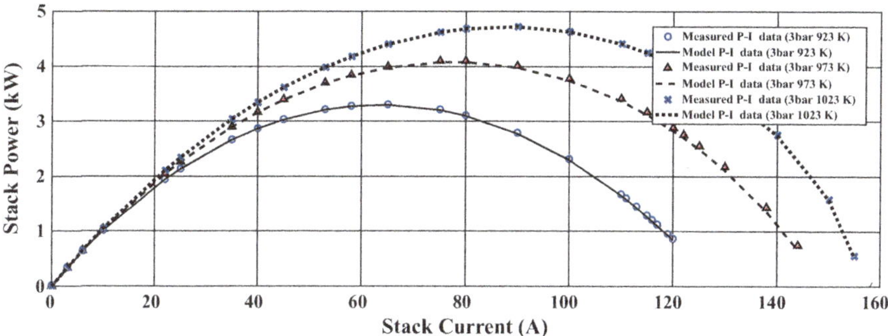

Fig. 5 Power (P-I) curve of SOFC 5-kW stack at different conditions

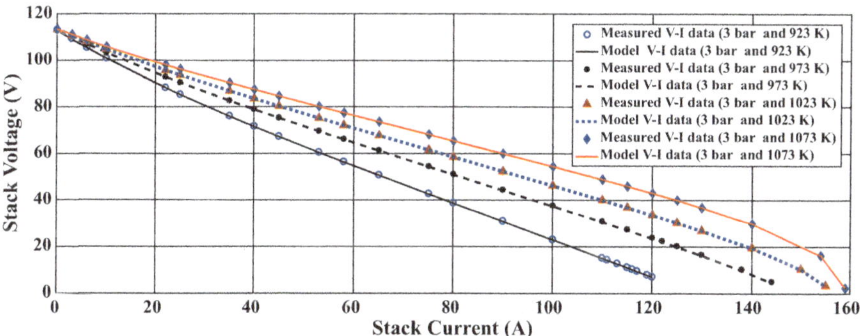

Fig. 6 A set of polarization (V-I) curve of SOFC 5-kW stack at four operating conditions

manner their social behavior, interact, and exchange experiences among them to reach the objective. COA is rapid, smooth, and steady in convergence process. In addition, COA differs from other algorithms, those which require more efforts to adjust the control variable of the algorithm. The simulation results prove robustness and high quality of performance of proposed COA to produce accurate parameters with good convergences. Also, the obtained estimated V-I and P-I curves are very close to the experimental data at different operating conditions. COA can be considered a very good candidate for promising optimization tool especially for nonlinear complex problems.

References

1. Xu, H., Chen, B., Tan, P., Cai, W., He, W., Farrusseng, D., et al. (2018). Modeling of all porous solid oxide fuel cells. *Applied Energy, 219*, 105–113.
2. Abdalla, A. M., Hossain, S., Azad, A. T., Petra, P. M. I., Begum, F., Eriksson, S. G., et al. (2018). Nanomaterials for solid oxide fuel cells: a review. *Renewable and Sustainable Energy Reviews, 82*(353), 353–368.

3. Ramadhani, F., Hussain, M. A., Mokhlis, H., & Hajimolana, S. (2017). Optimization strategies for Solid Oxide Fuel Cell (SOFC) application: a literature survey. *Renewable and Sustainable Energy Reviews, 76*, 460–484.
4. Buonomano, A., Calise, F., d'Accadia, M. D., Palombo, A., & Vicidomini, M. (2015). Hybrid solid oxide fuel cells-gas turbine systems for combined heat and power: a review. *Applied Energy, 156*, 32–85.
5. Roshandel, R., Golzar, F., & Astaneh, M. (2018). Technical, economic and environmental optimization of combined heat and power systems based on solid oxide fuel cell for a greenhouse case study. *Energy Conversion and Management, 164*, 144–156.
6. Oh, S. R., Sun, J., Dobbs, H., & King, J. (2014). Model predictive control for power and thermal management of an integrated solid oxide fuel cell and turbocharger system. *IEEE Transactions on Control Systems Technology, 22*(3), 911–920.
7. Vrečko, D., Nerat, M., Vrančić, D., Dolanc, G., Dolenc, B., Pregelj, B., et al. (2018). Feedforwardfeedback control of a solid oxide fuel cell power system. *International Journal of Hydrogen Energy, 43*, 6352–6363.
8. Sarmah, P., & Gogoi, T. K. (2017). Performance comparison of SOFC integrated combined power systems with three different bottoming steam turbine cycles. *Energy Conversion and Management, 132*, 91–101.
9. Zeng, H., Wang, Y., Shi, Y., & Cai, N. (2017). Biogas-fueled flame fuel cell for micro-combined heat and power system. *Energy Conversion and Management, 148*, 701–707.
10. Virkar, A., Williams, M. C., & Singhal, S. (2007). Concepts for ultra-high power density solid oxide fuel cells. *ECS Transactions, 5*(1), 401–421.
11. Zhu, L., Zhang, L., & Virkar, A. V. (2015). A parametric model for solid oxide fuel cells based on measurements made on cell materials and components. *Journal of Power Sources, 291*, 138–155.
12. Jie, Y., Xi, L., Jian, H. J., Li, J., Lei, Z., Jin, G. J., et al. (2011). Parameter optimization for tubular solid oxide fuel cell stack based on the dynamic model and an improved genetic algorithm. *International Journal of Hydrogen Energy, 36*(10), 6160–6174.
13. Jiang, J., Li, X., Deng, Z., Yang, J., Zhang, Y., & Li, J. (2013). Control-oriented dynamic model optimization of steam reformer with an improved optimization algorithm. *International Journal of Hydrogen Energy, 38*(26), 11288–11302.
14. Niu, Q., Zhang, L., & Li, K. (2014). A biogeography-based optimization algorithm with mutation strategies for model parameter estimation of solar and fuel cells. *Energy Conversion and Management, 86*, 1173–1185.
15. Askarzadeh, A., & Coelho, L. D. S. (2014). A backtracking search algorithm combined with Burger's chaotic map for parameter estimation of PEMFC electrochemical model. *International Journal of Hydrogen Energy, 39*(21), 11165–11174.
16. Gong, W., Cai, Z., Yang, J., Li, X., & Jian, L. (2014). Parameter identification of an SOFC model with an efficient, adaptive differential evolution algorithm. *International Journal of Hydrogen Energy, 39*(10), 5083–5096.
17. Bo, J., Ning, W., & Wang, L. (2014). Parameter identification for solid oxide fuel cells using cooperative barebone particle swarm optimization with hybrid learning. *International Journal of Hydrogen Energy, 39*(1), 532–542.
18. Ding, J., He, X., Jiang, B., & Wu, Y. (2015). Parameter identification for area-specific resistance of direct methanol fuel cell using Cuckoo search algorithm. *Bio-inspired Computation: Theory and Applications, 2015*, 107–112.
19. El-fergany, A. A., Kalogirou, S. A., & Christodoulides, P. (2018). Extracting optimal parameters of PEM fuel cells using Salp Swarm Optimizer. *Renewable Energy, 119*, 641–648.
20. El-Hay, E. A., El-Hameed, M. A., & El-Fergany, A. A. (2018). Steady-state and dynamic models of solid oxide fuel cells based on Satin Bowerbird Optimizer. *International Journal of Hydrogen Energy, 43*, 14751–14761.

21. Guojiang, X., Jing, Z., Dongyuan, S., & Yu, H. (2018). Parameter identification of solid oxide fuel cells with ranking teaching-learning based algorithm. *Energy Conversion and Management, 174*, 126–137.
22. Pierezan, J., & Coelho, S. (2018). Coyote optimization algorithm: a new metaheuristic for global optimization problems. In *Proceedings of the IEEE Congress on Evolutionary Computation (CEC)* (pp. 2633–2640). Rio de Janeiro: IEEE.
23. Parshant, S., & Gurtej, S. (2017). Solid oxide fuel cell connected to load. *International Journal of Engineering Management Research, 7*(2), 419–428.
24. Gebregergis, A., Pillay, P., Bhattacharyya, D., & Rengaswemy, R. (2009). Solid oxide fuel cell modeling. *IEEE Transactions on Industrial Electronics, 56*(1), 139–148.
25. Flavio, F., Chuahy, D. F., & Kokjohn, S. (2019). Solid oxide fuel cell and advanced combustion engine combined cycle: a pathway to 70% electrical efficiency. *Applied Energy, 235*, 391–408.
26. Ni, M., Leung, H., & Leung, C. (2007). Parametric study of solid oxide fuel cell performance. *Energy Conversion and Management, 48*(5), 1525–1535.
27. Strunz, K., Lan, T. (2017). Multi-physics transients modeling of solid oxide fuel cells: methodology of circuit equivalents and use in EMTP-type power system simulation. *IEEE Transactions on Energy Conversion*.
28. Wang, C., & Nehrir, H. (2007). A physically based dynamic model for solid oxide fuel cells. *IEEE Transactions on Energy Conversion, 22*(4), 887–897.
29. Ma, R., Gao, F., Breaz, E., Huangfu, Y., & Briois, P. (2018). Multi-dimensional reversible solid oxide fuel cell modeling for embedded applications. *IEEE Transactions on Energy Conversion, 33*, 692–701.
30. Lim, T. H., Song, R. H., Shin, D. R., Yang, J. I., Jung, H., Vinke, I. C., & Yang, S. S. (2008). Operating characteristics of a 5 kW class anode-supported planar SOFC stack for a fuel cell/gas turbine hybrid system. *International Journal of Hydrogen Energy, 33*, 1076–1083.
31. Sung, P., Young, L., & Kook, A. (2014). Performance analysis of an SOFC/HCCI engine hybrid system: system simulation and thermo-economic comparison. *International Journal of Hydrogen Energy, 39*, 1799–1810.
32. Abir, Y., Domenico, F., Hacen, D., Pierluigi, L., Khalifa, S., & Massimo, S. (2018). Electrochemical performance of solid oxide fuel cell: experimental study and calibrated model. *Energy, 142*, 932–943.
33. Tseronis, K., Bonis, I., Kookos, K., & Theodoropoulos, C. (2012). Parametric and transient analysis of non-isothermal, planar solid oxide fuel cells. *International Journal of Hydrogen Energy, 37*, 530–547.

Simulation Study of Terahertz Radiation Coupling Inside Field Effect Transistors

Marwa Mohamed and Nihal Ibrahim

Abstract Field effect transistors have been used lately for the detection of terahertz radiation beyond their cutoff frequency. The contacts are expected to play a significant role in detecting this high-frequency radiation. However, the effect of AC signal coupling between those contacts and the FET itself in this frequency range is not well known. In this work, a simulation study was conducted to extract the characteristic impedance of the contacts and the FET channel input, and these were used to extract the scattering coefficient. The results indicate that it is possible to reach high-power delivery rates in this frequency range. The results also indicated that the selection of the input terminal can affect the operation frequency range and that the gate potential can largely affect the overall signal coupling. And finally, the results indicate the possibility of operating the contacts FET coupling at an LC resonance condition subject to proper tuning of the system impedance.

Keywords FET · Terahertz radiation detection · Contacts coupling

1 Introduction

Recent research on the use of field effect transistors for very high-frequency (around THz) radiation detection beyond their transit time cutoff frequency, fs, has been the focal point of increasing attention [1]. However, it was noted that most of the past research efforts were spotlighting only on studying the nonlinear rectifying FET response [2, 3], while the antenna part of the detector still involves many unanswered questions.

M. Mohamed (✉)
Faculty of Engineering, Department of Communication and Electronics Engineering, Cairo University, Giza, Egypt

N. Ibrahim
Faculty of Engineering, Department of Engineering Mathematics and Physics, Cairo University, Giza, Egypt

© Springer Nature Switzerland AG 2020
M. H. Farouk, M. A. Hassanein (eds.), *Recent Advances in Engineering Mathematics and Physics*, https://doi.org/10.1007/978-3-030-39847-7_11

The detector's antenna part is the first stage in the operation of THz FET detectors, where the EM radiation is converted to electronic signal and is coupled to the transistor channel.

The antenna part has two main configurations that can be determined in the literature: the effective [4, 5] and the built-in antenna configurations [6, 7]. In the effective antenna configuration, on-chip metallization is considered as the effective antenna, and no separate antenna component is connected [4, 5]. This configuration has been used successfully to detect sub-THz/THz radiation with sensible response strength. Sakowicz et al. [8] studied this structure under 100 GHz and concluded that most of the detection process are from the on-chip and off-chip wires. However, as the radiation frequency increased, the wavelength is reduced to values close to the dimensions of the FET contacts (smaller than the dimensions of chip bonding and wiring). Therefore, the role of the chip contacts in detection is expected to increase for increasing radiation frequency [9], and more study is thus required.

This work is more focusing on the use of FET contacts and on-chip wiring in their planar stacking configuration as an effective THz antenna. This is done through studying the internal coupling between the contacts and the FET channel at high frequencies exceeding the cutoff frequency of the FET. The main motivation of this study is to understand the physics behind the inevitable coupling loss within the FET which was usually disregarded in favor of external antenna coupling. Optimizing the contact design for maximum coupling even in the absence of added antenna allows maximization of the FET response and gives the designer more control on its operational properties.

This is done through conducting several simulations of selected test structures to extract the characteristic impedance of the contacts of the FET on their own and when connected to on-chip wiring. These are followed by simulations of the FET devices itself to extract the characteristic impedance that opposes the AC signal entering the channel, a comparative approach between both sets of simulation for different coupling points. The coupling points in this context refer to the points at which coupling of the AC between the contacts and the semiconductor devices is considered. For example, two sets were considered: contact-to-body coupling (source to body) and source-to-gate coupling. The results extracted from these simulations are considered as a sample for typical devices.

The guidelines extracted from this study address consideration for proper contact design for operation beyond the cutoff of a FET in the THz regime. This adds to the potential of operation of FET devices as THz detector. The high-frequency contacts characteristics were simulated using the EDA tool (CST-MWS Microwave Studio), while the semiconductor device characteristics were simulated using SILVACO simulator.

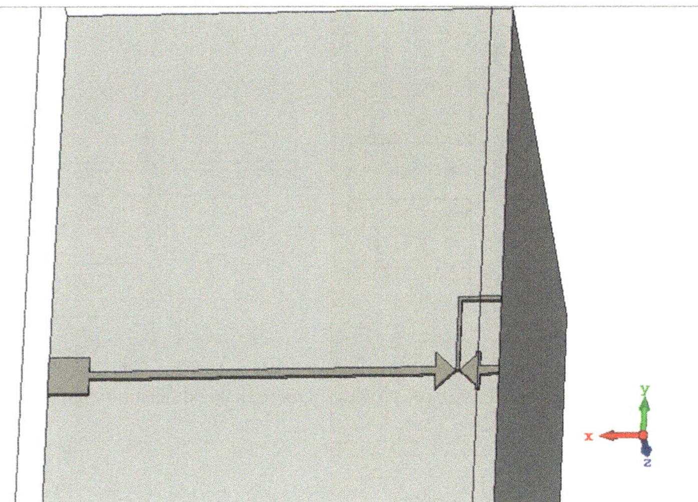

Fig. 1 3D diagram of the simulated test structure

2 Simulation Structure

The basic test structure is shown in Fig. 1. It is composed of three metal contacts representing those of a typical FET. The two outermost contacts represent the source and drain contacts. They are symmetric, and a bow-tie shape was selected for them in this simulation. The separation between the two contacts is 1 μm representing the channel, and the gate contact (Ga) occupies 0.8 μm of that distance. To represent the practical wiring connected to the contacts, two wiring extension were added to the contacts of lengths 5 μm (Sa) and 100 μm (Sw). The substrate is chosen to be lossy silicon, with refractive index $n = 3.6$, of surface 200 μm × 110 μm and depth of 150 μm.

A microwave port is added in two main configurations: the first is between each contact and the ground on the other side of the substrate (S_a or S_w to body), and the second is between the adjacent source-to-gate contacts (S-G). In the second configuration, the substrate is either grounded (Sa-G/grounded) or left floating (Sa-G/floating) (Fig. 2).

In this simulation the test structure is operated as a radiation transmitter instead of detector. According to the reciprocity theorem, the properties and parameters of a transmitting antenna are similar to those of a receiving antenna.

The simulation is set up for broadband simulation for the terahertz frequency range (0.5–3 THz). The basic simulation parameter considered here is the complex characteristic impedance ($Z = R + iX$).

High-frequency signal power delivered to the FET P_{in} channel depends on the scattering between the FET and the contacts$|S_{11}|$ as shown in Fig. 3, such that

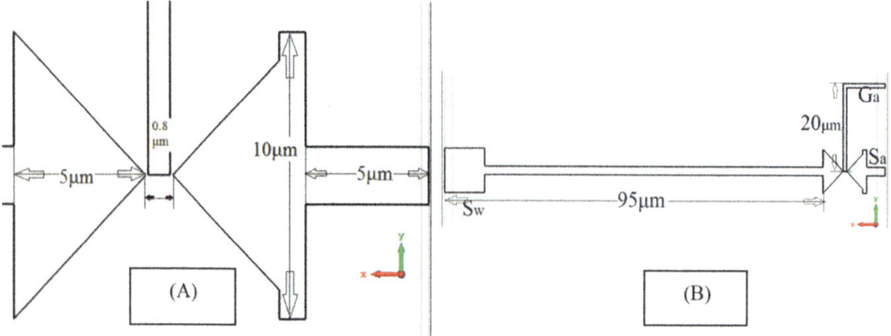

Fig. 2 Front view of the test structure showing the detailed shape and dimensions of the FET contacts and their dimensions. Figure (**a**) provides a more detailed diagram than the one shown in (**b**)

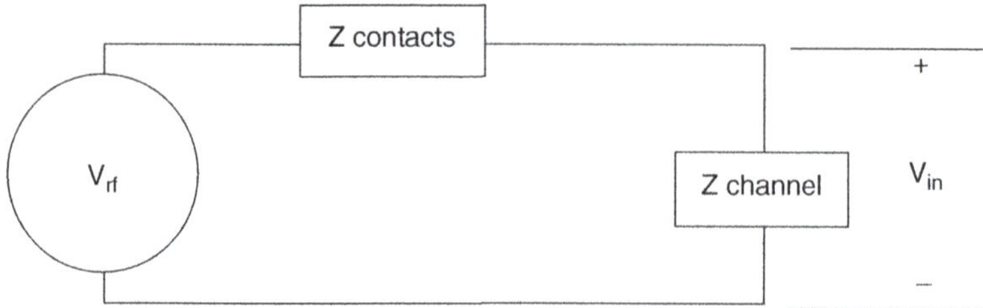

Fig. 3 Schematic of the tested structure (contacts are the effective antenna, and the effective signal is extracted by the FET channel)

$$P_{\text{in}} = \left(1 - |S_{11}|^2\right) P_{\text{rf}} \tag{1}$$

where P_{rf} is the RF power absorbed by the contacts. The scattering coefficient depends on the characteristic impedance of both the contacts and the FET such that

$$S_{11} = \frac{Z_{\text{FET}} - Z_{\text{contacts}}}{Z_{\text{FET}} + Z_{\text{contacts}}} \tag{2}$$

where Z_{contacts} is the contacts impedance and Z_{FET} is the FET channel input impedance.

The FET was modeled using Silvaco TCAD simulator with channel length of 1 μm. A standard FET structure was selected for this simulation as shown in Fig. 4. Using the selected simulator, we were able to extract the characteristic impedance between the source and ground which was placed on the other side of substrates as in the RF simulation mentioned above. The simulation was repeated for multiple gate potential values to account for the nonlinearity of the channel input characteristics. While the drain potential was kept constant (2 V), the gate potential was varied between 0 and 0.7 V. This allowed us to investigate the coupling of the AC signal to

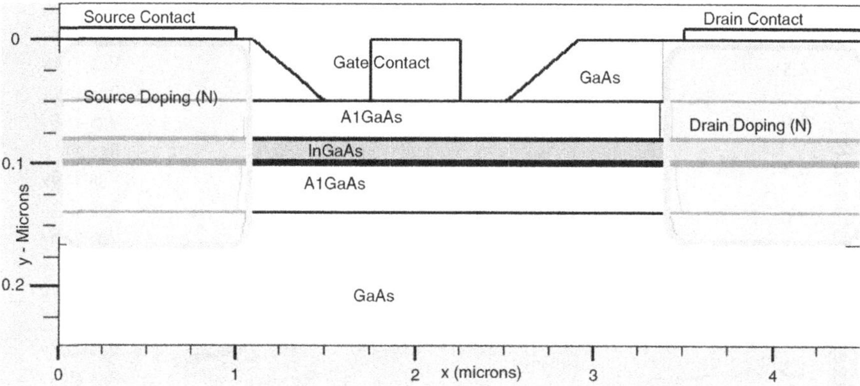

Fig. 4 Schematic diagram of the simulated FET structure

either the gates/source terminal or source-to-body terminal under practical operation conditions. The results presented later can affect our efforts in modeling and designing FET THz detectors.

3 Results and Discussion

Two simulation sets were conducted. The first set focused on gate-to source-coupling in both the contacts and the FET. The other set was focused on the source/body coupling. In the remainder of this section, results of each set are presented and discussed.

3.1 Gate-to-Source Coupling

In this configuration we focused on the AC signal coupling between the gate and source contacts. Through the microwave simulator, we extracted the characteristic impedance between these two contacts. Then, through the semiconductor device simulator, we extracted the impedance seen by the same AC signal as it is introduced to the channel (FET input impedance). Both floating and grounded bodies were considered in this simulation. The simulation results are presented in Fig. 5 (real part of the impedance R) and in Fig. 6 (imaginary part of the impedance, X).

The results show that as the frequency increases in the terahertz frequency scale, the resistances of the contacts as well as the resistance of the FET become closer in value. The dependence of the FET channel impedance on the applied bias allows control of the scattering coefficient as well. As a result, Fig. 5 shows that the gate

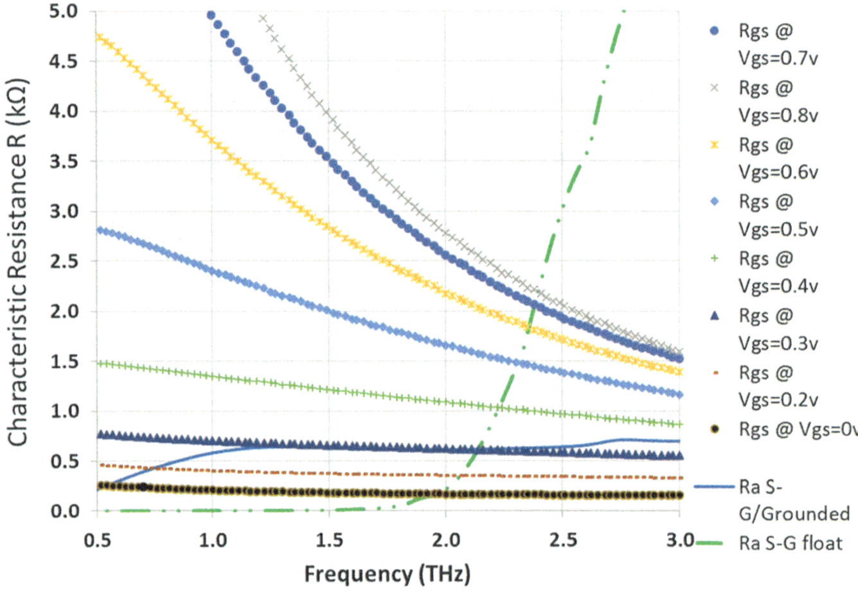

Fig. 5 Characteristic resistance between the gate and source contacts (solid and dashed lines) and between the gate/source terminals of the FET under varying gate/source potentials (marker lines)

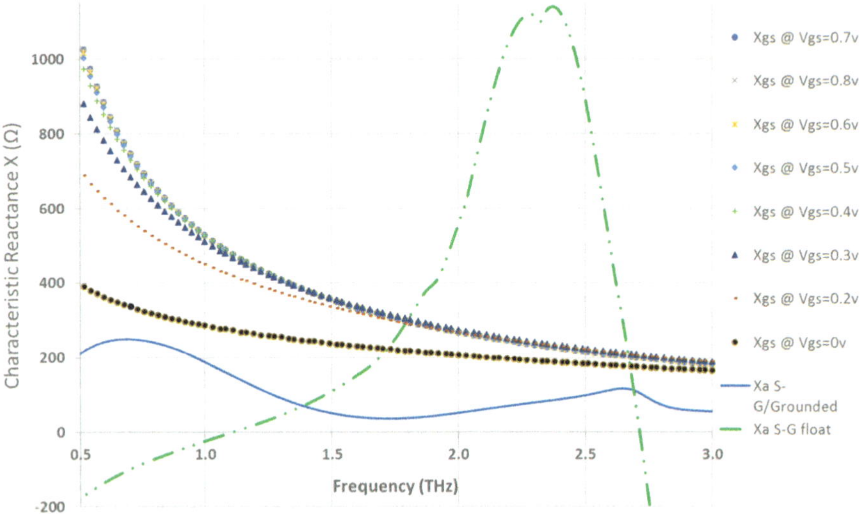

Fig. 6 Characteristic reactance between the gate and source contacts (solid and dashed lines) and between the gate/source terminals of the FET under varying gate/source potentials (marker lines)

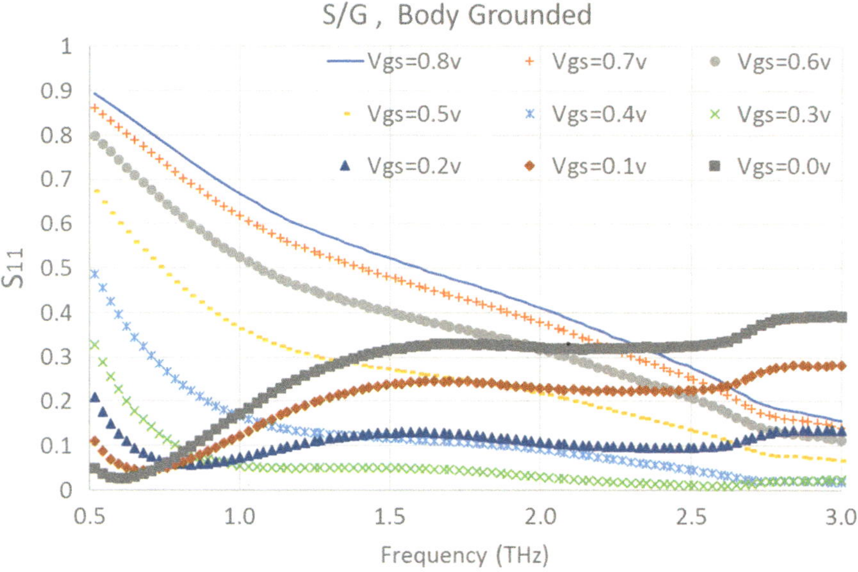

Fig. 7 Scattering coefficient of the coupling between gate/source contacts (with grounded body back metal plate) and the FET device subject to varying gate potential

voltage (~0.7 V) can help drive the FET contact to minimum scattering condition (R_a-R_{FET}~0).

The reduction of the FET reactance in high frequency is a typical sign of exceeding the FET cutoff frequency. However, this reactance matches the very small reactance of the small FET contacts, increasing the influence of signal radiation detection in this frequency range.

Figures 7 and 8 show the extracted scattering constants for the AC signal flow between the source and gate contacts in the two considered cases (grounded body and floating body (no back metal sheets)). In the first case represented by Fig. 7, the tunability provided by the gate potential is clearly seen. At around 0.3 V gate potential. The resistance of both the FET and contacts is almost equal along a broad frequency range, resulting in minimized scattering and maximized signal delivery to the FET.

In the second case represented by Fig. 8, the same potential tunability is seen; however, the response is not broadband as in the above case. This can be attributed to the characteristic impedance of the contacts which varies strongly with frequency in the terahertz range.

An interesting feature, however, can be seen in the floating contacts case (Fig. 7), that is, the characteristic inductive reactants below 1 THz and above 2.7 THz. This opens a window for engineering the contacts to provide resonant RLC characteristics. The main obstacle would be the high contacts resistance which may severely dampen this response.

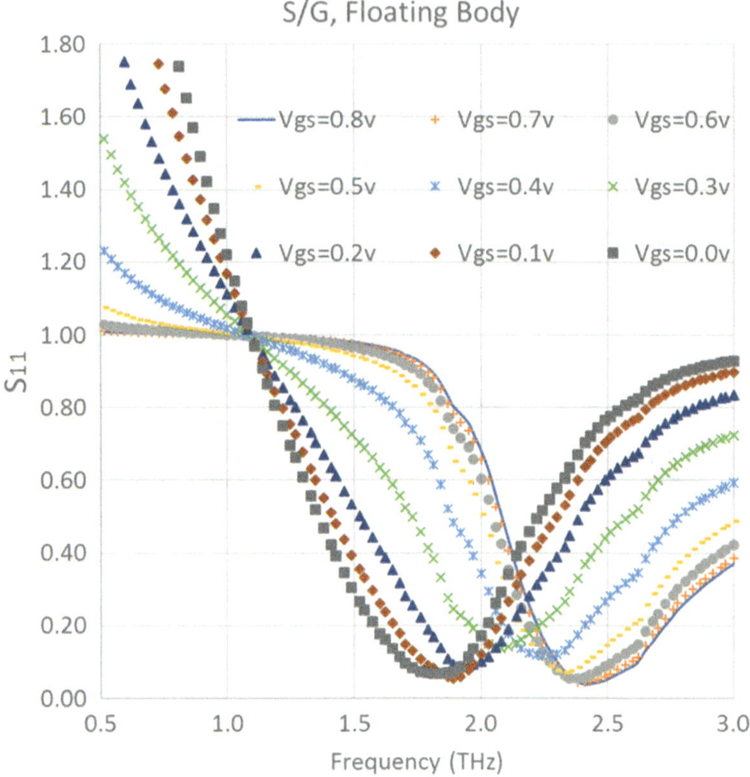

Fig. 8 Scattering coefficient of the coupling between gate/source contacts (floating body) and the FET device subject to varying gate potential

3.2 *Source-Body Coupling*

The second case considered in this simulation is the case of coupling AC incident radiation detected by the source-ground effective antenna, to the FET's source to body terminals. Two cases are considered, short and long on-chip wire connected to the source contacts. In this context, short refers to 5 μm wire strip, while long refers to and 95 μm wire strip. The characteristic resistance and reactance of both the contacts and the FET are shown in Figs. 9 and 10, respectively. The magnitude of the scattering coefficient of the long (Sw) and short (Sa) contacts interface with the FET are shown in Figs. 11 and 12, respectively.

The figures of the resistance and reactance clearly show Fabry–Perot ripples due to confinement of the radiation in the substrate. These ripples caused mostly narrow bands of low scatterings, resulting in almost lossless data transmission to the FET and providing superior response to the previous case. Increasing the length of the wire strip connected to the contact did not affect the location of the ripples in the characteristic impedance results, as they are related to the substrate depth. However,

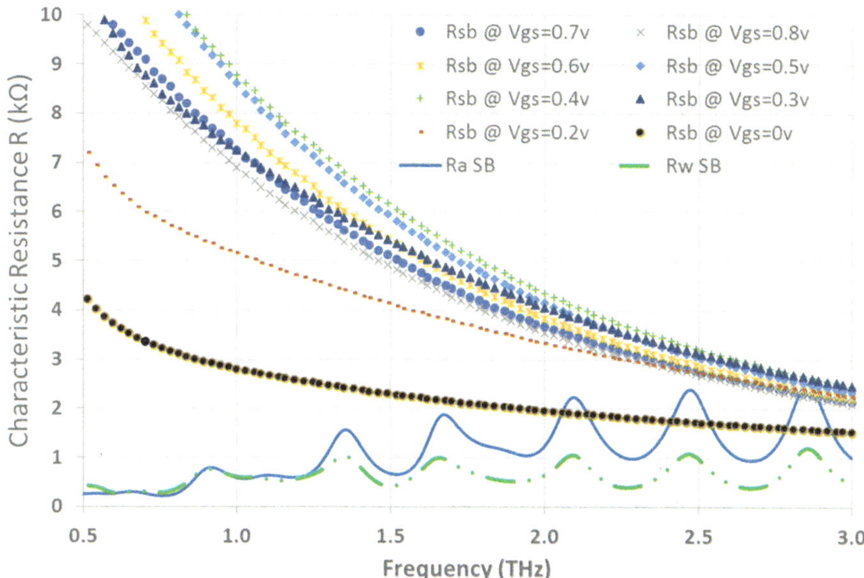

Fig. 9 Characteristic resistance between the source and gate contacts (solid and dashed lines, for short and long metal wires, respectively) and between the source/body terminals of the FET under varying gate/source potentials (marker lines)

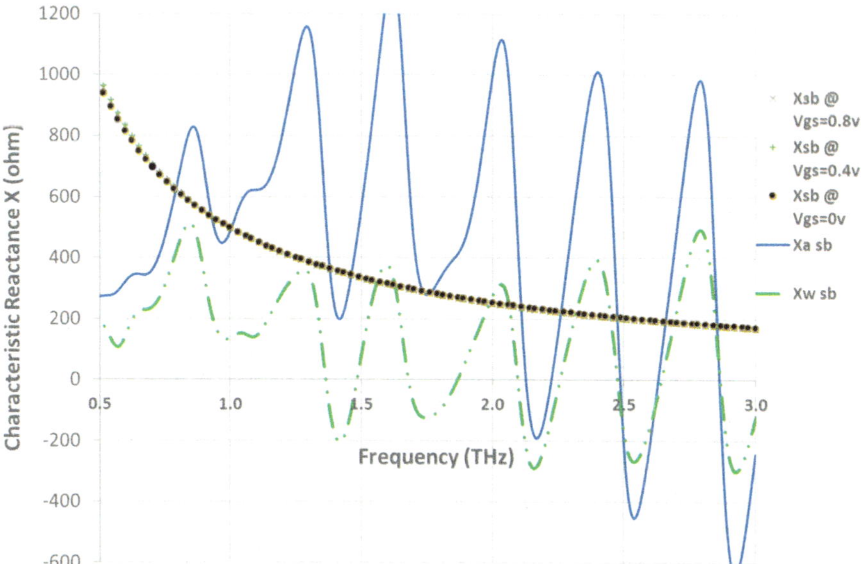

Fig. 10 Characteristic reactance between the source and gate contacts (solid and dashed lines, for short and long metal wires, respectively) and between the source/body terminals of the FET under varying gate/source potentials (marker lines)

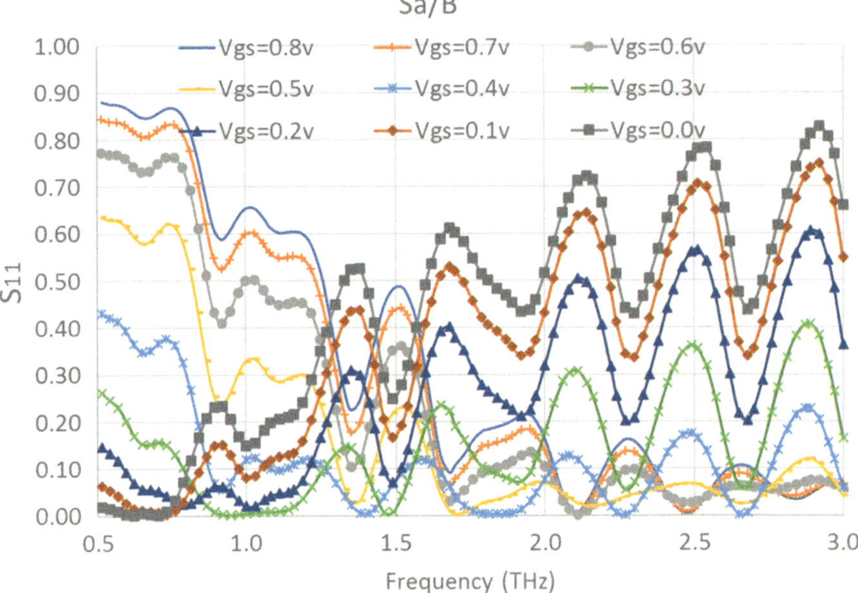

Fig. 11 Scattering coefficient of the coupling between source/body contacts (with short metal wire strip connected) and the FET device subject to varying gate potential

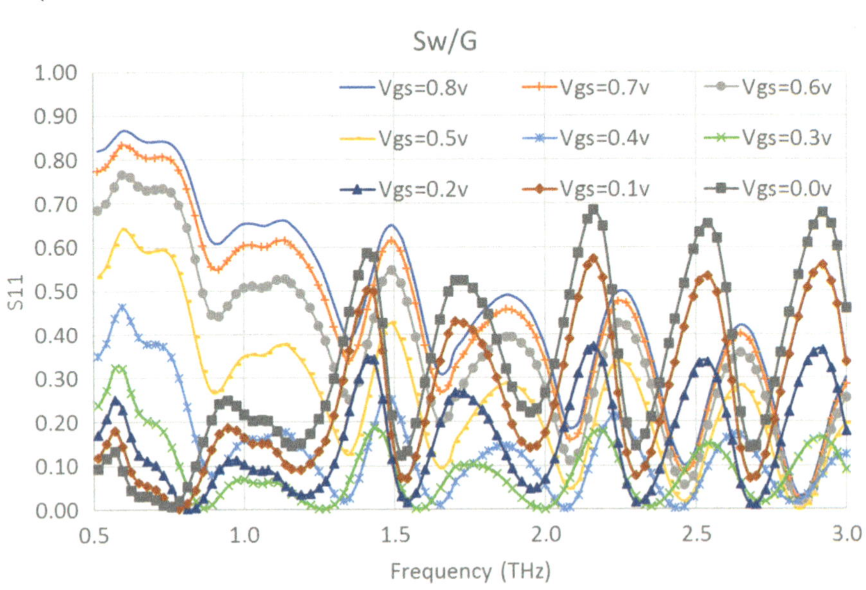

Fig. 12 Scattering coefficient of the coupling between source/body contacts (with long metal wire strip connected) and the FET device subject to varying gate potential

it caused reduction in the overall impedance, decreasing the scattering coefficient and increasing data transmission to the FET.

Meanwhile, despite the independence of the FET reactive impedance (capacitance) of the gate potential, the resistive FET impedance shows strong dependence. As a result, gate potential can be seen to provide response tunability as in the above cases. It is important to note that inductive response of the contacts can be seen repeatedly in the reactive characteristics of the source wires especially in the short wires case (Fig. 10). This opens a window for engineering the FET structure and contacts to reach RLC resonance conditions.

4 Conclusion

When FETs are operated at terahertz radiation detectors beyond their cutoff frequency, their contacts play an effective role in the detection process. In this range, internal scattering between the contacts and the FET channel gains increasing significance in the FET operation. For this reason, we conducted a simulation of the characteristic impedances of the contacts when operated as an antenna and the electrical impedance of the FET channel in this frequency range.

The results show that it is important to take into consideration the design of the contacts themselves as a key element in detecting and coupling the high-frequency signal to the FET channel. The results also show that the quality of the detected signal varies significantly depending on the effective input port. If broadband signal detection is targeted, then optimizing the FET design to extract gate source signal in the presence of grounded body plate is advised. However, for effective narrowband detection, it would be advised to adjust the design and operation conditions of the FET to extract the signal delivered by the source-body terminal.

Finally, our results indicated that proper adjustment of the contacts impedance relative to the FET impedance can strongly affect the operation of FET THz detectors. There are indications that the device may be driven to LC resonance when the FET contacts exhibit inductive characteristics, comparable to FET's small capacitive characteristics. Matching the resistive components may decrease the signal damping enough to drive the FET into resonance.

References

1. Knap, W., Nadar, S., & Videlier, H., et al. (2010). Field Effect transistors for fast terahertz detection and imaging. In *The 18th International Conference on Microwave Radar and Wireless Communications (MIKON), 2010, Vilnius, Lithuania*, pp. 1–3.
2. Veksler, D., Teppe, F., Dmitriev, A., et al. (2006). Detection of terahertz radiation in gated two-dimensional structures governed by dc current. *Physical Review B, 73*(12), 125328.
3. Ibrahim, N., Rafat, N., & Elnahwy, S. (2015). Drift transport model of field effect transistors in saturation beyond cutoff. *Journal of Physics D: Applied Physics, 48*(13), 135102.

4. Tauk, R., Teppe, F., Boubanga, S., et al. (2006). Plasma wave detection of terahertz radiation by silicon field effects transistors: Responsivity and noise equivalent power. *Applied Physics Letters, 89*(25), 253511–253513.
5. But, D., Drexler, C., Sakhno, M., et al. (2014). Nonlinear photoresponse of field effect transistors terahertz detectors at high irradiation intensities. *Journal of Applied Physics, 115*(16), 164514.
6. Kopyt, P., Marczewski, J., Kucharski, K., et al. (2011). Planar antennas for THz radiation detector based on a MOSFET. In *36th International Conference on Infrared, Millimeter and Terahertz Waves (IRMMW-THz), Houston, USA*, pp. 1–2.
7. Ravaro, M., Locatelli, M., Viti, L., et al. (2014). Detection of a 2.8 THz quantum cascade laser with a semiconductor nanowire field-effect transistor coupled to a bow-tie antenna. *Applied Physics Letters, 104*(8), 083116.
8. Sakowicz, M., Łusakowski, J., Karpierz, et al. (2008). Polarization sensitive detection of 100 GHz radiation by high mobility field-effect transistors. *Journal of Applied Physics, 104*(2), 024519.
9. Ibrahim, N., Rafat, N. H., & Elnahwy, S. E.-D. (2016). Simulation study for the use of transistor contacts for sub-terahertz radiation detection. *IET Microwaves, Antennas & Propagation, 10*(7), 784–790.

Second-Order Rectification of High-Frequency Radiation in Bipolar Junction Transistor

Ahmed M. Elsayed, Hassan M. Emam, Hussein S. Ahmed,
Yousof O. Moustafa, and Nihal Y. Ibrahim

Abstract Broadband rectification of high-frequency electromagnetic radiation has been observed in field effect transistors and heterostructure transistors. It was used for terahertz radiation detection and imaging. In this work we report experimental evidence of rectification in bipolar junction transistor beyond its cutoff frequency. We used discrete off-the-shelf NPN bipolar junction transistor to detect high-frequency electromagnetic radiation. The incident radiation was shown to exceed the transistor's cutoff frequency. Despite the complete decay of the AC component, we have shown that the DC characteristics of the transistor showed nonlinear dependence on the incident radiation. We suggest that this response is second-order rectification of the incident radiation. In this case, current models of second-order rectification need to be revised to account for the physics of bipolar transistors.

Keywords Rectification · Radiation detection · Bipolar junction transistors · Cutoff frequency

1 Introduction

Since the invention of the transistor by Shockley, Bardeen, and Brattain [1], the transistor's operation was divided into DC bias and AC amplification. The former was considered independent of any applied AC signal. Now, most of the transistors fall into two main categories which are bipolar junction transistors (BJT) and field effect transistors (FET). Many different variations within these categories have

A. M. Elsayed · H. M. Emam (✉) · H. S. Ahmed · Y. O. Moustafa
Faculty of Engineering, Department of Communications and Electronics Engineering, Cairo University, Giza, Egypt

N. Y. Ibrahim
Faculty of Engineering, Department of Engineering Mathematics and Physics, Cairo University, Giza, Egypt

© Springer Nature Switzerland AG 2020

M. H. Farouk, M. A. Hassanein (eds.), *Recent Advances in Engineering Mathematics and Physics*, https://doi.org/10.1007/978-3-030-39847-7_12

emerged such as HBTs [2], FinFET [3], Tunnel-FET [4], and graphene-FET [5], with improved characteristics, frequency response, and power consumption.

Recent work on terahertz radiation detection using FETs [6] shed light on a new mode of operation in field effect transistors (FET). In this mode, high-frequency signals can affect the DC response of the FET beyond its cutoff frequency through rectification of second-order terms of the incident signal. While this response is very small, it attracted a lot of attention because it can be used in applications such as high-speed, room-temperature THz imaging [7]. This effect was lately seen also in high-speed heterostructure bipolar transistor [8]. Multiple theories were proposed to model the physics of the buildup of second-order rectification (SOR) in these devices [9, 10]. However, these theories were mostly based on the physics of electron plasma in a two-dimensional electron gas in the FET channel and the transport of high-mobility charge carriers.

In this work, we report the detection of second-order rectification (SOR) in bipolar junction transistors beyond their cutoff frequency. To the best of the authors' knowledge, this is the first work reporting SOR detection in BJT. We used a commercially available BJT with low cutoff frequency and applied electromagnetic radiation with frequency that far exceeds the cutoff frequency of the transistor. We were able to detect a change in the DC characteristics of the BJT due to this applied radiation. These results call for a revised theory of the development of SOR that is independent of the physics of the FET channel and of high-mobility charge carriers.

In the following section, an overview of the experimental setup is presented followed by the results of the experiment. These results are then discussed in Sect. 3, and the conclusion is finally presented in Sect. 4.

2 Experimental Setup

To extract SOR signal in BJT transistors beyond the cutoff frequency, we used the commercially available small-signal NPN-BJT (2N3904) [11]. Its transition frequency is 25 MHz, with current gain between 30 and 300, and the maximum collector current is 200 mA.

The circuit configuration tested was a simple common collector configuration. Variable base bias was supplied through a 220 Ω resistance, and the output potential was measured directly from the emitter. Figure 1 shows a schematic of the transistor circuit described above. The DC characteristics of the transistor were extracted first in the absence of any incident radiation and are shown in Fig. 2. The results were registered using GWINSTEK GDS-2102A digital oscilloscope.

Next, high-frequency radiation was applied directly to the transistor through a 1mw-10GHz RF transmitter. As can be seen, the transmitter frequency far exceeds the cutoff frequency of the transistor.

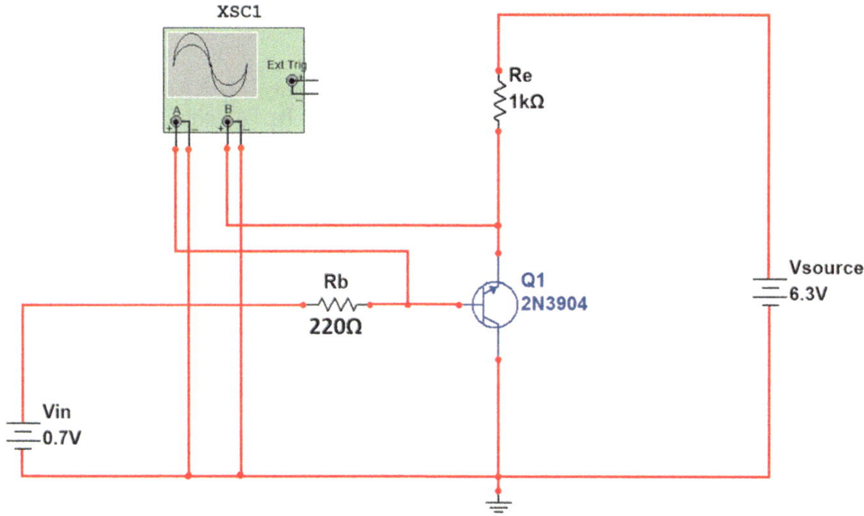

Fig. 1 Schematic circuit diagram for the experimental setup used. The input voltage is changed from 0 to 1 V throughout the experiment

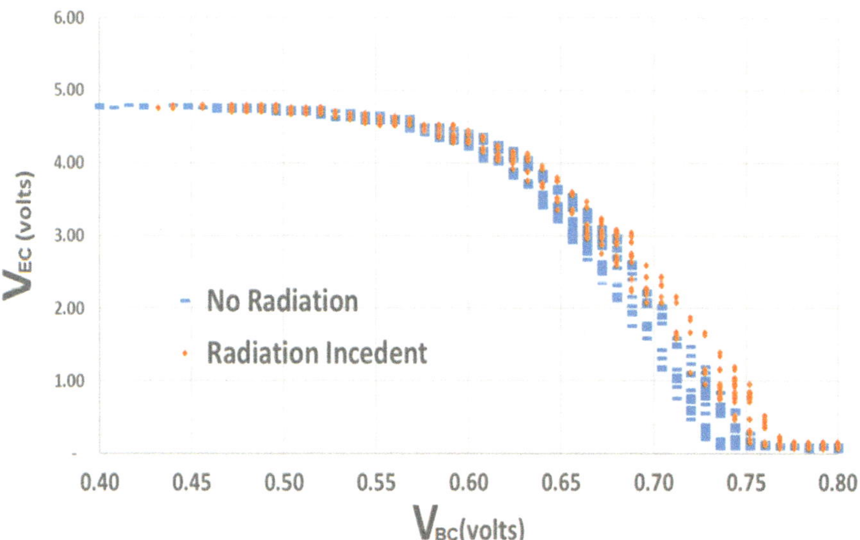

Fig. 2 Experimental (unprocessed) data of the DC emitter/collector potential at varying base potential. Blue dashes indicate the DC characteristics in the absence of incident radiation, while orange dots indicate the DC characteristics when the high-frequency radiation is incident on the transistor

3 Results and Discussion

The transistor used in this work is a small-signal transistor with a cutoff frequency that is much lower than that of the applied radiation. Its DC characteristics in the absence of incident radiation are shown in Figs. 2 and 3. When the same DC characteristics were extracted in the presence of incident high-frequency radiation, small deviation from the DC characteristics was detected around the saturation potential of the transistor (Figs. 2 and 3). This change in the DC characteristics, due to incident AC high-frequency radiation, can be related to nonlinear rectification of higher harmonics of the incident radiations generated within the transistor. The figures show maximum V_{EC} of 4.75 V instead of the applied 6.3 V, which can be attributed to the existence of a light-emitting diode connected to the battery.

It must be mentioned that the results showed an increase in the noise generated around the saturation point potential. This can be seen in Fig. 2 where the original measured data were plotted; however, the deviation from the average DC response is clear.

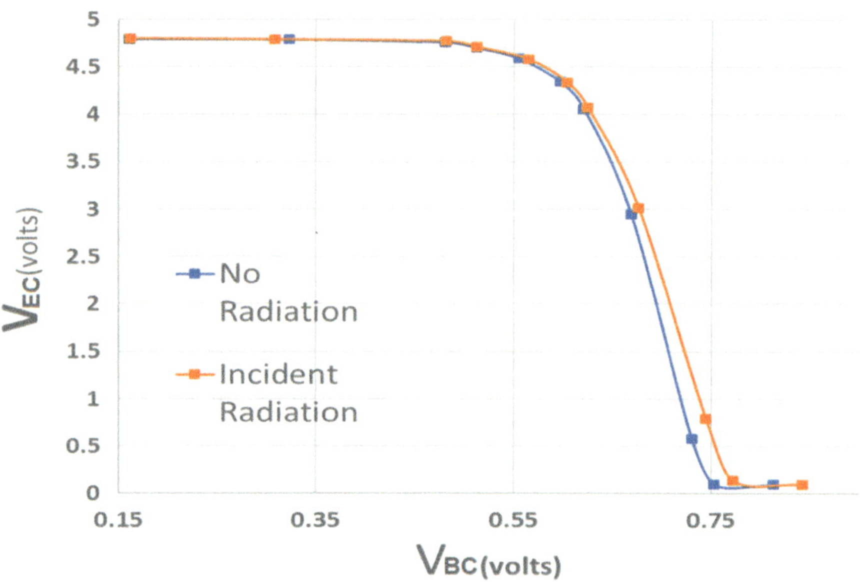

Fig. 3 Average experimental DC transistor characteristics of the DC emitter/collector potential at varying base potential. Blue dashes indicate the DC characteristics in the absence of incident radiation, while orange dots indicate the DC characteristics when the high-frequency radiation is incident on the transistor

4 Conclusion

As mentioned above, the results given here are, to the best of the authors' knowledge, the first indication of SOR in silicon BJTs. In silicon BJTs, there is no 2DEG, populated by high-mobility charges. Therefore, available theories explaining the emergence of SOR in other devices cannot be applied to explain the results presented here. However, recent work [12] have indicated that the SOR in FETs may be explained by the nonlinearity of terminal boundary conditions of the transistor instead of in the FET's channel. This may be a starting point to explain and model the SOR seen in this experiment.

The detection of SOR is believed to be a promising candidate of ultrahigh-frequency detection such as terahertz radiation. It provides a strong candidate for low-cost, high-speed, room-temperature candidate for terahertz radiation imaging which is a field of high-demand applications. Proper understanding and modeling of the SOR would allow increasing the SOR magnitude, therefore increasing the responsivity of the sensing element (Fig. 4).

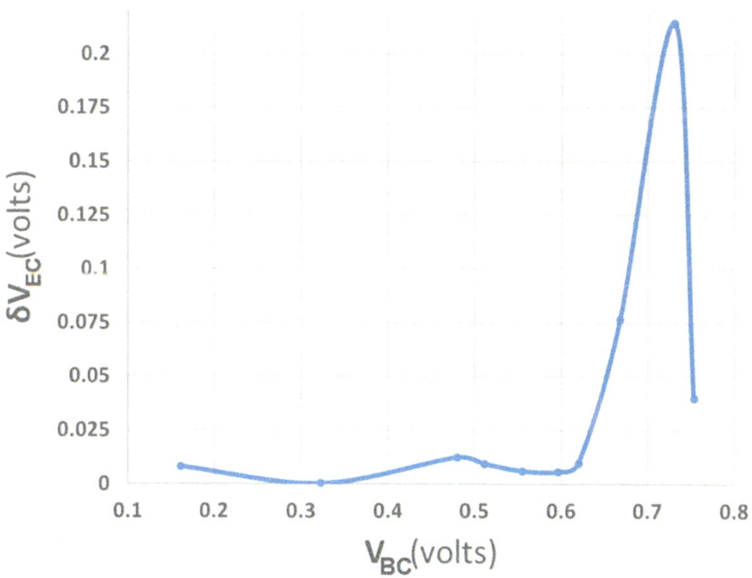

Fig. 4 SOR of the transistor represented as ΔV_{EC} which is the change in the DC emitter potential when high-frequency radiation is incident on the transistor

References

1. Shockley, W., Bardeen, J., & Brattain, W. (1948). The electronic theory of the transistor. *Science, 108*, 1948.
2. SiGe HBT technology: a new contender for Si-based RF and microwave circuit applications - IEEE Journals & Magazine. Retrieved July 28, 2019, from https://ieeexplore.ieee.org/abstract/document/668665
3. Yeung, C. W., Zhang, C., Xu, P., Bu, H., & Cheng, K. (2019). *Fin field-effect transistor for input/output device integrated with nanosheet field-effect transistor*. US10297667B1.
4. Lin, C. H., Tang, C. J., Chen, H. Y., & Hsieh, S. W. (2019). *Tunneling field effect transistor and method for fabricating the same*. US20190131453A1.
5. Di Lecce, V., Grassi, R., Gnudi, A., Gnani, E., Reggiani, S., & Baccarani, G. (2013). Graphene Base transistors: A simulation study of DC and small-signal operation. *IEEE Transactions On Electron Devices, 60*(10), 3584–3591.
6. Elkhatib, T. A., Kachorovskii, V. Y., Stillman, W. J., Rumyantsev, S., Zhang, X.-C., & Shur, M. S. (2011). Terahertz response of field-effect transistors in saturation regime. *Applied Physics Letters, 98*(24), 243505–243503.
7. Al Hadi, R., et al. (2012). A 1 k-pixel video camera for 0.7–1.1 terahertz imaging applications in 65-nm CMOS. *IEEE Journal of Solid-State Circuits, 47*(12), 2999–3012.
8. Coquillat, D., et al. (2016). High-speed room temperature terahertz detectors based on InP double heterojunction bipolar transistors. *International Journal of High Speed Electronics and Systems, 25*(03n04), 1640011.
9. Veksler, D., Teppe, F., Dmitriev, A. P., Kachorovskii, V. Y., Knap, W., & Shur, M. S. (2006). Detection of terahertz radiation in gated two-dimensional structures governed by dc current. *Physical Review B, 73*(12), 125328.
10. Preu, S., Kim, S., Verma, R., Burke, P. G., Sherwin, M. S., & Gossard, A. C. (2012). An improved model for non-resonant terahertz detection in field-effect transistors. *Journal of Applied Physics, 111*(2), 024502–024509.
11. "2N3904: NPN Bipolar Transistor, TO-92." [Online]. Retrieved Jul 28, 2019, from https://www.onsemi.com/PowerSolutions/product.do?id=2N3904
12. Ayoub, A. B., Ibrahim, N. Y., & Elnahwy, S. E. A. (2020). 2nd order non-quasi-static, compact model of field effect transistor revealing terminal rectification beyond their cutoff frequency. *IET Circuits, Devices & Systems*.

Structural and Optical Properties of Thin Layers of Zinc Oxide Under the Effect of Microdroplet Doping

N. Hamzaoui, M. Ghamnia, and A. Boukhachem

Abstract Zinc oxide (ZnO) is a binary material, a semiconductor with a large direct gap (3.3 eV). With their good optoelectronic properties, thin films find several applications such as solar cells, gas sensors, sensors piezoelectric, etc. In this work, thin films of aluminum-doped zinc oxide and lithium-doped zinc oxide are obtained using a simple process called microdroplet. The synthesized samples were characterized by X-ray diffraction (XRD), atomic force microscope (AFM) and UV-visible spectroscopy. X-ray diffraction (XRD) analysis reveals that, thanks to the incorporation of aluminum or lithium, all the thin films prepared are well crystallized in the structure of wurtzite with the preferential orientation (002) in the direction parallel to the C axis. The morphology of the surface of the doped ZnO layers is studied by AFM atomic force microscopy which showed less rough images as well as the surface topography measured between 4 and 10 nm. The optical study by UV-visible spectroscopy revealed a high transparency of the order of 80% and a visible domain emission linked to the doped ZnO compound.

Keywords Zinc oxide · Microdroplet · X-ray diffraction

N. Hamzaoui (✉)
Ecole Supérieure en Génie Electrique et Energétique d'Oran, Oran, Algeria

Laboratoire LSMC, Département de Physique, Faculté des Sciences Exactes et Appliquées, Université d'Oran 1 Ahmed Ben Bella, Oran, Algeria

M. Ghamnia
Laboratoire LSMC, Département de Physique, Faculté des Sciences Exactes et Appliquées, Université d'Oran 1 Ahmed Ben Bella, Oran, Algeria

A. Boukhachem
Unité de Physique des Dispositifs à Semi-Conducteurs, Faculté des Sciences de Tunis, Université de Tunis El Manar, Tunis, Tunisia

M. H. Farouk, M. A. Hassanein (eds.), *Recent Advances in Engineering Mathematics and Physics*, https://doi.org/10.1007/978-3-030-39847-7_13

1   Introduction

Zinc oxide is a metal oxide that is considered as one of the most important II-V semiconductor materials as it is characterized by its wide band gap energy of 3.3 eV [1] and n-type conductivity. Doping metal oxides was used to improve and modify the characteristics of the oxides to enhance finally the performance of various electronic devices. Many methods were used to obtain ZnO thin films. Thin films of aluminum-doped zinc oxide (ZnO/Al) and lithium-doped zinc oxide (ZnO/Li) were prepared by a chemical simple microdroplet method inexpensive.

The present work aims to study the effect of aluminum or lithium doping on the structural, morphological, and optical properties of thin ZnO films.

2   Experimental

The thin layers of ZnO/Al and ZnO/Li were synthesized by the microdroplet method in aqueous solution. The experimental device includes a heating plate for the glass substrates which were cleaned in an ultrasonic bath. A prepared solution is deposited in the form of drops; each drop is characterized by a diameter of the orifice of the pipette. The precursor solution was prepared from 5×10^{-2} M zinc dihydrate acetate [Zn $(CH_3COO)_2$, $2H_2O$] in a beaker containing 20 mL of methanol as the solvent. The aluminum doping was carried out by adding aluminum nitrate nonahydrate (Al $(NO_3)_3$, $9H_2O$) as an aluminum precursor in the initial solution. In another beaker we proceeded to dissolve controlled quantities of hydrated lithium chloride (LiCl, H_2O) in a solution of zinc acetate dissolved in methanol under the same conditions. The solutions were deposited on heated glass substrates between 100 and 200 °C. The number of drops deposited on substrates is 20 drops.

X-ray diffraction measurements of thin films of ZnO/Al or ZnO/Li were characterized with a copper source diffractometer with CuK. Optical analyses, in the visible UV-spectrum, were recorded with a Specord 50 plus dual-beam spectrophotometer apparatus in a 300–1100 nm wavelength range. Topography and roughness of the surface were measured by Bruker's Edge Dimension Atomic Force Microscope in tapping mode using tips of antimony and cantilever doped silicon material between 115 and 135 μm in length.

3   Results and Discussion

3.1   Structural Properties

The structural study by using X-ray diffraction was realized out to identify the phases of the pure ZnO/Al films and ZnO/Li and some other parameters. Figure 1

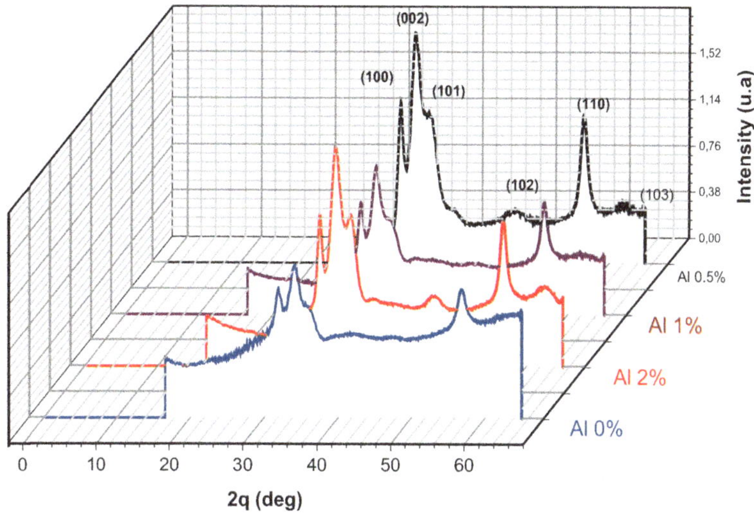

Fig. 1 X-ray diffraction spectra of ZnO thin films prepared with different rates Al doping

Table 1 Average values of grain size of aluminum-doped ZnO thin films	Dopant Al (%)	D (nm)
	0.0	88.95
	0.5	88.99
	1.0	88.98
	2.0	89.02

shows the X-ray diffraction spectra of thin films of pure ZnO and all thin films of ZnO/Al.

The spectra of the aluminum-doped zinc oxide films show defined peaks of (103), (102), (101), (100), and (002) principal orientation, corresponding to the hexagonal structure of wurtzite, in accordance with JCPDS N0 36-1451. The size of the crystallites and the total width at the maximum half (FWHM) of the peaks (hkl) are deduced with the Debye-Scherrer relation:

$$D = \frac{K\lambda}{\beta_{\frac{1}{2}} \cos\theta} \tag{1}$$

where $k = 0.90$ is the Scherrer constant, $\beta_{1/2}$ is the full width at half maximum of (002) peak, and $\lambda = 1,5406$ Å is the wavelength of Cu Ka radiation. The calculated crystallite size values are shown in Table 1.

Peak intensity is observed to increase for (002), (101), and (102) with increasing aluminum concentration and then decrease with the decrease of the latter. These results are similar with many of the results of other researchers [2, 3]. We conclude that the atomic concentration of the dopant influences the growth kinetics of the thin films.

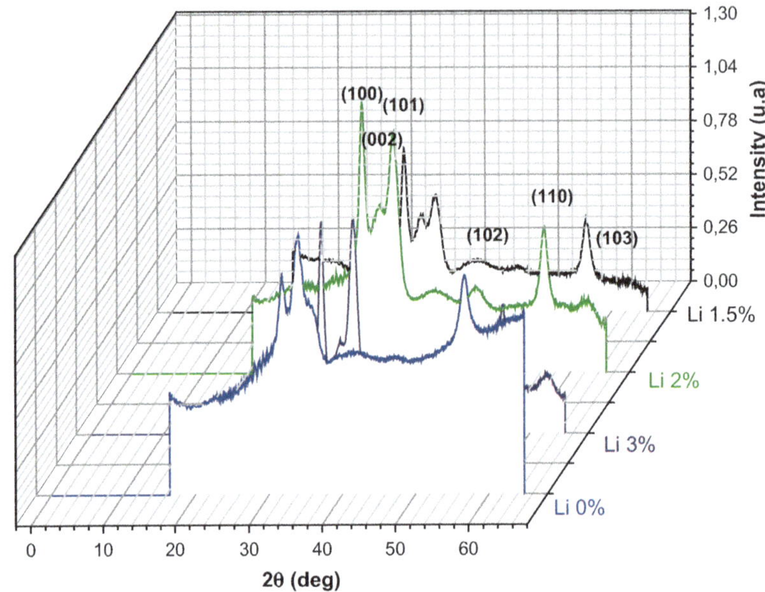

Fig. 2 X-ray diffraction spectra of thin films of ZnO prepared with different rates of Li doping

Table 2 Average values of grain size, dislocation density, and microstrain of lithium-doped ZnO thin films	Dopant Li (%)	D (nm)	ε (10^{-4})	δ (10^5 line/m^2)
	1.5	89.02	40.63	12.61
	2	89.05	40.62	12.60
	3	89.09	40.60	12.59

As in the case of ZnO/Al, the ZnO/Li films show a similar X-ray diffraction pattern, or we can see the peaks relative to the different planes corresponding to the polycrystalline structure in Fig. 2.

In addition, the orientation of the peak, representing the plane (002), has a lower intensity in the pure ZnO film than in all ZnO/Li films. This indicates that the incorporation of lithium has changed the crystallinity of the films. From the results presented in Table 2, it can be seen that the position of the peak (002) shifted gradually to higher angles with increasing percentage of lithium.

Lithium atoms take interstitial sites rather than replace Zn site; they deform network parameters [4]. Thereby, the microstrain (ε), developed in these thin films, was estimated with this relation:

$$\varepsilon = \frac{\beta_{\frac{1}{2}}}{4 \tan \theta} \tag{2}$$

Also, the dislocation density (δ) was obtained by the relation below:

$$\delta = \frac{1}{D^2} \tag{3}$$

The microstrain decreased from 40.63×10^{-4} to 40.60×10^{-4} when the concentration of lithium increased. In addition, the dislocation density decreases from 12.61 to 12.59 lines/m^2 with the increase of the lithium doping rate.

3.2 Morphological Analysis

Atomic Force Microscopy images of various thin films pure ZnO are shown in Fig. 3 and ZnO/Al, ZnO/Li with different doping levels are shown in Figs. 4, 5, 6, 7, 8, 9, and 10.

The images were processed with WSXM 5.0 software (Windows Scanning X Microscope) [5]. The AFM images in Figs. 3, 4, 5, 6, 7, 8, 9, and 10a, c show the morphology of pure ZnO films, ZnO/Al and ZnO/Li in 2D and 3D. The surface of the films is formed of small round and interconnected grains which form a more or less smooth surface morphology. The profile shown in Figs. 3, 4, 5, 6, 7, 8, 9, and 10b shows the variation in Z height; the profile depends on the doping concentration and the nature of the dopant.

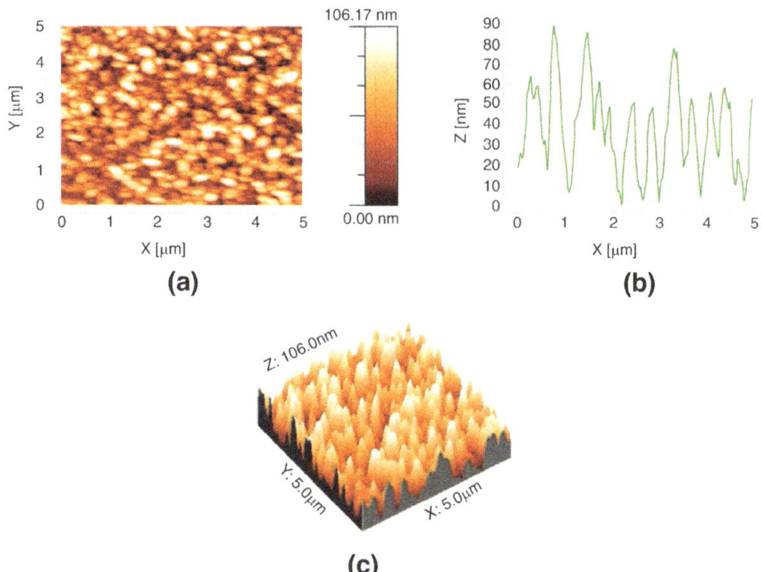

Fig. 3 AFM images of ZnO pure (**a**) 2D image, (**b**) approximate profile, and (**c**) 3D image of a thin film

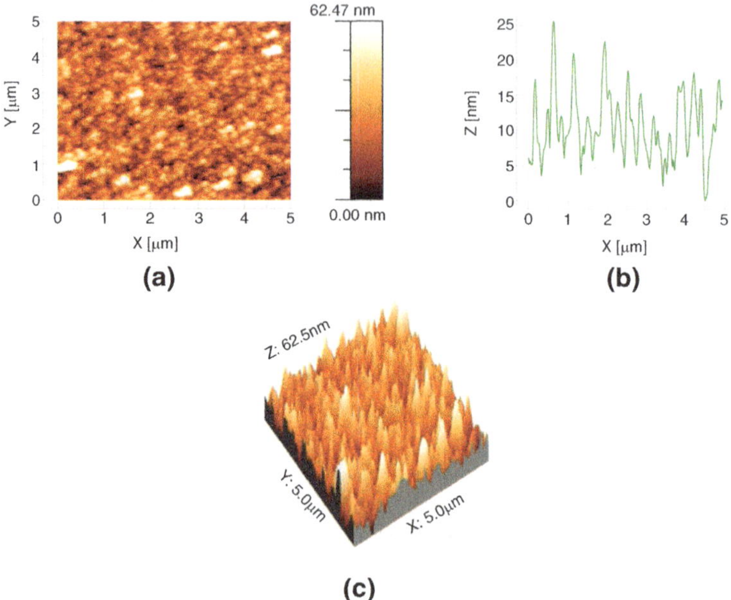

Fig. 4 AFM images of Al-doped ZnO (**a**) 2D image, (**b**) approximate profile, and (**c**) 3D image of a thin film for a doping rate of 0.5%

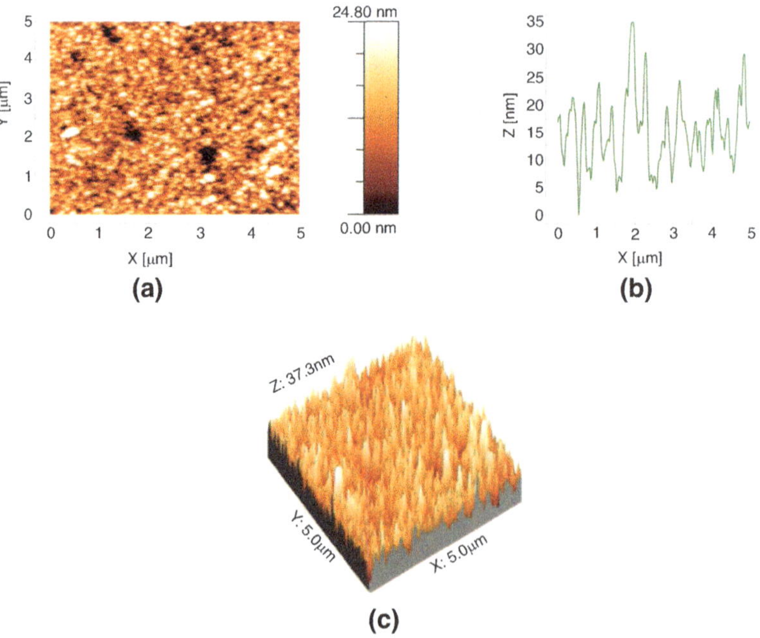

Fig. 5 AFM images of Al-doped ZnO (**a**) 2D image, (**b**) approximate profile, and (**c**) 3D image of a thin film for a doping rate of 1%

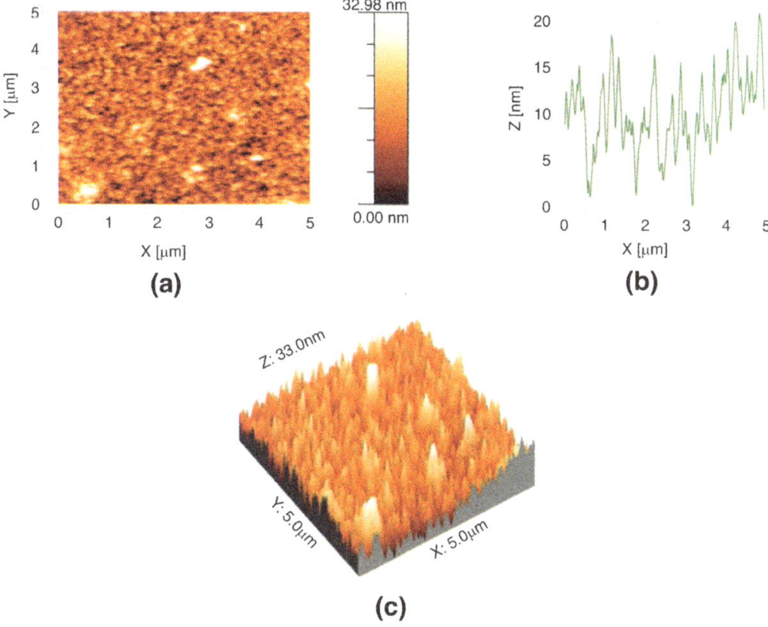

Fig. 6 AFM images of Al-doped ZnO (**a**) 2D image, (**b**) approximate profile, and (**c**) 3D image of a thin film for a doping rate of 1.5%

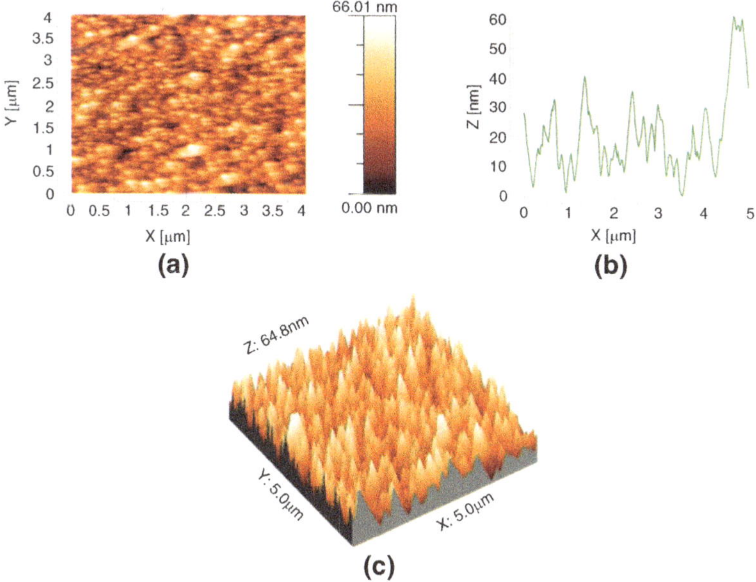

Fig. 7 AFM images of Al-doped ZnO (**a**) 2D image, (**b**) approximate profile, and (**c**) 3D image of a thin film for a doping rate of 2%

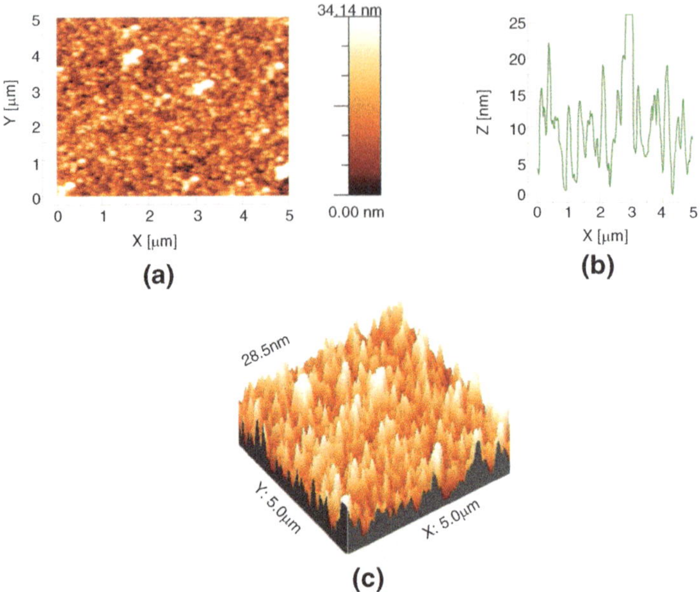

Fig. 8 AFM images of Li-doped ZnO (**a**) 2D image, (**b**) approximate profile, and (**c**) 3D image of a thin film for a doping rate of 1.5%

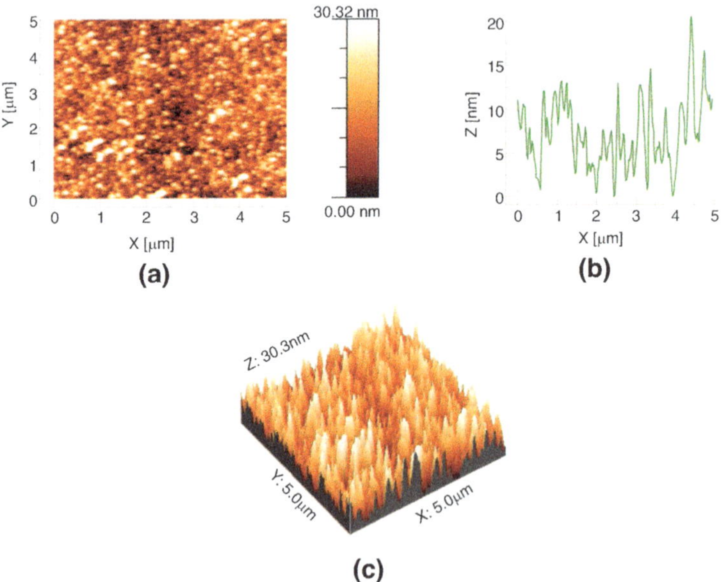

Fig. 9 AFM images of Li-doped ZnO (**a**) 2D image, (**b**) approximate profile, and (**c**) 3D image of a thin film for a doping rate of 2%

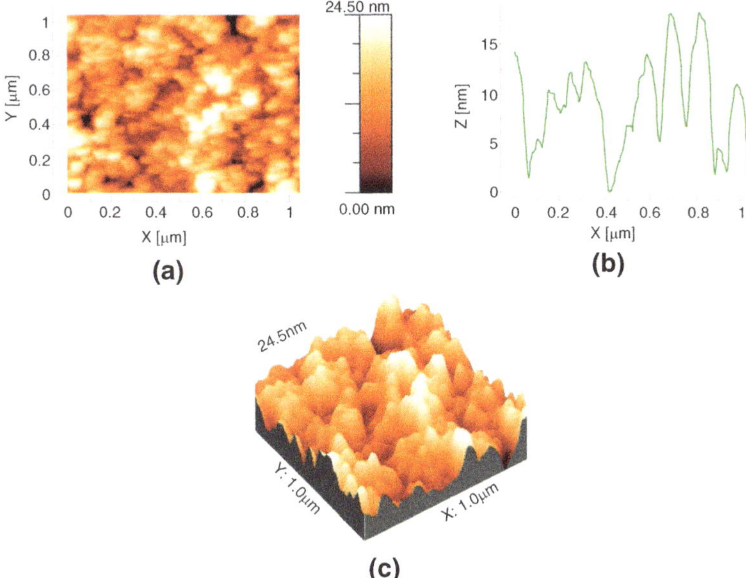

Fig. 10 AFM images of Li-doped ZnO (**a**) 2D image, (**b**) approximate profile, and (**c**) 3D image of a thin film for a doping rate of 3%

3.3 Optical Properties

The spectra of transmission and optical absorption of ZnO/Al and ZnO/Li in a 300–1100 nm wavelength range have been shown in Figs. 11 and 12.

These films are characterized by high transparency in the visible range with average transmittance ranging from 60 to 85%. Transmission is maximum for pure ZnO. Also, a very low absorptivity in this range of wavelength [6]. From the transmission spectra, we can determine the absorption coefficient (α) and thereby band gap energy:

$$\alpha = \frac{1}{d} \ln \left(\frac{1}{T} \right) \tag{4}$$

Furthermore, the absorption coefficient can be given by

$$(\alpha h\nu)^2 = A(h\nu - Eg) \tag{5}$$

where A is a constant, hν corresponds to the photon energy, and Eg represents the optical band gap energy. Obviously, the Eg decrease from 3.25 to 3.20 eV, respectively, for 0.5% and 2% (Fig. 13). For ZnO/Li the gap is 3.38 to 3.34 eV 1.5 to 3% (Fig. 14). This result can be explained by ZnO crystal imperfection which can be related to the influence of various factors, such as thickness and grain size.

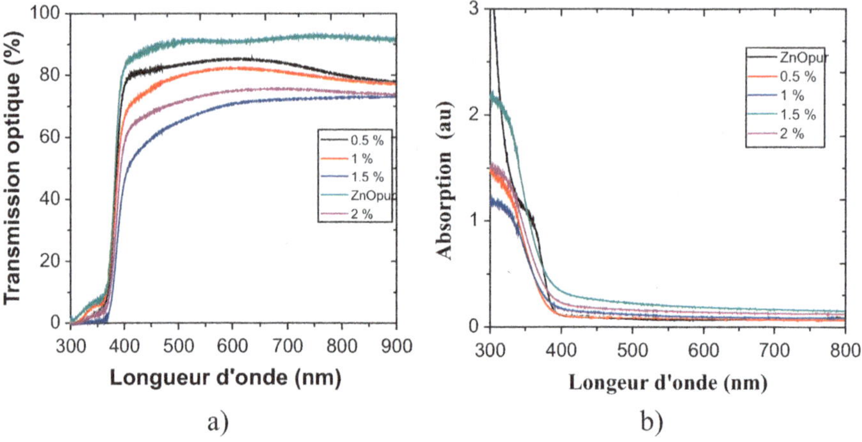

Fig. 11 Transmission (**a**) and absorption (**b**) spectrum of Al-doped ZnO samples

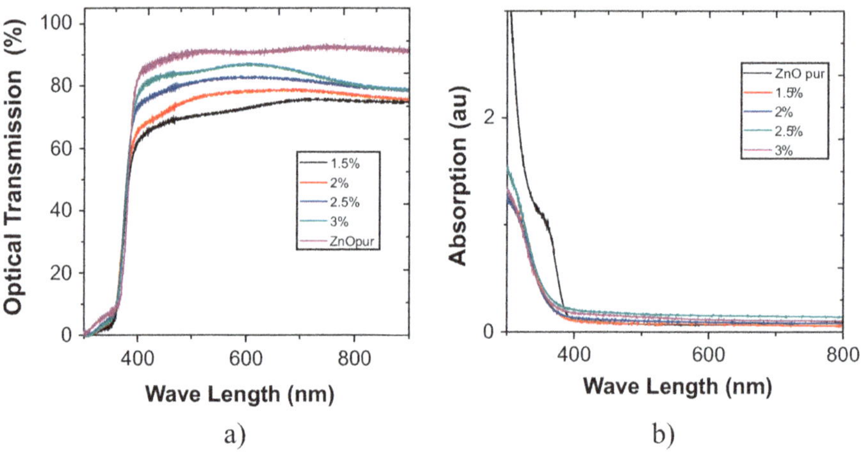

Fig. 12 Transmission (**a**) and absorption (**b**) spectrum of Li-doped ZnO samples

4   Conclusion

Thin layers of aluminum- or lithium-doped ZnO have been synthesized with different.

percentages of doping on glass substrates using the microdroplet technique. In conclusion, we studied the effect of aluminum lithium doping on the structure and morphological and optical properties. X-ray diffraction study shows that all prepared ZnO/Al and ZnO/Li films are in hexagonal polycrystalline wurtzite phase (with preferential orientation of the C axis). The peak (002) is shifted to a Bragg angle of 34°. The morphological study of different thin films of ZnO/Al and ZnO/ Li shows

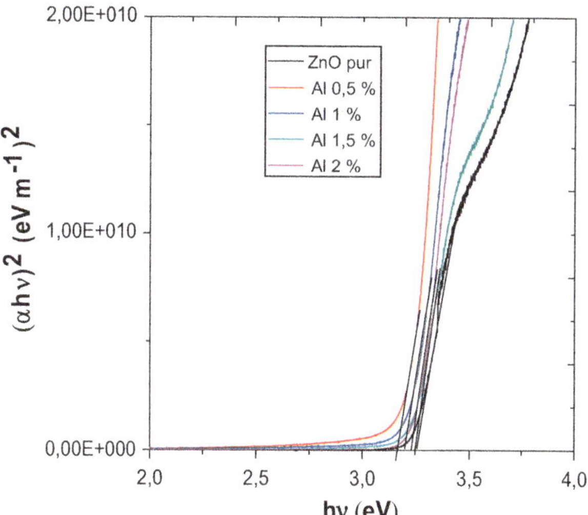

Fig. 13 Plot of $(\alpha h\upsilon)^2$ versus photon energy $(h\upsilon)$ for thins films of Al-doped ZnO

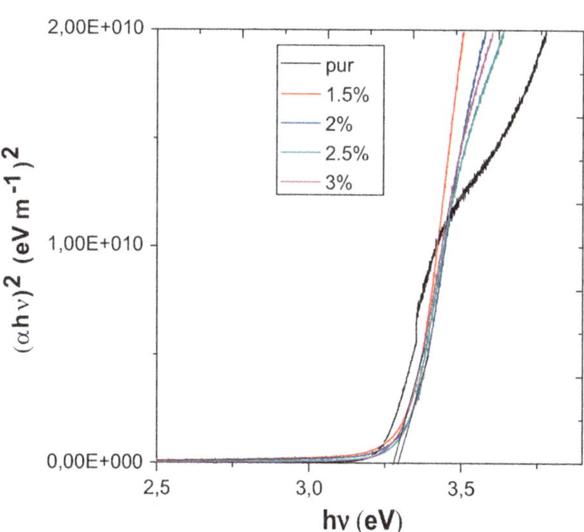

Fig. 14 Plot of $(\alpha h\upsilon)^2$ versus photon energy $(h\upsilon)$ for thins films of Li-doped ZnO

the surface going from a rough morphology to a more or less smooth morphology. Doping has reduced the roughness of surfaces. The different profiles of the Z height show this behavior well. The transmittance varies from 60 to 85% when the doping rate varies. The variation of the ZnO/Al gap is from 3.25 to 3.20 eV and for thin ZnO/Li films, from 3.38 to 3.34 eV.

 N. Hamzaoui et al.

References

1. Manoharan, C., Pavithra, G., Dhanapandian, S., Dhamodaran, P., & Shanthi, B. (2015). *Spectrochimica Acta, Part A: Molecular and Biomolecular Spectroscopy, 141*, 292–299.
2. Kayani, Z. N., Nazir, F., Riaz, S., & Naseem, S. (2015). *Superlattices and Microstructures, 82*, 472–482.
3. Viswanatha, R., ArthobaNayak, Y., Venkatesha, T. G., & Vidyasagar, C. C. (2013). Synthesis, characterization and optical properties of Sn- ZnO nanoparticle. *Nanoscience and Nanotechnology: An International Journal, 3*(8), 16–20.
4. Zeng, Y.-J., Ye, Z.-Z., Xu, W.-Z., Chen, L.-L., Li, D.-Y., Zhu, L.-P., Zhao, B.-H., & Hu, Y.-L. (2005). *Journal of Crystal Growth, 283*, 180–184.
5. Horcas, I., Fermandez, R., Gomez-Rodriguez, J. M., Colchero, J., Gomez-Herrero, J., & Baro, A. M. (2007). *Rev Sci Intrum, 78*, 013702.
6. Wang, J., Meng, L., Qi, Y., Li, M., Shi, G., & Liu, M. (2009). The Al-doping contents dependence of the crystal growth and energy band structure in Al:ZnO thin films. *Journal of Crystal Growth, 311*, 2305–2308.

Part IV
Electromagnetic Waves

Design Optimization for High-Gain Quad Array of Helical Antennas for Satellite Applications

Maha Maged ⓘ, Mohammed El-Telbany ⓘ,
and Abdelrahman El-Akhdar ⓘ

Abstract This chapter presents a design of a high-gain quad array of helical antennas placed inside a truncated horn ground plane for satellite applications. The antennas within the truncated horn produce a circularly polarized and low side-lobe level. The gain in radiating direction is about 19:6dBi in the operating frequency range, which is more than 3dBi that of quad array using a square ground plane and lower side lobes. The design of the truncated horn is obtained by optimization process using genetic algorithms.

Keywords Helical antenna · Array · Ground plane · Optimization · Genetic algorithms

1 Introduction

In order to establish a wireless communication system for space applications, it is requisite to use antennas for radiating and receiving the radio waves [1]. Helical antennas have been widely used in space communications because of its circularly polarized, reasonably wide frequency band and relatively simple structure characteristics.

Using arrays of the helical antennas increases the gain compared to a single helical antenna of the same axial length. Moreover, by adjusting the geometry and shape of the ground conductor, various performances can be achieved [2–4]. Increasing the

M. Maged (✉)
Egyptian Space Agency, Cairo, Egypt
e-mail: maha_maged@narss.sci.eg

M. El-Telbany
Sinai University, North Sinai, Egypt
e-mail: mahammed.elsaid@su.edu.eg

A. El-Akhdar
Technical Research and Development Center, Cairo, Egypt
e-mail: a.m.el-akhdar@ieee.org

© Springer Nature Switzerland AG 2020
M. H. Farouk, M. A. Hassanein (eds.), *Recent Advances in Engineering Mathematics and Physics*, https://doi.org/10.1007/978-3-030-39847-7_14

gain of the helical antenna without changing the number of turns and its length is desirable to simplify the antenna fabrication, and the effect of gain boosting is strong [5]. Shaping the ground conductor has been considered as a useful method to enhance the gain and improve the radiation characteristics of helical antennas [6]. The ground plane of the helical antenna, with either cylindrical or truncated conical cup at the base of the helix, as shown in Fig. 1, improves the gain, front-to-back ratio, and side-lobe level, thereby reducing the susceptibility of the antenna to multipath [2, 6–8].

Djordjevic et al. [9] employ the search algorithms for optimizing the truncated-cone reflector. Olcan et al. [10] optimize both a helical antenna and a truncated-cone reflector simultaneously for maximum gain using particle swarm optimization (PSO). In this chapter, we present a design of a high-gain quad array of uniform helical antennas with an incorporated truncated horn where the quad array fits into the truncated horn. Optimally designed truncated horn reflector for quad-helical antennas improves the performance and gives better properties than helical antennas of the same size without this reflector. To find the optimal design parameters of the reflector for quad array of helical antennas, the genetic algorithm (GA) (optimization algorithm) is used, where the derivative of the cost function is unknown or not easily computed [11, 12]. The rest of the chapter is organized as follows. Section 2 defines the geometry of the considered quad array of helical antennas and presents the corresponding ground plane model. Section 3 describes the optimization algorithm and lists the optimal parameters of the horn ground plane. Section 4 presents the simulation and optimization results by FEKO software [13] of the quad array with horn ground plane. Section 5 concludes the chapter.

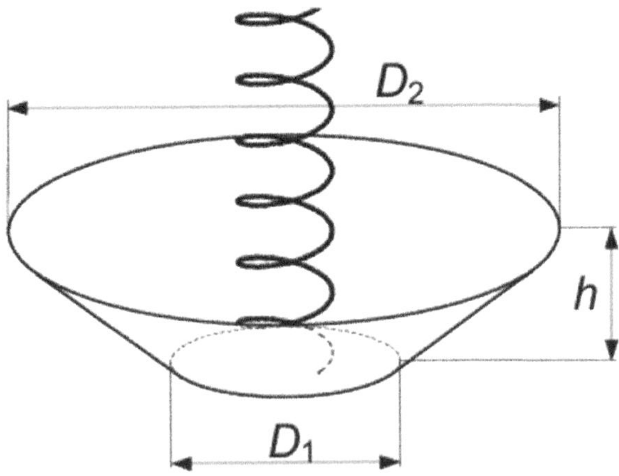

Fig. 1 Helical antenna with truncated-cone reflector [7]

2 Quad Array of Helical Antenna Geometry

2.1 Basic Helical Antenna Geometry

The geometry of an axial mode helical antenna is illustrated in Fig. 2. The geometrical parameters of a helical antenna are the turn number N, the diameter of one coil D, the spacing between each turn S, the wire length of one coil Lo, the antenna length L, the pitch angle α, and the circumference C, as shown in Fig. 2. The designed antenna consists of a four-turn conductor wound in the form of helix. The dimensions of the antenna are $C = 1.18\ \lambda = (3.9\ \text{cm})$, pitch angle $\alpha = 13°$, diameter of the helix $D = 1.25$ cm, and length of the helix $L = 3.5$ cm, where $\lambda = 3.3$ cm is the wavelength at $f = 9$ GHz.

2.2 Array of Helical Antenna Geometry

The designed quad array of helical antennas, which operates in the axial mode for the frequency range from 8.4 to 9.4 GHz, is shown in Fig. 3. The antennas are placed inside a truncated horn ground plane (a reflector) of height side h with the flare angle Φ.

The feeding points of the antennas are located at vertices of a square of side r (Fig. 3b). The center of the array coincides with the center of the truncated horn reflector. The 1×4 Wilkinson power divider with an equal RF power distribution ratio has been used for feeding network. This power divider provides high isolation between output ports and matches the antenna to 50 Ω.

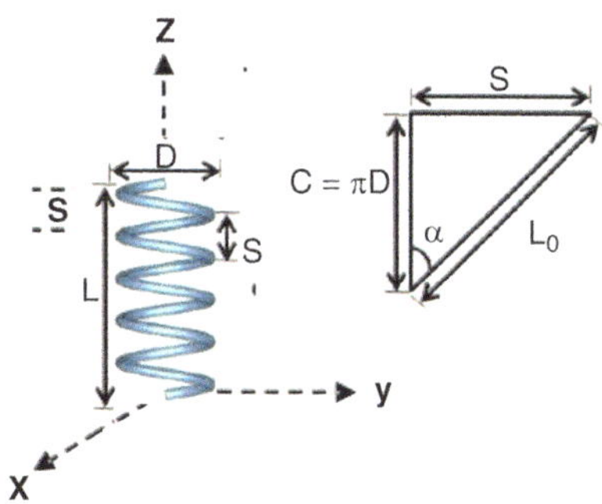

Fig. 2 Helical antenna geometry

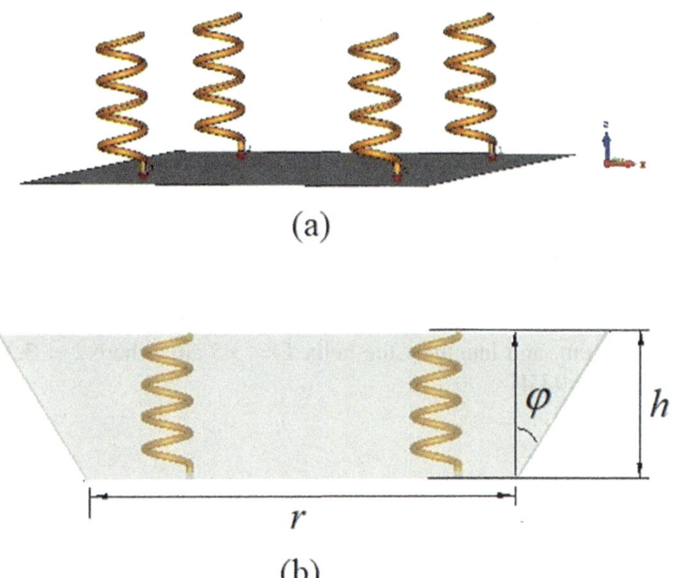

Fig. 3 Quad array helical antennas (**a**) with square ground plane and (**b**) with truncated horn ground plane

3   Optimization Strategy

Optimizing the design ground plane is accomplished by genetic algorithm (GA) [11, 12]. GA is a stochastic search method based on the Darwinian principles and concepts of natural selection and evolution. The GA provides optimal solutions by successively creating populations that improve over many generations. Selecting, mating, and mutating the previous population each creates a new generation. This process continues until the population converges to a single optimal solution. The state variables of the ground plane, which are changed during the optimization, are three real numbers (r, h, and ϕ) put into chromosome. The height (h) is changed within the limits $0.5\,\lambda \leq h \leq 1.5\,\lambda$, the ground width ($r$) is changed within the limits $2.5\,\lambda \leq h \leq 4\,\lambda$, and the angle ($\phi$) is changed from 25 to 50°. The optimization goal is to maximize the antenna gain (G) in the axial direction at the operating frequency $f_0 = 9$ GHz. Mathematically, this can be represented by the following constraint optimization problem:

$$\max \text{fitness}(r, h, \phi) = G \tag{1}$$

These parameters (r, h, and ϕ) are changed during the optimization to reach the maximum of the global objective function fitness. The GA is used by FEKO with the following control parameters in the optimization process: number of iteration = 20, population size $= 20$, crossover probability $= 0.5$, and creep mutation probability $= 0.5$, where population of individuals is randomly generated.

4 Simulation Results

The designed quad array helical antennas with truncated horn ground plane are simulated and optimized using FEKO software [13]. The optimum values for , h, and ϕ are then determined to be 3.6 λ, 1.2 λ, and 40.9°, respectively. A comparison of the gains for the antennas at $f = 9$ GHz is shown in Fig. 4. It can be observed that throughout the operational band, the optimized truncated horn demonstrates sufficiently high gains and low side lobes in the entire frequency range of interest.

Figure 5 demonstrates the simulated input reflection coefficients for the array with optimized parameters. It is seen that the value of $|S_{11}|$ is less than 10 dB from 8.4 to 9.4 GHz. Figure 6 shows that the simulated results represent that the quad array with truncated horn exhibits a good axial ratio (AR) of less than 3 at the operating frequency. The simulated gains against the frequency for the quad array helical antennas with square ground plane and with truncated horn ground plane are

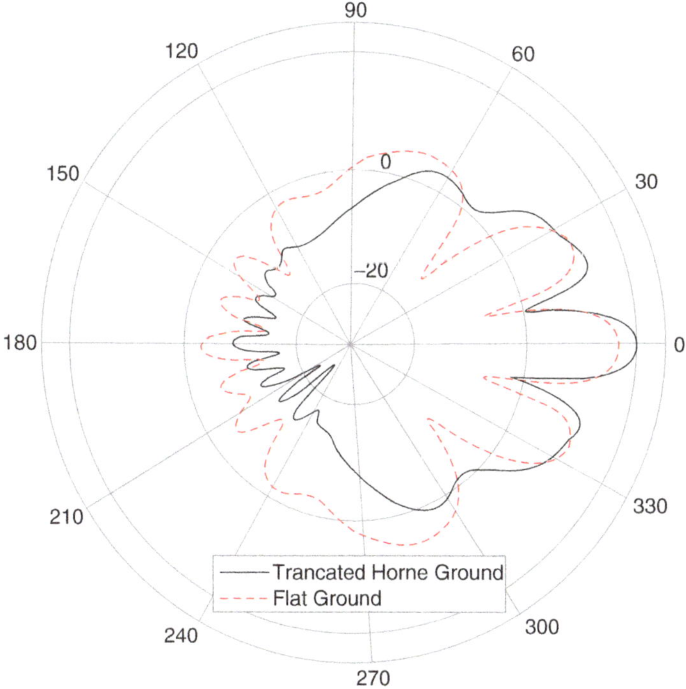

Fig. 4 A comparison between the resulting gain radiation pattern of the square flat and truncated horn ground plane at $f = 9$GHz

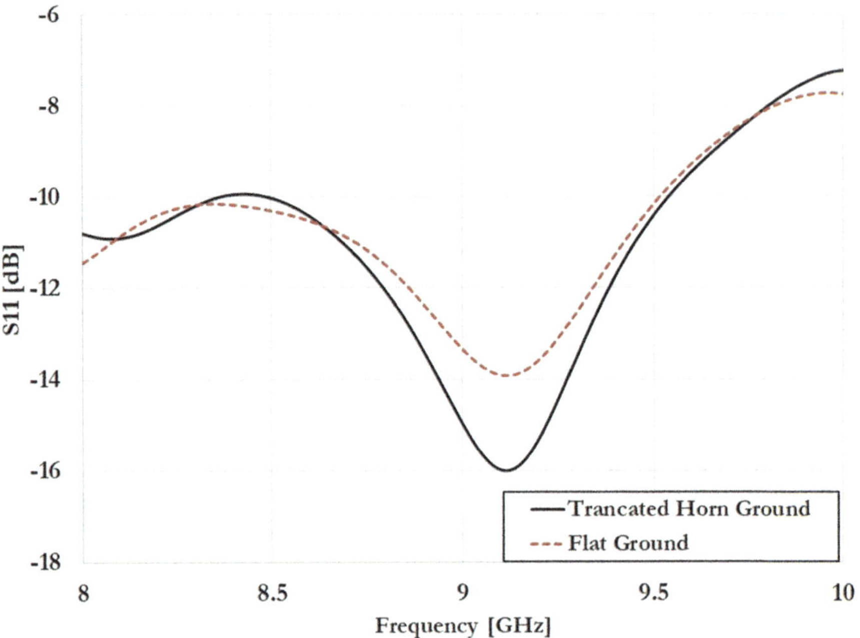

Fig. 5 The return loss at the input port of quad array antenna with the square flat and truncated horn ground plane

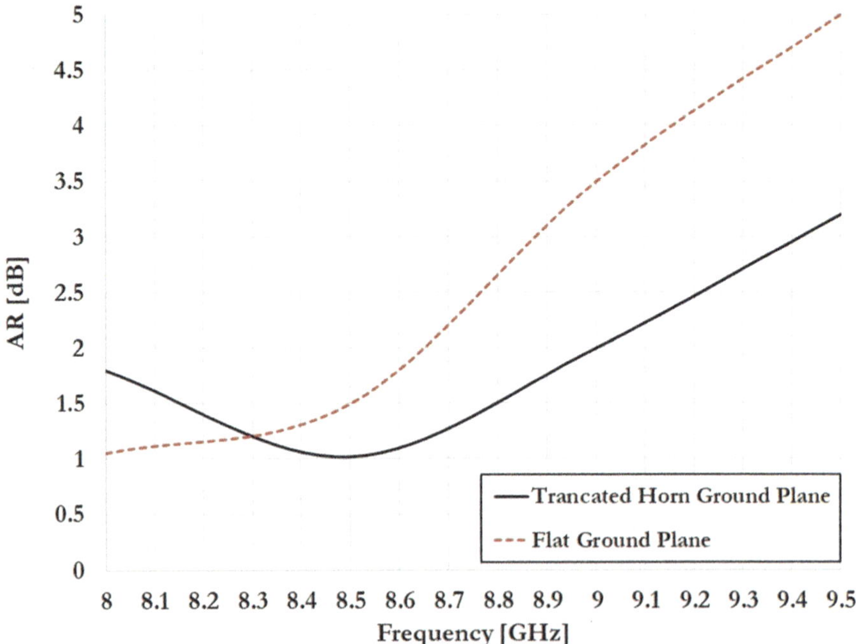

Fig. 6 The axial ratio of quad array antenna with the square flat and truncated horn ground plane

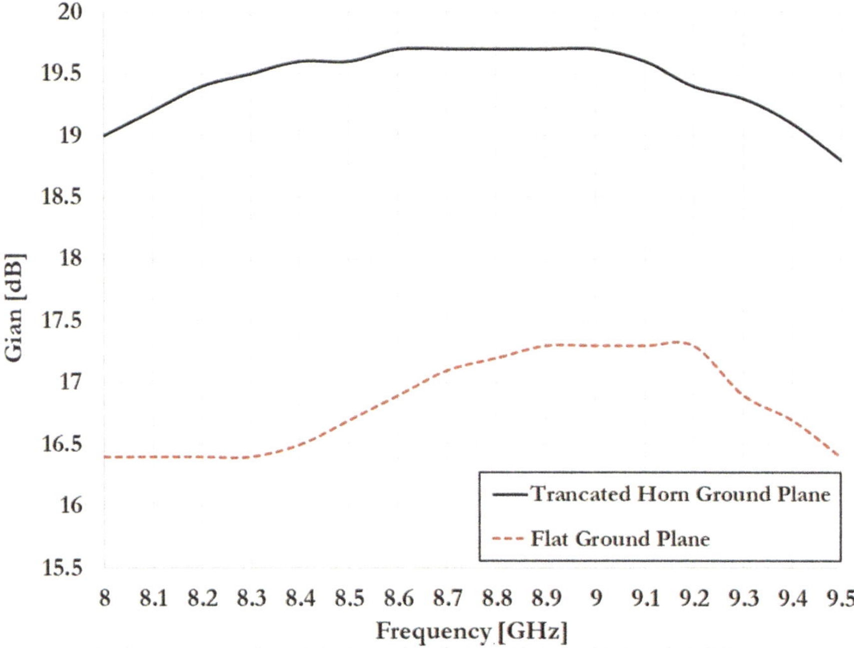

Fig. 7 Gain bandwidth comparison between helical antennas with square ground plane and with truncated horn ground plane

presented in Fig. 7. It is obvious that the truncated horn ground plane significantly improves the gain bandwidth in comparison with a traditional square ground plane.

5 Conclusions

In this chapter, the effects of the optimization process on truncated horn ground plane of an axial mode quad array helical antenna on its radiation characteristics, namely its gain, bandwidth, and side lobe level (SLL), have been studied. Specifically, the gain of the quad array helical antenna is a relatively complex function of diameter, height, and angle of the truncated horn ground plane. Thus, the maximum gain has to be achieved through optimization, especially when the size and mass limits are concerned. The results show that the truncated horn ground plane has a significant effect on the gain and bandwidth of the antenna. In the future, we plan to use multi-objective cost function to maximizing the performance of the ground plane design.

References

1. Balanis, C. (2016). *Antenna theory: Analysis and design* (4th ed.). Hoboken: Wiley.
2. Djordjevic, A., Zajic, A., & Ilic, M. (2006). Enhancing the gain of helical antennas by shaping the ground conductor. *IEEE Antennas and Wireless Propagation Letters, 5,* 138–140. https://doi.org/10.1109/LAWP.2006.873946
3. Djordjevic, A., Zajic, A., Ilic, M., & Stuber, G. (2006). Optimization of helical antennas. *IEEE Antennas & Propagation Magazine, 48*(6), 107–115. https://doi.org/10.1109/MAP.2006.323359
4. Liu, L., Li, Y., Zhang, Z., & Feng, Z. (2014). Circularly polarized patch-helix hybrid antenna with small ground. *IEEE Antennas and Wireless Propagation Letters, 13,* 361–364. https://doi.org/10.1109/LAWP.2014.2306494
5. Cai, R., Lin, S., Wang, L., Wang, J., Lu, Y., Huang, G., Qiu, J., & Wang, J. (2010). Design and experiment of a high gain axial-mode helical antenna. In: *IEEE 12th International Conference on Communication Technology.* https://doi.org/10.1109/ICCT.2010.5688887
6. Sadeghkia, F., & Horestani, A. (2018). Design guidelines for helicone antennas with improved gain. *Microwave and Optical Technology Letters, 61*(4), 1–6. https://doi.org/10.1002/mop.31663
7. Djordjevic, A., Ilic, M., Zajic, A., Olcan, D., & Nikolic M. (2007). Why Does Reflector Enhance The Gain of Helical Antennas? In: *The 2nd European Conference on Antennas and Propagation, EuCAP 2007.* https://doi.org/10.1049/ic.2007.0966
8. Sadeghkia, F., Mahmoodi, M., Hashemi-Meneh, H., & Ghayoomeh, J. (2014). Helical antenna over different ground planes. In: *The 8th European Conference on Antennas and Propagation (EuCAP 2014).* https://doi.org/10.1109/EuCAP.2014.6902243
9. Djordjevic, A., Olcan, D., Ilic, M., & Zajic, A. (2006). Design of Optimal Ground Conductor for The Helical Antenna. In *Proceedings of the 50th ETRAN Conference.*
10. Olcan, D., Zajic, A., Ilic, M., & Djordjevic, A. (2006). On The Optimal Dimensions of Helical Antenna with Truncated-Cone Reflector. In: *Proceedings of the 1st European Conference on Antennas and Propagation.* https://doi.org/10.1109/EUCAP.2006.4584946
11. Goldberg, D. (1989). *Genetic algorithms in search, optimization and machine learning.* Boston: Addison-Wesley Longman Publishing.
12. Whiele, D., & Michielssen, E. (1997). Genetic algorithm optimization applied to electromagnetics: A review. *IEEE Transactions on Antennas and Propagation, 45*(3), 343–353. https://doi.org/10.1109/8.558650
13. EM Software & Systems. (2011). *FEKO—Field computations involving bodies of arbitrary shape.* Stellenbosch: S.A. (Pty) Ltd. Retrieved from http://www.feko.info

C-Band SIW Slot Synthetic Aperture Radar Antenna for Remote Sensing Applications

Maha Maged ⓘ, Ahmed Ali Yousef, Haitham Akah ⓘ,
and Essam El-Diwany

Abstract This chapter presents a substrate integrated waveguide (SIW) slot array antenna designed for synthetic aperture radar (SAR) to be mounted on small satellites. This prototype single-layer dual-polarized C-band antenna was designed and simulated with the concern of remote sensing applications. The simulation results showed that the designed antenna has excellent radiation characteristics such as bandwidth, gain, and sidelobe level (SLL), with values 290 MHz, 17.9 dB, and −13.6 dB, respectively.

Keywords Substrate integrated waveguide (SIW) · Slot antenna array · Synthetic aperture radar · Remote sensing

1 Introduction

Nowadays, synthetic aperture radar (SAR) imaging has been extensively used in remote sensing applications. The SAR imaging provides unique information independent of cloud coverage and during night time. Recently, a variety of antennas with multifrequency and polarimetric have been widely studied which provides additional information of monitored area [1–4]. These antennas have been developed using microstrip patch radiating elements [1]. Many SAR satellite programs such as RadarSAT-1, X-SAR, ERS-1/2, and TerraSAR-X have adopted waveguide slot array antennas. In [5–7], a waveguide passive SAR antenna with slot array is

Supported by Egyptian Space Agency.

M. Maged (✉) · A. A. Yousef · H. Akah
Egyptian Space Agency, Cairo, Egypt
e-mail: maha_maged@narss.sci.eg; ahmed.ali@narss.sci.eg; haitham_akah@narss.sci.eg

E. El-Diwany
Electronics Research Institute, Cairo, Egypt
e-mail: essamdiwany@eri.sci.eg

M. H. Farouk, M. A. Hassanein (eds.), *Recent Advances in Engineering Mathematics and Physics*, https://doi.org/10.1007/978-3-030-39847-7_15

constructed with deployable panels and one fixed panel on the satellite center body. Substrate integrated waveguide (SIW) has emerged as a new concept for millimeter-wave (mm-wave) antennas and integrated circuits and systems due to advantages like high gain, efficiency, and low profile [8, 9]. Using SIW cavity, slots are widely used as radiating elements in SIW array antennas [10, 11, 12–16]. Using small satellites significantly required low profile, lightweight, and low cost of SAR antenna [17]. Due to these requirements, the SIW slot array antenna is designed to operate at the C-band and mounted on low Earth orbit (LEO) small satellite which has additional advantages such as a dual polarization, compact size, lightweight, low cost, and easier to integrate with other circuits. In this contribution, a design of a dual-polarized slot array antenna to enable a proof-of-concept SAR antenna for C-band remote sensing is presented. However, dual-polarized slot array antenna fed by SIW is rare because SIW feeding network occupies a large space, making it a challenge to feed the antenna array in dual polarization [18–20]. For the aim of improving the transmission efficiency and compacting the size, a series feeding network with multiple power-division vias for the dual-polarized SAR antenna is proposed. The rest of the chapter is organized as follows. Section 2.1 describes the basic principles of SIW design. Section 2.2 defines the geometry of the proposed SAR antenna. Section 3 presents the simulation and optimization results by CST Microwave Studio of the SAR antenna. Section 4 concludes the chapter.

2 Design of Dual-Polarized SAR Antenna

2.1 SIW Design Principles

SIW is a quasi-rectangular waveguide formed by periodic via-hole of two rows of metallized via-holes or grooves connected with two metal layers, and TE_{10} is the dominant mode (see Fig. 1).

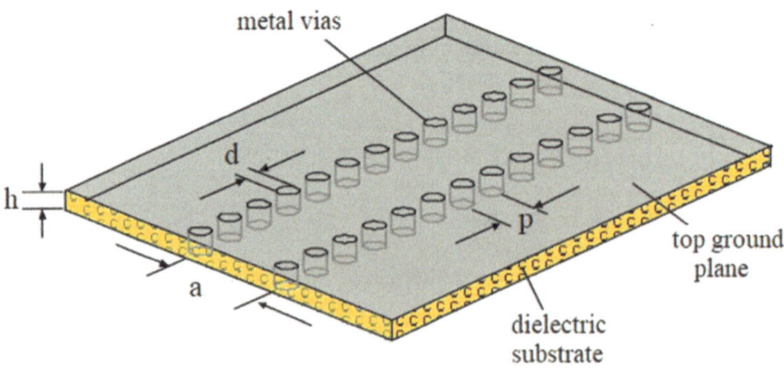

Fig. 1 Substrate integrated waveguide geometry

The design relations are given in [8], where the cutoff frequency in SIW structures are calculated as follows:

$$f_c = \frac{c}{2a_e\sqrt{\mu\varepsilon}} \tag{1}$$

where c is speed of light, a_e is the equivalent width of SIW, and μ and ε are the permittivity and permeability of the substrate, respectively. The equivalent width (a_e) is given in Eq. (2):

$$a_e = a - 1.08\frac{d^2}{p} + 0.1\frac{d^2}{a} \tag{2}$$

where a is the width of SIW, d is the diameter of metalized via-holes, and p is the pitch between adjacent via-holes.

2.2 *SIW-Based SAR Antenna Design*

As shown in Fig. 2, the size of the antenna is 1×4.5 m^2, which is divided into three identical panels that satisfied the design goals [21–23]. Figure 3 illustrates the structure of central and terminal panels, where the proposed panel is composed of radiating pair waveguides and connected with dual-feeding networks.

The SAR antenna is designed to be dual-polarized (horizontal and vertical) with two orthogonal slots in adjacent waveguides. In our designed prototype, each radiating waveguide has five slots of interdigital structure for 45° dual linear polarization arrays and is connected with two series feeding power networks [10, 24]. The unit-cell structure of this antenna is shown in Fig. 4. Each radiation slot in the unit-cell antenna array consists of one 45° inclined slot with three reflection-canceling tuning vias placed around the slot. The radiation cells are feeding in a series way to achieve a compact and simple feeding network, as shown in Fig. 5. The series feeding power divider uses *adaptive vias* to have impedance matching, compact size, and low sidelobe levels for the antenna [9, 25]. The proposed prototype antenna panel will be composed of four radiating pair waveguides and is designed for the frequency band 5.3 GHz. The antenna array

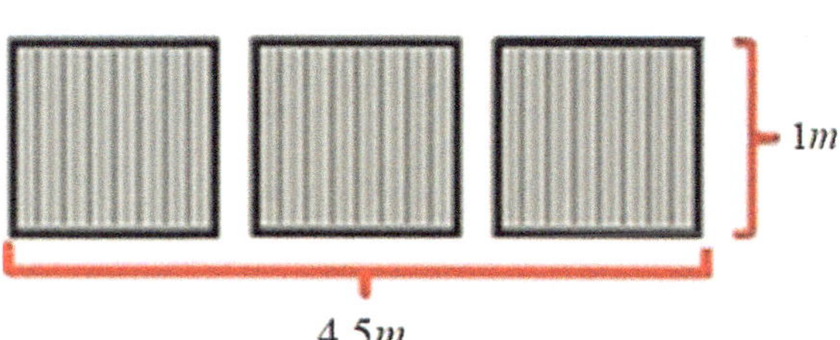

Fig. 2 SAR antenna panel configuration

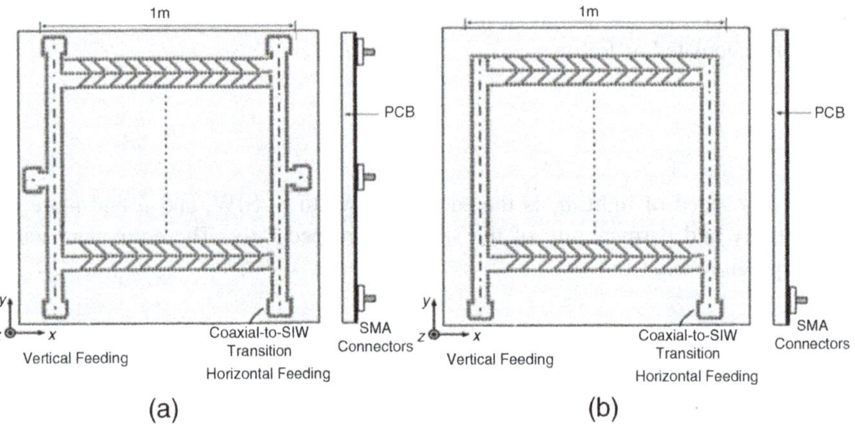

Fig. 3 The panel front and side view of (**a**) main panel and (**b**) terminal panel

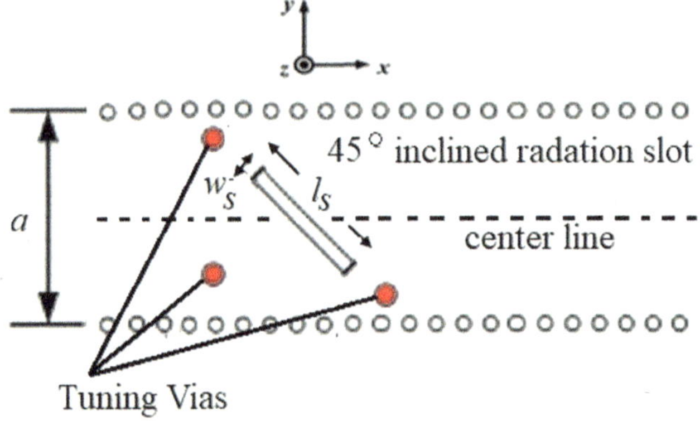

Fig. 4 Configuration of one cell of radiation slot

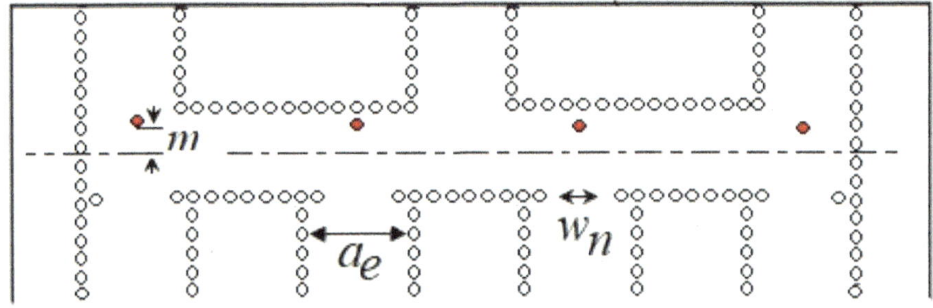

Fig. 5 Structure of multi-way power divider

feeding network is realized by multi-way feeding power divider that distributes the input power into each branch of the slot array. As shown in Fig. 5, the power divider consists of several cadenced T-junctions. In order to divide the power of SIW feeding network into each branch of the planar array in such a way to obtain an equal radiation pattern, tuning vias are added to the feeding network, where two vias are indented to forming an inductive window of width w_n. A matching via is placed at a distance m from the center of the feeding SIW. The basic SIW parameters are $d = 0.5$ mm, $p = 0.8$ mm, $a = 23$ mm, $w_s = 1$ mm, $w_n = 20$ mm, and height $h = 1.524$ mm. Other parameters such as l_s and m should be designed separately for every element to cancel the reflection and support a uniform power distribution.

3   Simulation Results

The prototype antenna panel was designed, simulated, and adjusted using the electromagnetic numerical software CST Studio Suite [26]. Simulated reflection coefficients $|S_{11}|$ of antenna panel are shown in Fig. 6. The simulated radiation patterns of E-plane and H-plane are shown in Fig. 7. The antenna maximum gain is 17.26 dB at the resonant frequency with a very stable 3 dB gain bandwidth of 290 MHz. The radiation pattern indicates that low sidelobe level is -13.6dB in E-plane. The simulated E-field distribution of the series feeding network of radiators at the center frequency is shown in Fig. 8. It shows ideal in-phase feeding and distribution of the input power into each branch.

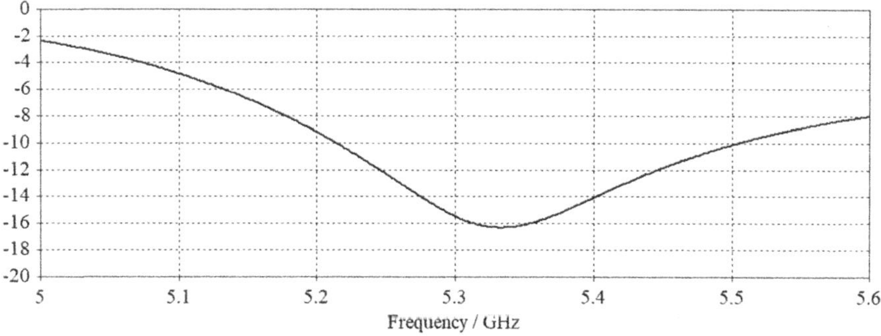

Fig. 6 The return loss of the proposed antenna

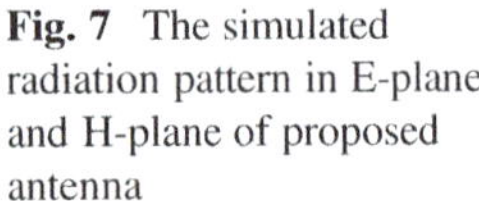

Fig. 7 The simulated radiation pattern in E-plane and H-plane of proposed antenna

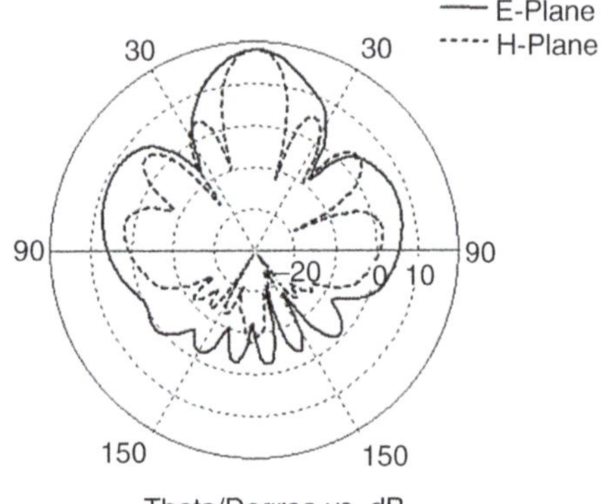

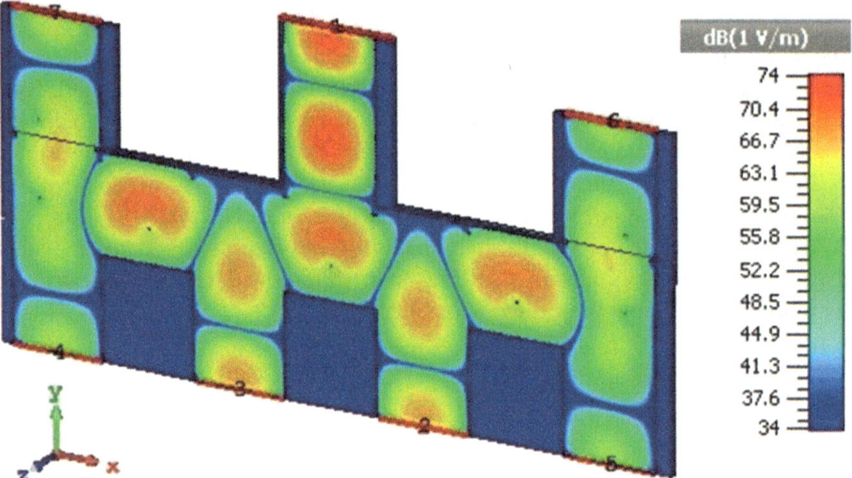

Fig. 8 Simulated E-field distribution of the feeding network

4 Conclusions

In this chapter, a design of dual-polarized single-layer SIW slot array antenna is presented with application to C-band SAR system. The radiating and series feeding networks were integrated into a single-layer PCB with the SIW technique. The series feeding network is a more efficient approach to improve the transmission efficiency as well as reduce the size of SAR antenna. The simulation result is presented which almost satisfies the requirement. This antenna panel is a first step of completing the SAR antenna with three panels. Future work is to realize such SAR antenna and evaluate its performance. Moreover, we can use the soft-coupled type series-to-series coupling slots in the feeding SIW [13, 14].

Acknowledgments The authors would like to thank Prof. *Fatma M. Hefnawi* for her inspiring discussion on SAR antennas.

References

1. Pokuls, R., Uher, J., & Pozar, D. (1998). Dual-frequency and dual-polarization microstrip antennas for SAR applications. *IEEE Transactions on Antennas and Propagation, 46*(9), 1289–1296. https://doi.org/10.1109/8.719972
2. Di Bari, R., Brown, T., Gao, S., Notter, M., Hall, D., & Underwood, C. (2011). Dual- polarized printed S-band radar Array antenna for spacecraft applications. *IEEE Antennas and Wireless Propagation Letters, 10*, 987–990. https://doi.org/10.1109/LAWP.2011.2167951
3. Qu, X., Zhong, S., Zhang, Y., & Wang, W. (2007). Design of an S/X dual-band dual-polarized microstrip antenna array for SAR applications. *IET Microwaves, Antennas & Propagation, 1*(2), 513–517. https://doi.org/10.1049/iet-map:20060232
4. Vetharatnam, G., & Koo, V. (2009). Compact L- & C-band SAR antenna. *Progress in Electromagnetics Research C, 8*, 105–114. https://doi.org/10.2528/PIERL09031805
5. Saito, H., Akbar, P., Watanabe, H., Ravindra, V., Hirokawa, J., Ura, K., & Budhaditya, P. (2017). Compact X-band synthetic aperture radar for 100kg class satellite. *IEICE Transactions on Communications, E100.B*(9), 1653–1660. https://doi.org/10.1587/transcom.2016PFI0008
6. Hirako, K., Shirasaka, S., Obata, T., Nakasuka, S., Saito, H., Nakamura, S., & Tohara, T. (2018). Development of small satellite for X-band compact synthetic aperture radar. *Journal of Physics: Conference Series, IOP Publishing, 1130*. https://doi.org/10.1088/1742-6596/1130/1/012013
7. Pyne, B., Akbar, P., Ravindra, V., Saito, H., Hirokawa, J., & Fukami, T. (2018). Slot-array antenna feeder network for space-borne X-band synthetic aperture radar. *IEEE Transactions on Antennas and Propagation, 66*(7), 3463–3474. https://doi.org/10.1109/TAP.2018.2829805
8. Bozzi, M., Georgiadis, A., & Wu, K. (2011). Review of substrate-integrated waveguide circuits and antennas. *IET Microwaves Antennas & Propagation, 5*(8), 909–920. https://doi.org/10.1049/iet-map.2010.0463
9. Cheng, Y. (2016). *Substrate integrated antennas and arrays*. Boca Raton: CRC Press.
10. Park, S., Okajima, Y., Hirokawa, J., & Ando, M. (2005). A slotted Post-Wall waveguide array with interdigital structure for 45° linear and dual polarization. *IEEE Transactions on Antennas and Propagation, 53*(9), 2865–2871. https://doi.org/10.1109/TAP.2005.854554
11. Chen, P., Hong, W., Kuai, Z., & Xu, J. (2009). A substrate integrated waveguide circular polarized Slot radiator and its linear Array. *IEEE Antennas and Wireless Propagation Letters, 8*, 120–123. https://doi.org/10.1109/LAWP.2008.2011062
12. Zhang, Q., & Lu, Y. (2008). Design of 45-degree linearly polarized substrate integrated waveguide-fed slot array antennas. *International Journal of Infrared and Millimeter Waves, 29*(11), 1019–1027. https://doi.org/10.1007/s10762-008-9398-x
13. Kim, D., Chung, W., Park, C., Lee, S., & Nam, S. (2011). Design of a 45° -inclined SIW resonant series slot array antenna for Ka-band. *IEEE Antennas and Wireless Propagation Letters, 10*, 318–321. https://doi.org/10.1109/LAWP.2011.2141105
14. Kim, D., Chung, W., Park, C., Lee, S., Nam, S., & Series Slot, A. (2012). Array antenna for 45° -inclined linear polarization with SIW technology. *IEEE Transactions on Antennas and Propagation, 60*(4). https://doi.org/10.1109/TAP.2012.2186270
15. Zhou, M., Cheng, Y., & Huang, W. (2016). Substrate integrated slot array antenna with required radiation pattern envelope. *International Journal of Antennas and Propagation*. https://doi.org/10.1155/2016/4031029

16. Xu, J., Hong, W., Chen, P., & Wu, K. (2009). Design and implementation of low sidelobe substrate integrated waveguide longitudinal slot array antennas. *IET Microwaves, Antennas & Propagation, 3*(5), 790–797. https://doi.org/10.1049/iet-map.2008.0157

17. Urata, A., Tetuko, J., Santosa, C., & Viscor, T. (2018). Development of an L-band SAR microsatellite antenna for earth observation. *Aerospace, 8*(128), 1–16. https://doi.org/10.3390/aerospace5040128

18. Chen, Z., Liu, H., Yu, J., & Chen, X. (2018). High gain, broadband and dual-polarized substrate integrated waveguide cavity-backed slot antenna array for 60 GHz band. *IEEE Access, 6,* 31012–31022. https://doi.org/10.1109/ACCESS.2018.2845917

19. Chu, H., Li, P., & Guo, Y. (2019). A beam-shaping feeding network in series configuration for antenna array with cosecant-square pattern and low sidelobes. *IEEE Antennas and Wireless Propagation Letters, 4*, 18. https://doi.org/10.1109/LAWP.2019.2901948

20. Ali, M. Sharma, K. & Yadav, R. (2019). Empirical design formulae for series-fed substrate integrated waveguides power divider. *International Journal of RF and Microwave Computer-Aided Engineering*. https://doi.org/10.1002/mmce.21859

21. Ulaby, F., Moore, R., & Fung, A. (1982). *Microwave remote sensing: Active and passive* (Vol. II). Reading, MA: Addison-Wesley.

22. Curlander, J., & McDonough, R. (1991). *Synthetic aperture radar: Systems and signal processing*. Hoboken: Wiley-Interscience.

23. Skolnik, M. (2008). *Radar handbook* (3rd ed.). New York: McGraw-Hill Education.

24. Hirokawa, J., & Ando, M. (2000). 45° linearly polarized post-wall waveguide-fed parallel plate slot arrays. *IEE Proceedings Microwaves, Antennas Propagation, 147*(6), 515–519. https://doi.org/10.1049/ip-map:20000795

25. Raida, Z., Puskely, J., & Mikulasek, T. (2013). Design of a compact wideband antenna array for microwave imaging applications. *Radio Engineering, 22*(4), 1224–1232.

26. CST Studio Suite. (2016). *Computer simulation technology, Copyright 1998–2016, Darmstadt*. Retrieved from https://www.cst.com/products/cstmws

Ultra-Wideband Compact T-Junction with Optimized V Cut for Millimeter Wave Applications

Islam Afifi ⓘ, Abduladeem Beltayib, and Abdel Razik Sebak

Abstract In this chapter, an ultra-wideband printed ridge gap T-junction is presented. A T-junction is an essential component in feeding antenna arrays and power division networks. The printed ridge gap technology is selected as it has a low profile, supports quasi-TEM mode, and has very low losses. The design depends on using one quarter wave transformer and optimized V cut for wideband matching where the genetic algorithm is used for the optimization. The device has a compact size and a wide bandwidth from 25.18 to 40.52 GHz (46.7%).

Keywords Millimeter wave components · Printed ridge gap · Power divider · Wideband

1 Introduction

The growth of the communication systems and its applications makes it necessary to search for high-frequency bands that have wideband to accommodate the large demand for high data rates and the increased number of users and connected devices. The millimeter wave band is a good candidate for future communication systems as it can provide a wide bandwidth. The conventional technologies for the manufacturing of the millimeter wave components have lots of problems. In the microstrip technology, there are high radiation and dielectric losses. On the other hand, the rectangular waveguide (RWG) becomes very expensive as the dimension goes small and needs electrical contact between all its parts. Recently, there are two new technologies to implement components for the millimeter wave band, the substrate

I. Afifi (✉)
Concordia University, Montreal, Canada

Cairo University, Giza, Egypt
e-mail: i_afifi@encs.concordia.ca

A. Beltayib · A. R. Sebak
Concordia University, Montreal, Canada

© Springer Nature Switzerland AG 2020
M. H. Farouk, M. A. Hassanein (eds.), *Recent Advances in Engineering Mathematics and Physics*, https://doi.org/10.1007/978-3-030-39847-7_16

integrated waveguide (SIW) and the ridge gap waveguide (RGW). The substrate integrated waveguide has the same concept of filled rectangular waveguide and is compatible with PCB technology which is cheap. However, it has a high dielectric loss and the propagating mode is TE_{10} which causes signal distortion for wideband signals. The ridge gap waveguide has the advantage of low losses as the wave is mainly propagating in an air gap region, and there are no radiation losses as it is a closed structure. Therefore, in this chapter, the printed ridge gap technology (PRGW) is chosen to implement the proposed T-junction.

The T-junction is an important component in microwave and millimeter wave circuits. It has been built in many technologies such as microstrip, RWG, SIW, and RGW. In [1], a microstrip T-junction of unequal power ratio and tapered matching sections is presented where a bandwidth from 1.7 to 5 GHz is achieved with -10 dB matching. However, its size is very large. In the RWG technology, a T-junction is designed in [2] with the addition of a load to reduce the coupling between output ports when used as a power combiner. It shows a matching level less than -17 dB and isolation better than 20 dB over the frequency range of 27–33 GHz (20%). Another RWG E-plane T-junction is in [3], in which the input is rotated to make the structure planar. The achieved bandwidth is from 26.3 to 39.4 GHz (39.88%) with reflection coefficient lower than -20 dB. In [4], a ridge waveguide T-junction is designed where a bandwidth from 27.5 to 34.9 GHz is achieved (23.7%) with a reflection coefficient below -20 dB. A double-ridge waveguide T-junction is in [5] with a bandwidth from 2 to 4 GHz (66.67%) and a matching better than -20 dB. In the SIW technology, a T-junction is designed in [6] where a matching bandwidth from 58.97 to 61.2 GHz (3.71%) is obtained (S11 < -20 dB). In [7], an SIW T-junction fed by a rectangular waveguide is presented that has a bandwidth from 28.60 to 30.25 GHz (5.61%) with -10 dB matching level. In the grove waveguide technology, a T-junction is designed, and a bandwidth from 51 to 70 GHz (31.4%) is achieved with a return loss better than 15 dB [8]. Another grove gap waveguide T-junction is in [9] where a bandwidth from 70 to 100 GHz (35.29%) is achieved with a reflection coefficient less than -20 dB. Regarding the RGW technology, a T-junction is presented in [10] where a bandwidth from 57 to 64 GHz (11.57%) is achieved with a matching level less than -15 dB. Another one is in [11] where a two-section Chebyshev quarter wave transformer is used to achieve a bandwidth of 62.1% around 15.3 GHz with a matching level below -17 dB. However, it is large in size as two quarter wave transformers are used and also it is built in the metal RGW which is expensive.

The proposed T-junction is compact with a wideband. Moreover, it is implemented in the PRGW which makes it cheap and has low losses.

2 Theory

The ridge gap waveguide concept is first introduced by Kildal in 2009 [12]. He states that no wave can propagate in an air gap region between a perfect electric conductor (PEC) and a perfect magnetic conductor (PMC) surfaces when the air gap height is

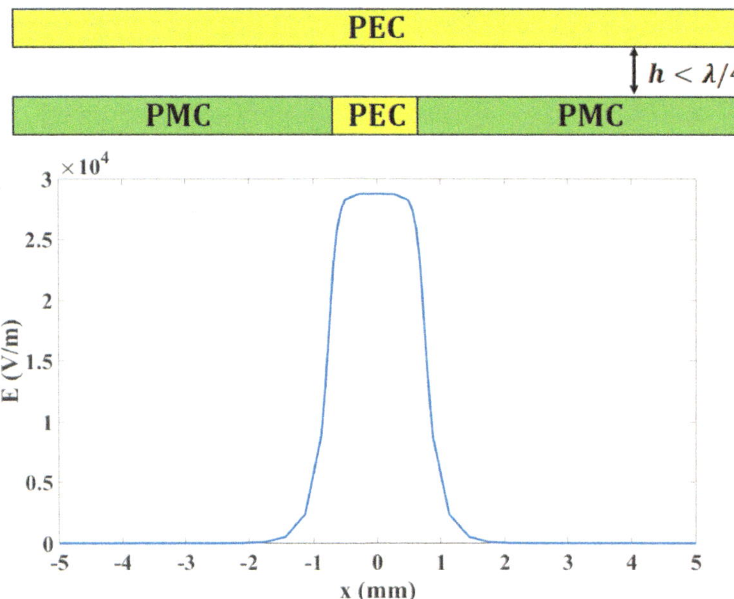

Fig. 1 Ridge gap concept using ideal boundary conditions

smaller than 0.25 λ_0. If a PEC strip is placed on the PMC surface as shown in Fig. 1, a wave can propagate between the PEC strip and the PEC surface. The wave is decaying outside the strip region as shown from the electric field distribution in Fig. 1 (the E-field distribution at 30 GHz is presented, and the gap is 0.289 mm which is much smaller than $\lambda_0/4$ to ensure the band gap region). It is clear that the field is almost flat in the strip region and exponential decaying outside the strip region.

Since natural PMC does not exist, an artificial magnetic conductor (AMC) is used instead. The common way to construct an AMC surface is to have a bed of nails with a length of $\lambda_0/4$ connected to a ground plane [13]. The produced surface has a very high impedance that behaves as a PMC.

In this chapter, the printed ridge gap waveguide is used to implement the proposed T-junction. It has the same working principle as the RGW, but it is built using the PCB technology rather than the CNC technology (CNC is used to build metal RGW structures). This reduces the cost of fabrication and facilitates its integration with other PCB devices. The unit cell of the PRGW is shown in Fig. 2 with its associated band gap. It consists of a mushroom-shaped structure that acts as an AMC surface and a PEC surface; they are separated by an air gap. The mushroom-shaped structure is implemented on RT6002 material ($\varepsilon_r = 2.94$, tan $\delta = 00012$, and thickness (h_{sub}) of 0.762 mm) with upper circle diameter (d_{cap}) of 1.35 mm and connected to a ground plane by a via of diameter $d_{\text{via}} = 0.4$ mm. The unit cell size (a) is 1.6 mm, and the air gap height ($h_{\text{air gap}}$) is 0.289 mm. The band gap of the unit cell is from 24.7 to 47.6 GHz. Figure 3 shows the dispersion diagram

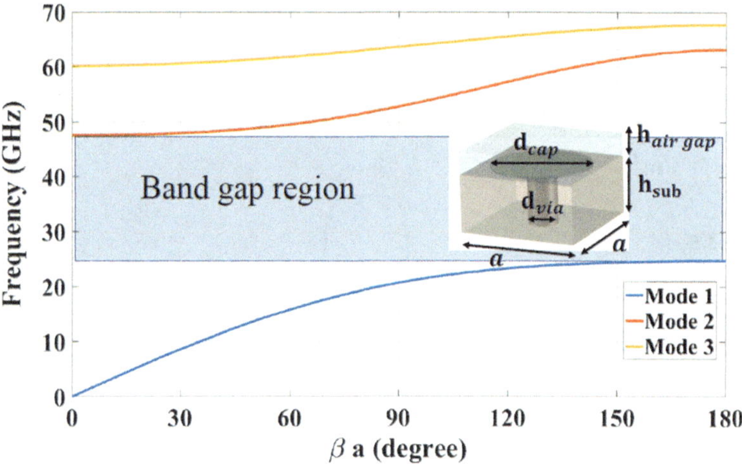

Fig. 2 The PRGW band gap unit cell and its dispersion diagram

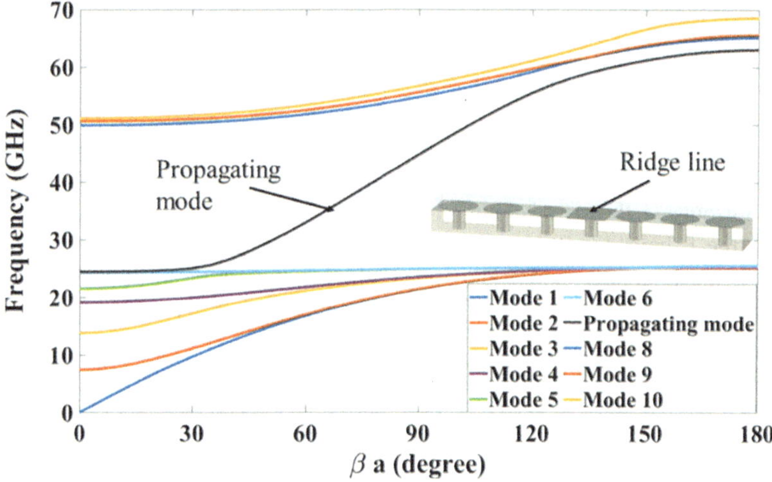

Fig. 3 The dispersion diagram of a PRGW line surrounded by three band gap unit cells from each side

of a ridge line surrounded by three-unit cells from each side. It is clear that only one mode exists in the band gap region, which is the desired propagation mode of the structure (the appearance of extra modes is due to the truncation in the periodicity of the structure in the direction normal to the line). The electric field distribution on the ridge line is shown in Fig. 4 at different frequencies where it is obvious that the field is confined along the line in the band gap region (at 26 and 40 GHz) and is distributed in the structure outside the band gap region (at 22 and 54 GHz).

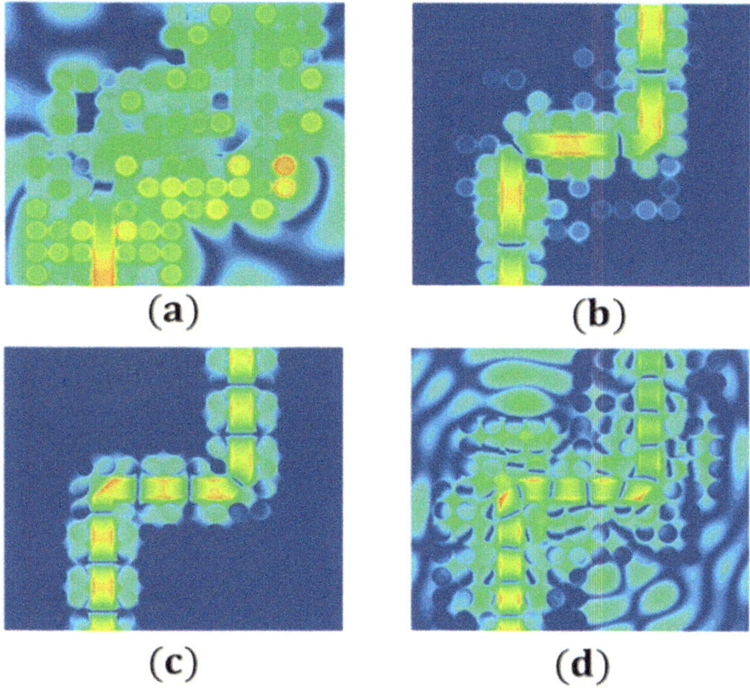

Fig. 4 Electric field distribution in the middle of the air gap region at different frequencies. (a) 22 GHz, (b) 26 GHz, (c) 40 GHz, and (d) 54 GHz

After the concept of the printed ridge gap waveguide has been illustrated, the design of the proposed T-junction is explained here. The T-junction is a three-port component that is reciprocal, lossless, and not all port matched. The S-matrix of the ideal T-junction is [14].

$$S = \begin{bmatrix} 0 & \dfrac{1}{\sqrt{2}} & \dfrac{1}{\sqrt{2}} \\ \dfrac{1}{\sqrt{2}} & -\dfrac{1}{2} & \dfrac{1}{2} \\ \dfrac{1}{\sqrt{2}} & \dfrac{1}{2} & -\dfrac{1}{2} \end{bmatrix} \tag{1}$$

The T-junction is usually used for dividing the input power equally between two output ports. This is done by having the line impedances of the output port equal to each other (Z_0) and the line impedance of the input port half of it ($Z_0/2$). In order to have all the input and output ports with the same impedance, a quarter wave transformer is used between the input line and the junction with an impedance equal to $Z_0/\sqrt{2}$.

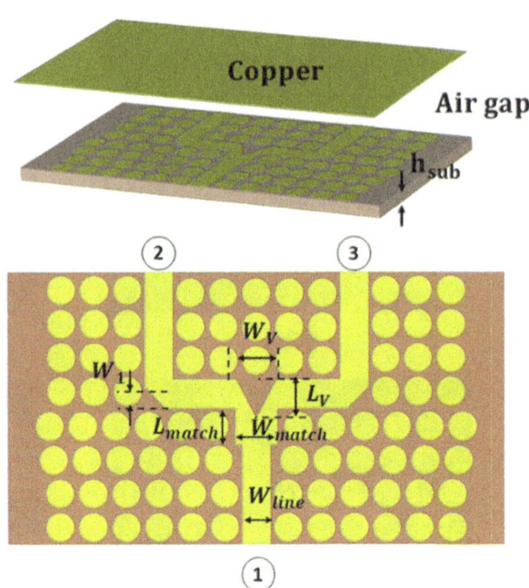

Fig. 5 Structure of the PRGW T-junction

Table 1 Dimensions of the T-junction

Parameter	W_1	W_v	W_{match}	W_{line}	L_{match}	L_v
Value (mm)	1.016	1.994	2.019	1.37	1.717	1.839

3 Simulation and Results

The structure of the proposed PRGW T-junction is illustrated in Fig. 5. The design depends on using one quarter wave transformer and an optimized V cut to have a wideband performance. The dimensions of the design are listed in Table 1. The structure has been simulated and optimized using CST studio, where the embedded genetic algorithm is used to achieve the optimum dimensions that satisfy a wideband performance with matching level less than -15 dB. The achieved bandwidth is from 25.18 to 40.52 GHz (46.7%), and the output power is -3.425 ± 0.225 dB for each output port. The s-parameters are shown in Fig. 6 ($S_{21} = S_{31}$ from the symmetry of the device). It covers the whole Ka-band (26.5–40 GHz) that makes it suitable to be used in MMW communications and MMW imaging systems.

4 Conclusion

A brief introduction to the printed ridge gap technology is introduced. After that, an ultra-wideband PRGW T-junction is designed. The design is optimized using the genetic algorithm to have a wide bandwidth with a good matching level. The proposed design has a 47% fractional bandwidth and covers the whole Ka-band.

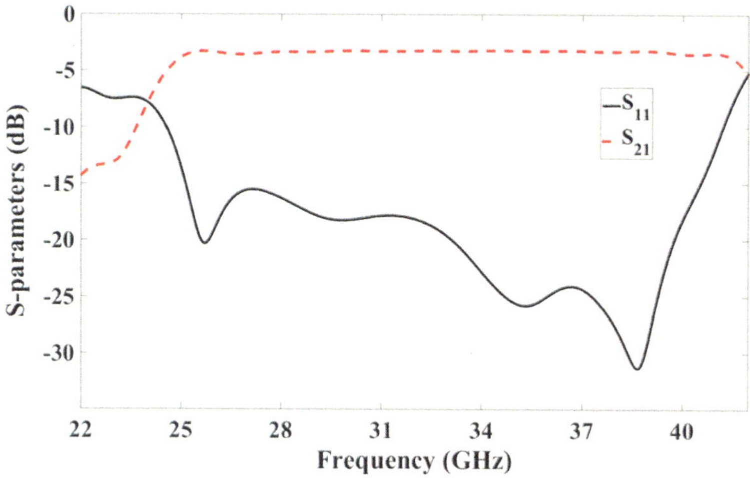

Fig. 6 S-parameter of the PRGW T-junction

References

1. Sangam, R. S., & Kshetrimayum, R. S. (2019). An Improved T-Junction Power Divider Using Linearly Tapered Microstrip Lines. In: *2019 URSI Asia-Pacific Radio Science Conference (AP-RASC), New Delhi, India*, pp. 1–4. https://doi.org/10.23919/URSIAPRASC.2019.8738357
2. Xu, Z., Xu, J., Cui, Y., & Qian, C. (2015). A novel rectangular waveguide T-junction for power combining application. *IEEE Microwave and Wireless Components Letters, 25*(8), 529–531. https://doi.org/10.1109/LMWC.2015.2440775
3. Zhao, P., Wang, Q., Zhang, F., & He, X. (2017). A novel T-junction waveguide power divider with anti-phases and broad bandwidth. In: *2017 IEEE International Symposium on Antennas and Propagation and USNC/URSI National Radio Science Meeting, San Diego, CA*, pp. 741–742. https://doi.org/10.1109/APUSNCURSINRSM.2017.8072413
4. You, Q., Huang, J., & Wang, D. (2016). Excitation of parallel plate waveguide by an array of ridge waveguide T-junction. In: *2016 IEEE MTT-S International Microwave Workshop Series on Advanced Materials and Processes for RF and THz Applications (IMWS-AMP), Chengdu*, pp. 1–3. https://doi.org/10.1109/IMWSAMP.2016.7588430
5. Ruiz, J., Soto, P., Boria, V. E., & Bias, A. A. S. (2015). Compensated double-ridge waveguide E-plane and H-plane T-junctions. In: *2015 IEEE 15th Mediterranean Microwave Symposium (MMS), Lecce*, pp. 1–4. https://doi.org/10.1109/MMS.2015.7375462
6. Kumar, K. B., & Shanmuganantham, T. (2017). 3-Port T-junction SIW power divider for60GHz applications. In: *2017 IEEE International Conference on Antenna Innovations and Modern Technologies for Ground, Aircraft and Satellite Applications (iAIM), Bangalore.* pp. 1–4. https://doi.org/10.1109/IAIM.2017.8402630
7. Abdel-Wahab, W., Al-Saedi, H., & Safavi-Naeini, S. (2017). E-plane RWG-to-SIW Transition Power Splitter for Ka-Band SATCOM Application. In: *2017 IEEE International Symposium on Antennas and Propagation and USNC/URSI National Radio Science Meeting, San Diego, CA.* pp. 2251–2252. https://doi.org/10.1109/APUSNCURSINRSM.2017.8073168
8. Farahbakhsh, A., Zarifi, D., & Zaman, A. U. (2017). 60-GHz groove gap waveguide based wideband H -plane power dividers and transitions: For use in high-gain slot Array antenna.

IEEE Transactions on Microwave Theory and Techniques, 65(11), 4111–4121. https://doi.org/10.1109/TMTT.2017.2699680

9. Liu, J., Zaman, A. U., & Yang, J. (2018). Design of a wideband array antenna prototype with gap waveguide for w-band wireless links. In: *12th European Conference on Antennas and Propagation (EuCAP 2018), London*, pp. 1–4. https://doi.org/10.1049/cp.2018.1215

10. Gupta, S., Briqech, Z., & Sebak, A. R. (2017). Analysis of 60 GHz ridge gap waveguide-based junctions, bends and losses. In: *IEEE International Conference on Antenna Innovations and Modern Technologies for Ground, Aircraft and Satellite Applications (iAIM), Bangalore*, pp. 1–4. https://doi.org/10.1109/IAIM.2017.8402635

11. Shams, S. I., & Kishk, A. A. (2016). Wide band power divider based on Ridge gap waveguide. In: *2016 17th International Symposium on Antenna Technology and Applied Electromagnetics (ANTEM), Montreal, QC*. pp. 1–2. https://doi.org/10.1109/ANTEM.2016.7550164

12. Kildal, P. (2009). Three metamaterial-based gap waveguides between parallel metal plates for mm/submm waves. In: *2009 3rd European Conference on Antennas and Propagation, Berlin*, pp. 28–32.

13. Polemi, A., Maci, S., & Kildal, P. (2011). Dispersion characteristics of a metamaterial-based parallel-plate ridge gap waveguide realized by bed of nails. *IEEE Transactions on Antennas and Propagation, 59*(3), 904–913. https://doi.org/10.1109/TAP.2010.2103006

14. Pozar, D. M. (2011). *Microwave engineering* (4th ed.). Hoboken: Wiley.

Excitation of the First High-Order Mode in Ridge Gap Waveguide

Abduladeem Beltayib, Islam Afifi ⓘD, and Abdel Razik Sebak

Abstract In this chapter, a method to excite the first higher-order mode of the ridge gap waveguide (RGW) is introduced. The fundamental mode of the RGW is well studied in a lot of papers, while the higher-order modes are ignored. However, they have some good properties that make them suitable for specific applications. The higher-order modes can exist by having a ridgeline. In the proposed work, the excitation of the first higher-order modes is achieved by having an L transition and vias that convert the fundamental quasi-TEM mode of the RGW to the first higher-order mode (first odd mode) and suppress the even modes. The proposed transition has a simple configuration and is designed by using metallic (RGW) technology. The proposed structure is simulated using a full wave simulator (CST Microwave Studio).

Keywords Ridge gap waveguide · High-order mode · Millimeter wave

1 Introduction

Recently, ridge gap waveguide (RGW) is considered one of the attractive guiding structures especially in millimeter and submillimeter bands [1, 2]. The main concept of RGW is coming initially from the soft and hard surfaces presented by Kildal in 1988 [3]. In 2009, the RGW is presented for the first time and used as a guiding structure. In the practical, RGW consists of two parts: the upper one is a PEC plate and the bottom is a PEC plate with an added bed of nails of length about $\lambda/4$ (acts as artificial magnetic conductor, AMC). A ridgeline is inserted between the beds of

A. Beltayib (✉) · A. R. Sebak
Concordia University, Montreal, Canada
e-mail: A_beltay@encs.concordia.ca

I. Afifi
Concordia University, Montreal, Canada

Cairo University, Giza, Egypt

© Springer Nature Switzerland AG 2020
M. H. Farouk, M. A. Hassanein (eds.), *Recent Advances in Engineering Mathematics and Physics*, https://doi.org/10.1007/978-3-030-39847-7_17

nails and connected to the bottom plate, so a wave can propagate between it and the upper PEC plate. The distance between the upper plate and the ridgeline should be less than $\lambda/4$ at the operation frequency to prevent the propagation of any waves between the parallel plates. In this case, there is only quasi-TEM mode that propagates in the gap between the upper PEC and the ridgeline [4, 5]. Based on the boundary condition of RGW, the dominant mode is the quasi-TEM mode. As well known, the quasi-TEM mode has a low signal distortion that can carry a wideband signal. Moreover, there is no need to ensure electrical contacts as the signal leakage is blocked by the periodic cells [6, 7]. Most of the designed RGW components are based on Q-TEM as a fundamental mode. However, there is no sufficient research about designing RGW based on higher-order modes. In [8], we have introduced a differential feeding for a cavity slot antenna based on the use of the first higher-order mode of the RGW, in which the first higher-order mode is excited by using a magic tee. In this chapter, a new technique to generate the odd mode RGW as a fundamental mode is introduced.

2 Ridge Gap Waveguide Analysis

In this section, the design of the band gap unit cell used to build the RGW is presented. After that, the RGW structure with the ridgeline and three cells from each side is presented in order to illustrate the existence of modes on the ridgeline. Finally, a comparison between the RGW modes and the microstrip line modes is used to establish the equations for the RGW higher-order modes.

2.1 Unit Cell

The dimensions of the unit cell are selected to have an operating bandwidth between 25 and 40 GHz. The dispersion diagram of the unit cell is obtained using CST Studio. The unit cell structure is illustrated in Fig. 1a, where the cells are formed by a metallic square pin with width $w_{\mathrm{p}} = 1$ mm, height $d = 2.5$ mm, period $a = 2.5$ mm, and air gap height $h = 1$ mm. The dispersion diagram of the unit cell is shown in Fig. 1b. It can be seen that there is a band gap from 22 to 40 GHz.

2.2 RGW Section

The complete RGW is implemented using a metal ridgeline surrounded by three square pins as shown in Fig. 2. The ridge width is selected to allow the existence of the fundamental (quasi-TEM) and the higher-order modes over the operating bandwidth.

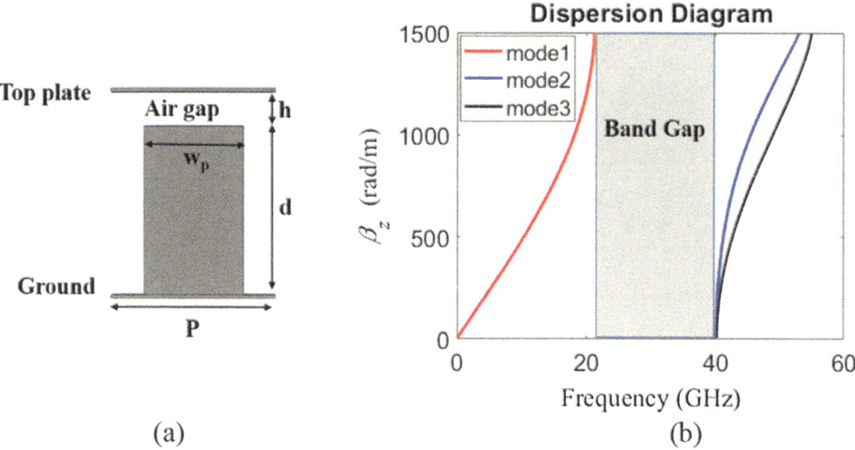

Fig. 1 The unit cell and its dispersion diagram. (**a**) The structure of the unit cell. (**b**) Dispersion diagram

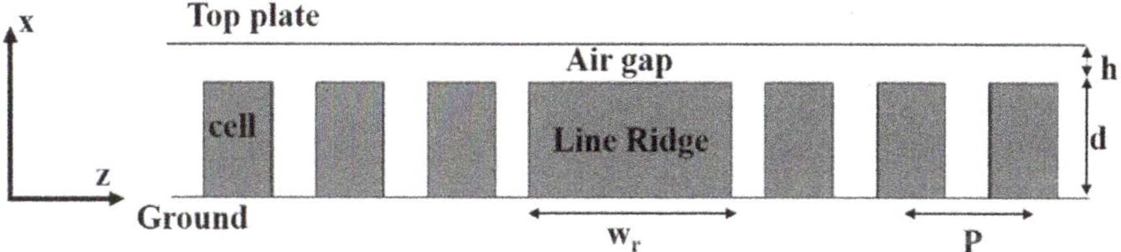

Fig. 2 The configuration of wide RGW. The line ridge width $w_r = 10$ mm

The periodic boundary condition is used along the y-axis, and PMC boundary conditions are used along the x-axis. A perfect electric boundary condition has been utilized to the upper and bottom of the RGW. The dispersion diagram of the RGW is shown in Fig. 3. The modes on the air gap along the ridge are quasi-TEM mode, first odd mode TE10, and the second even mode TE20, as illustrated in Fig. 3.

2.3 General Equations of the RGW Modes

The ridge gap waveguide is a line ridge surrounded by PEC/PMC boundaries which confine the field on the ridgeline and make it most even outside the ridgeline. This is similar to the field distribution of the microstrip line. The fundamental mode in both microstrip and RGW is Q-TEM.

In order to verify the similarity between the microstrip and the RGW, a comparison between the electric field distributions in the ridge gap and the microstrip structures is shown in the following Fig. 4, for the first three modes, where in the

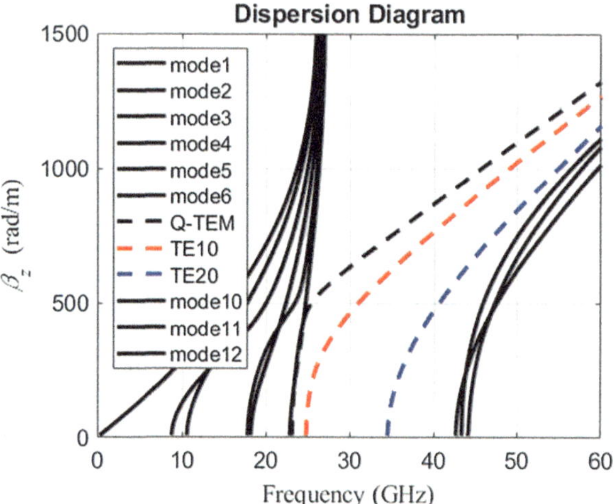

Fig. 3 The dispersion curves of RGW obtained from CST

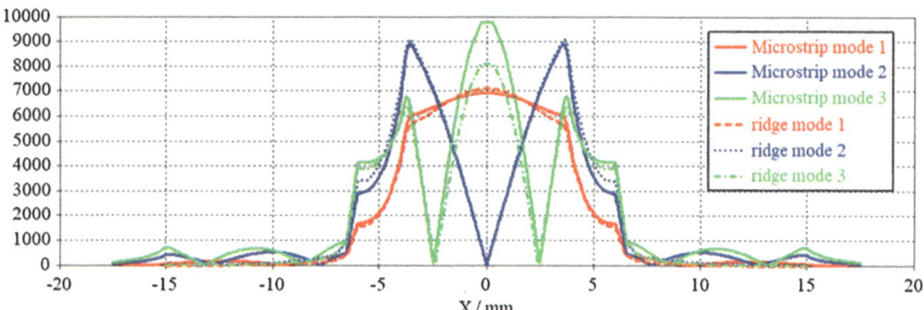

Fig. 4 A comparison between the electric field distributions in the RGW and the microstrip structures

microstrip case we use 1 mm substrate with $\varepsilon_r = 1.0$ and linewidth $= 7.5$ mm (same as the RGW case where the linewidth is 7.5 mm and the air gap $= 1$ mm). It is clear that the behavior is the same for both the ridge and the microstrip within the first three modes, and hence they can be represented by the same equation. From this point, we can write the expression for the electric field component on the ridge of the structure in Fig. 2 as follows:

$$E_y = \sum_{m=0,2,4}^{\infty} E_m \cos\left(k_{\mathrm{x}m}x\right)\mathrm{e}^{-jk_{\mathrm{z}m}z} + \sum_{m=1,3,5}^{\infty} E_m \sin\left(k_{\mathrm{x}m}x\right)\mathrm{e}^{-jk_{\mathrm{z}m}z} \tag{1}$$

where $x = 0$ is taken at the center of the ridge.

Figure 4 shows the magnitude of the electric field at the middle of the air gap for the ridge gap waveguide and at the middle of the substrate for the microstrip line at 30 GHz. Moreover, the height between the ridgeline and the upper PEC in the ridge gap waveguide is 1 mm so the higher-order mode with a y variation starts after

Fig. 5 Cross-sectional field distribution of the first three modes

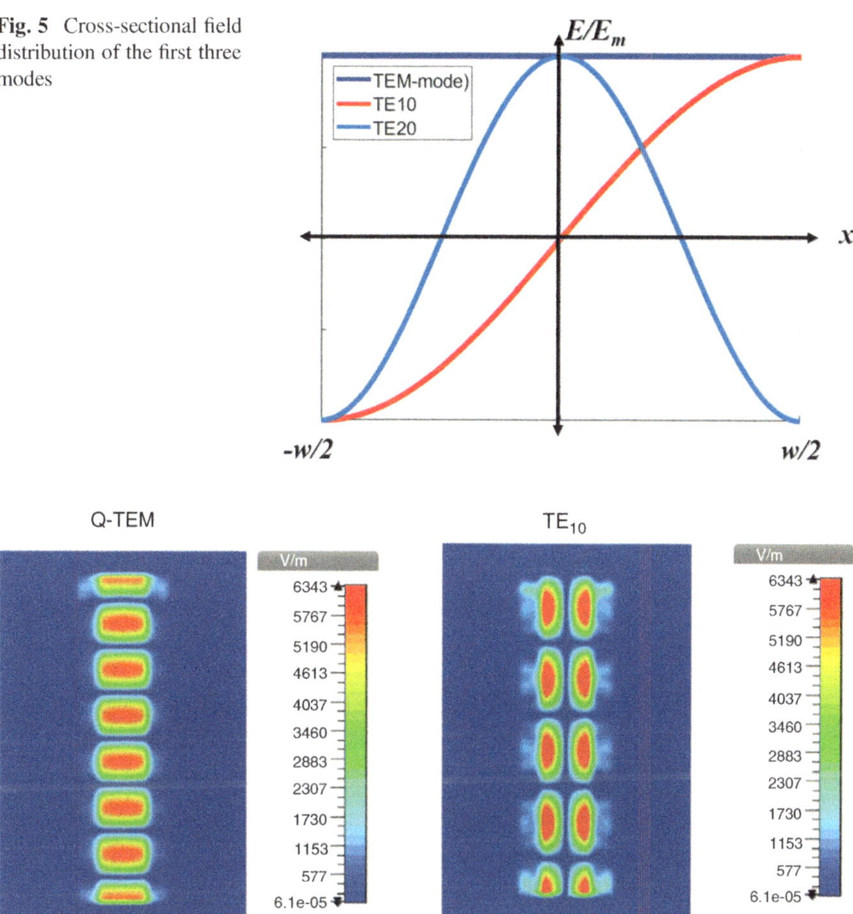

Fig. 6 Electric field distribution of the fundamental mode (Q-TEM) and first odd mode (TE10). (**a**) Electric field distribution of Q-TEM mode and (**b**) electric field distribution of TE10 mode

150 GHz which is far from the working frequency band (the field is constant along y-direction). Based on Eq. (1), the electric field distribution for the first three modes in the air gap along the x-axis can be estimated by the curves shown in Fig. 5. Using CST, the electric field distribution for the first two modes of a wide RGW is shown in Fig. 6. It very is clear that fundamental mode (quasi-TEM) has even symmetry, while the first higher mode (TE10) has odd symmetry.

3   Odd Mode RGW Transition Design

The odd mode RGW transition is designed based on transforming the regular Q-TEM mode to the desired odd mode. This is achieved by using an L-shape transition. The L-shape provides a 180° phase delay between the two edges of the wide ridgeline. After that, a row of metallic pins in the middle of the wide ridgeline is used to suppress the Q-TEM and the TE20 modes. The top view of the proposed structure and all dimensions are shown in Fig. 7 and Table 1, respectively. The top plate is hidden to show the internal details.

Matching is achieved by using a quarter wave transformer between the input port which supports only the quasi-TEM mode and the output port that supports the odd mode. The equation of the strip line can be used to have the starting dimension of the ridgeline [3]:

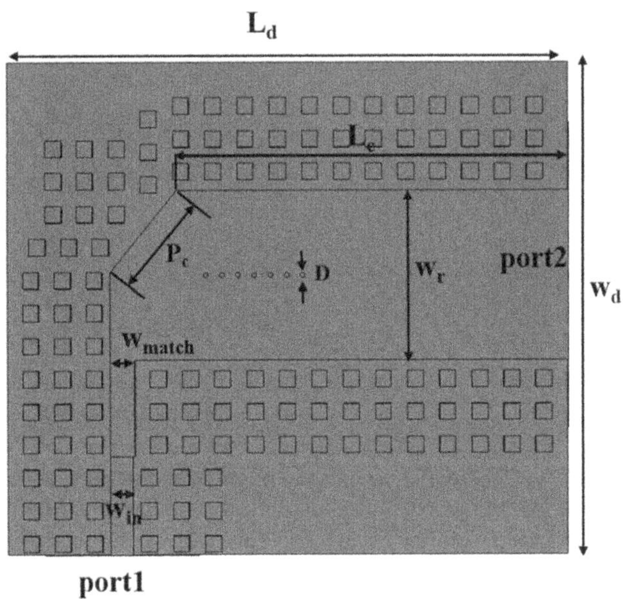

Fig. 7 Top view of odd mode RGW configuration with hidden top plate

Table 1 Parameters of the transition

Parameter	Value (mm)	Parameter	Value (mm)
P_c	6.5	W_r	10
W_{match}	1.5	W_d	30
W_{in}	1.4	L_d	25
L_c	14.3	D	0.3

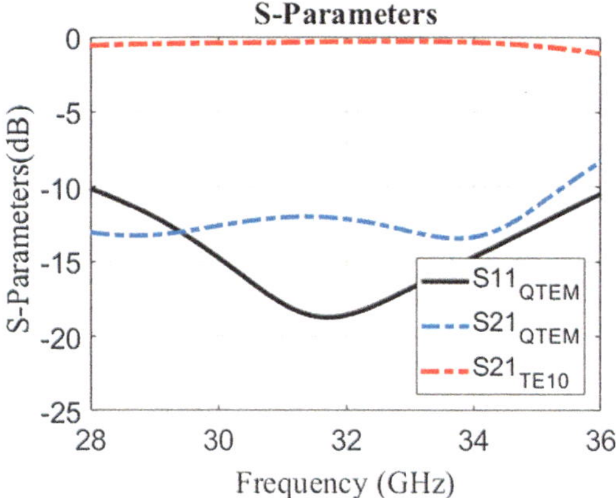

Fig. 8 The magnitude of the proposed RGW dual-mode S-parameters

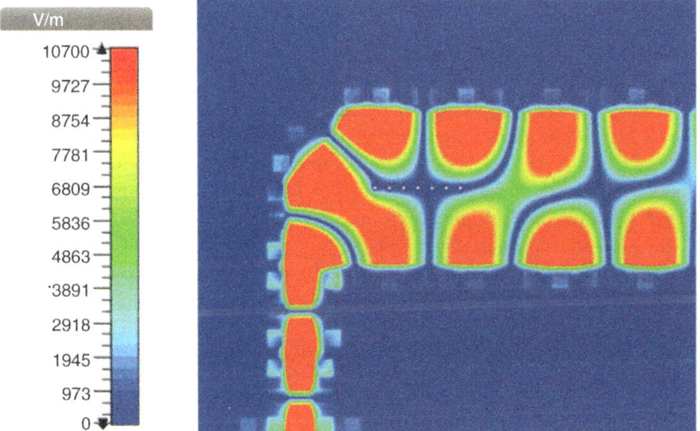

Fig. 9 The field distribution for the proposed odd mode RGW transition at 32 GHz

$$Z_{\mathrm{RGW}} = \frac{\eta_0}{2} \left(\frac{W_{\mathrm{r}}}{2h} + 0.0441 \right)^{-1} \tag{2}$$

where η_0 is the intrinsic impedance of the air.

An optimization process is carried out to have a wide bandwidth matching. The s-parameters of the proposed odd mode transition are shown in Fig. 8. The reflection coefficient of this structure is less than -10 dB over 8 GHz from 28 to 36 GHz, where more than 93% of the input power has been transformed into the odd mode at the output port. Figure 9 illustrates the electric field distribution on the proposed transition at 32 GHz. It is clear that the quasi-TEM mode at the input is efficiently converted to the odd mode (TE10) at the output port.

4 Conclusion

This chapter has introduced a new and simple structure to excite the first higher-order mode of the RGW as a fundamental mode instead of the conventional Q-TEM mode. The structure is fed by a quasi-TEM mode at the input port, and then it is transformed into the odd mode at the wide ridgeline. The whole structure has achieved a matching level of less than -10 dB from 28 to 36 GHz. With the proposed waveguide, the odd mode can be used which has a great potential for millimeter-wave devices that need the use of higher-order modes.

References

1. Suárez, F. C., Méndez, D. N., & Baquero-Escudero, M. (2013). Rotman lens with Ridge Gap Waveguide technology for millimeter wave applications. In: *2013 7th European Conference on Antennas and Propagation (EuCAP), Gothenburg*. pp. 4006–4009.
2. Pourahmadazar, J., Farahani, M., & Denidni, T. (2018). Printed Ridge Gap Waveguide Rotman Lens for Millimetre-wave Applications. In: *2018 18th International Symposium on Antenna Technology and Applied Electromagnetics (ANTEM), Waterloo, ON*. pp. 1–2.
3. Kildal, P.-S., Zaman, A. U., Raja-Iglesias, E., Alfonso, E., & Nogueira, A. V. (2011). Design and experimental verification of ridge gap waveguide in bed of nails for parallel-plate mode suppression. *IET Microwaves, Antennas & Propagation, 5*(3), 262–270.
4. Kildal, P. S., Alfonso, E., Valero-Nogueira, A., & Rajo-Iglesias, E. (2009). Local metamaterial-based waveguides in gaps between parallel metal plates. *IEEE Antennas and Wireless Propagation Letters, 8*, 84–87.
5. Kildal, P. S. (2009). Three metamaterial-based gap waveguides between parallel metal plates for mm/submm waves. In: *Proceedings of the Third European Conference on Antennas and Propagation, EuCAP*. pp. 28–32.
6. Shams, I., & Kishk, A. A. (2016). Wide band power divider based on Ridge gap waveguide. In: *2016 17th International Symposium on Antenna Technology and Applied Electromagnetics (ANTEM)*, Montreal, QC. pp. 1–2.
7. Sharifi Sorkherizi, M., & Kishk, A. A. (2017). Self-packaged, low-loss, planar bandpass filters for millimeter-wave application based on printed gap waveguide technology. *IEEE Transactions on Components, Packaging and Manufacturing Technology, 7*(9), 1419–1431.
8. Beltayib, A., Afifi, I., & Sebak, A. (2019). 4×4 -element cavity slot antenna differentially-fed by odd mode ridge gap waveguide. *IEEE Access, 7*, 48185–48195.

Double Via Row Radial Four-Way Power Combiner with Improved Isolation Performance

Abdelrahman El-Akhdar (iD), Maha Maged (iD), and Mohammed El-Telbany (iD)

Abstract In this chapter, a radial four-way combiner based on double via row substrate integrated waveguide (SIW) with SMA connector input and output ports is designed and simulated. The proposed combiner has an advantage of reduced leakage obtained by using double via row configuration and improved isolation by optimizing the isolation wall length. The proposed radial power combiner can be easily fabricated using traditional PCB fabrication facility. The characteristics of the simulated combiner are evaluated. The return loss at the center port is higher than -33 dB for a 2% bandwidth around the center design frequency. The maximal loss is less than 1 dB within the operating frequency band. The impedance of each port is 50 Ω SMA connectors as a matching system.

Keywords Substrate integrated waveguide (SIW) · Radial power combiner · Double via rows

1 Introduction

Today, the military and space industries also require high-performance microwave technology in combination with excellent reliability, as well as the lowest size and weight possible. This also refers to radar and satellite communication, where the main objective is to generate high power output at microwave and millimeter wave frequencies. In radar and wireless communication systems, robust, compact, and

A. El-Akhdar
Technical Research and Development Center, Cairo, Egypt
e-mail: a.m.el-akhdar@ieee.org

M. Maged (✉)
Egyptian Space Agency, Cairo, Egypt
e-mail: maha_maged@narss.sci.eg

M. El-Telbany
Sinai University, North Sinai, Egypt
e-mail: mahammed.elsaid@su.edu.eg

© Springer Nature Switzerland AG 2020

M. H. Farouk, M. A. Hassanein (eds.), *Recent Advances in Engineering Mathematics and Physics*, https://doi.org/10.1007/978-3-030-39847-7_18

high-performance amplifiers are crucial components [1]. The size and power capacity of solid-state power devices are significantly reduced, however, and single solid control devices are difficult to meet the requirements of an entire system. Therefore, a number of high-efficiency hybrid techniques have been developed to achieve higher output power than is possible in the transmission system from a single device [2–7]. The radial n-way combiners are efficient and compact structures that sum the power of n inputs in one single step and deliver the total power to the central port. Such radial combiners have high efficiencies since power does not have to pass through several stages of combiners and are therefore an interesting alternative to frequently used combining networks [4, 8, 9]. Typically, cavity-based radial power combiners are used in narrow-band applications and low bandwidth insertion loss [10]. The cavity-based combiners are realized using waveguide and substrate integrated waveguide (SIW) [11]. The advantages of SIW, such as low-loss, low-cost, and high-density integration, make SIW appropriate for the design of power combiners. Different SIW combing designs have been published in recent years in the literature. A compact radial cavity power combiner based on the SIW that excites the TE_{10} mode through the magnetic field is presented in [8]. Liu et al. present several Ku-band compact power combiners/dividers based on half-mode SIW (HMSIW) technique [12]. A four-way power combiner/divider is realized by integrating the microstrip lines to the open wall of the HMSIW [17]. Other high-performance combiners/dividers based on the different methods of implementation are reported in [13–16]. In this chapter, a novel radial four-way combiner is designed in the X-band utilizing the advantages of double via row SIW structure technology [18]. We demonstrate experimentally the results, which show high port isolation, low loss, high efficiency, wide band, and small lateral size.

2 Design of SIW Power Combiner

2.1 SIW Design Principles

SIW is a quasi-rectangular waveguide formed by periodic via hole of two rows of metallized via holes or grooves connected with two metal layers, and TE_{10} is the dominant mode. The cutoff frequency in SIW structures is calculated as follows:

$$f_c = \frac{c}{2a_e\sqrt{\mu\varepsilon}} \tag{1}$$

where c represents the speed of light, a_e is the equivalent width of SIW, and μ and ε are the permittivity and permeability of the substrate, respectively. In our design, we focused on the X-band applications, and the combiner has been designed to resonate at frequency of 8 GHz. The equivalent width (a_e) is given in Eq. (2) based on the *double* via *row* configuration according to [18] where inner and outer rows are staggered to reduce the leakage as possible:

$$a_e = a - 1.08\frac{d^2}{p} + 0.1\frac{d^2}{a} \qquad (2)$$

where a is the SIW width, d is the diameter of metalized via holes, row-to-row separation $D = 2d$, and p is the pitch between adjacent via holes. The distance between the two via rows and the via-to-via separation are optimized to provide a reduced sidewall leakage in addition to an improved power handling capability [18]. However, multiple rows may be used to form the via wall.

2.2 Radial Power Combiner Design

Figure 1 shows the structure of SIW radial cavity four-way power combiner. The SIW radial cavity is combined in a planar substrate with arrays of metallic double via rows to realize circumferential walls. The peripheral SIW cavity ports are four identical equispaced metallic input via holes. The four sectors are separated by metallic single via sidewalls of length L. The SIW cavity is axially symmetric with respect to the central port which can be viewed as an equivalent traditional radial cavity. The central probe and peripheral probes are achieved by metallic via hole and located on the opposite sides of the SIW radial cavity. The diameter of the radial d of SIW cavity is selected according to operating frequency range of interest [8, 19]. Thus, the radius is

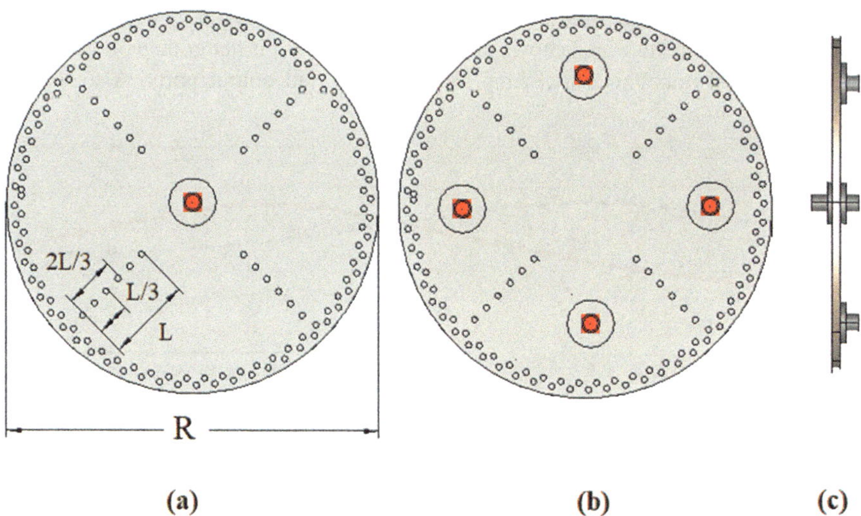

Fig. 1 SIW-based radial combiner with isolation wall (**a**) front side, (**b**) back side, and (**c**) side view section

$$R \approx \frac{Na_e}{2\pi} \tag{3}$$

where N is the number of sectors. Similarly, in order to achieve a low insertion loss and high isolation over the bandwidth, an optimization process of the physical structure has to be applied carefully [20].

3 Simulation Results

According to the presented analysis in Sect. 2, the radial power combiner is designed, simulated, and optimized using CST Microwave Studio. The optimization process is carried out by altering the isolation wall length using $L/3$, $2L/3$, and L, as shown in Fig. 1a. A parametric study of the isolation wall length L for $L/3 = 5.61$ mm, $2L/3 = 11.22$ mm, $L = 16.43$ mm, and no walls to observe the effect on return and insertion losses. The S-parameter as a result of the numeric simulation is shown in Figs. 2 and 3. It is important to mention that the maximum length of the isolation wall L is limited to the input matching; otherwise, the input return loss will be sufficiently high that degrades the overall circuit performance. The optimum results occurred at the isolation wall length (L), as shown in Fig. 2. The dielectric and conductor losses were included in the simulation. The final dimensions of the structure are $L = 16.843$ mm, $R = 63$ mm, $h = 1.53$ mm, $d = 1$ mm, $p = 3.15$ mm, and $D = 2$ mm.

As shown in Figs. 2 and 3, the insertion and return losses are improved when the isolation wall length equals (L) compared to other wall lengths ($L/3$, $2L/3$, and using no isolation wall). The simulated return loss of the double via row radial power combiner presented in Fig. 3 shows that the return loss S_{11} is better than 30 dB which indicates good impedance matching at both input and output ports. Also, Fig. 4

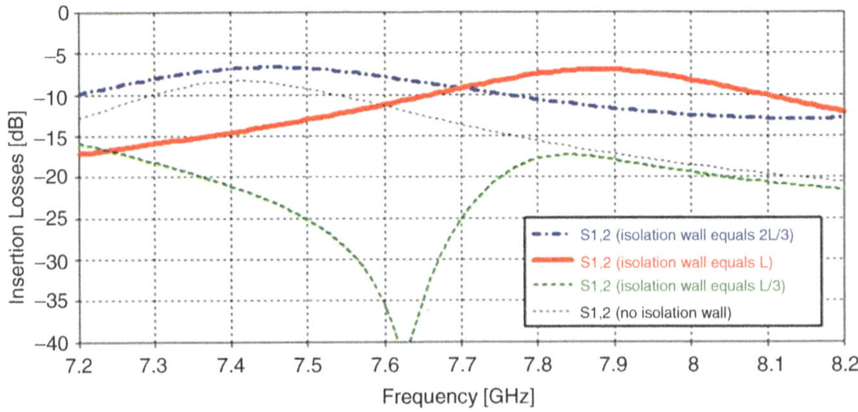

Fig. 2 The effect of the isolation wall length (L) on the combiner insertion loss

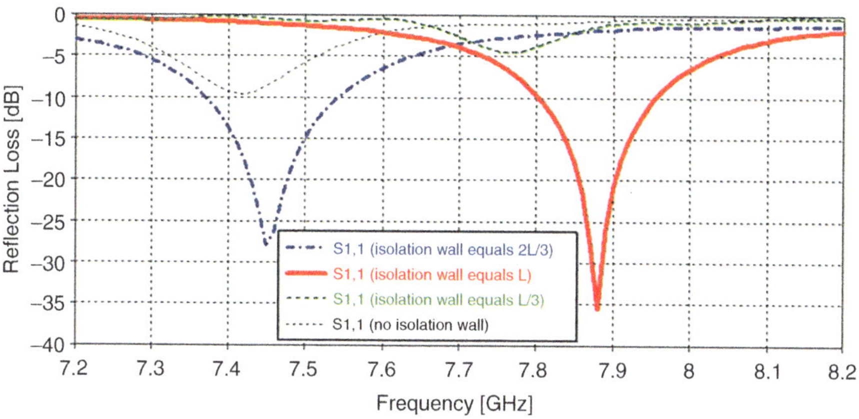

Fig. 3 The effect of the isolation wall length (L) on the combiner input return loss

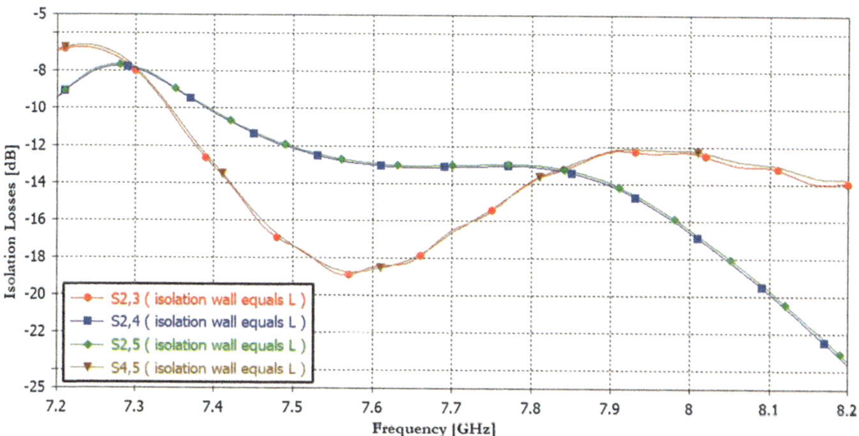

Fig. 4 The isolation between output ports of the proposed SIW radial power combiner

shows that the isolation losses (S_{23}, S_{24}, and S_{25}) are better than 13 ± 0.5 dB over the all operating band, it can be noticed that the isolation losses are improved by the use of sidewalls.

The maximum measured amplitude imbalance of the insertion loss, as shown in Fig. 5, is ± 0.5 dB.

The total insertion loss is calculated by substituting the measured values for S_{j1}; $j = 2; 3; \cdots; n + 1$ into

$$\text{losses} = -10 \log_{10} \left(\sum_{j=2}^{n+1} |S_{j1}|^2 \right) \tag{4}$$

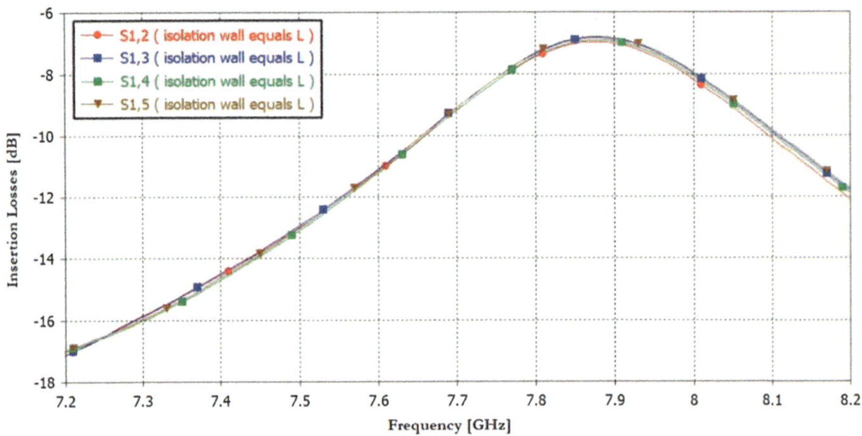

Fig. 5 Simulated insertion loss of the proposed SIW radial power combiner

As shown in Fig. 5, it is clear that maximum insertion loss of the proposed SIW radial power combiner with isolation sidewall length (L) is less than 1 dB in the operating band.

4   Conclusions

In this chapter, a four-way radial power combiner is designed and analyzed using double via row SIW structure technique. Through simulation, good improvement in isolation between adjacent peripheral ports can be obtained due to the metallic isolated walls, which were inserted symmetrically in the radial line section. The double via row configuration improved the results because it provides a reduction in leakage and improved power handling capability for the proposed radial power combiner [18]. In future work, the proposed radial power combiner will be fabricated using traditional PCB fabrication facility. Moreover, the optimization of sidewall length L will be established using evolutionary algorithms.

References

1. Cripps, S. (2006). *RF power amplifiers for wireless communications* (2nd ed.). Norwood: Artech House.
2. Russell, K. (1979). Microwave power combining techniques. *IEEE Transaction on Microwave Theory and Techniques, 27*, 472–478. https://doi.org/10.1109/TMTT.1979.1129651
3. Chang, K., & Sun, C. (1983). Millimeter-wave power-combining techniques. *IEEE Transaction on Microwave Theory and Techniques, 31*(2), 91–107. https://doi.org/10.1109/TMTT.1983.1131443

4. Song, K., Fan, Y., & Zhang, Y. (2006). Radial cavity power divider based on substrate integrated waveguide technology. *Electronics Letters, 42*(19), 1100–1101. https://doi.org/10.1049/el:20062012

5. Song, K., Fan, Y., & Zhang, Y. (2007). Design of low-profile millimeter-wave substrate integrated waveguide power divider/combiner. *International Journal of Infrared and Millimeter Waves, 28*(6), 473{478. https://doi.org/10.1007/s10762-007-9229-5

6. Song, K., Fan, Y., & He, Z. (2008). Broadband radial waveguide spatial combiner. *IEEE Microwave and Wireless Components Letters, 18*, 73–75. https://doi.org/10.1109/LMWC.2007.911984

7. Ding, J., Wang, Q., Zhang, Y., Wu, L., & Sun, X. (2015). High-efficiency millimeter-wave spatial power combining structure. *Electronics Letters, 51*(5), 397–399. https://doi.org/10.1049/el.2014.3695

8. Song, K., Fan, Y., & Zhang, Y. (2008). Eight-way substrate integrated waveguide power divider with low insertion loss. *IEEE Transactions on Microwave Theory and Techniques, 56*(6), 1473. https://doi.org/10.1109/TMTT.2008.923897

9. Song, K., Fan, Y., & Xue, Q. (2010). Millimeter-wave power amplifier based on coaxial waveguide power-combining circuits. *IEEE Microwave and Wireless Components Letters, 20*(1), 46–48. https://doi.org/10.1109/LMWC.2009.2035965

10. Ghanadi, M. (2012). *A new compact broadband radial power combiner. PhD. Thesis.* Berlin: Technical University of Berlin. https://doi.org/10.14279/depositonce-3096

11. Bozzi, M., Georgiadis, A., & Wu, K. (2011). Review of substrate-integrated waveguide circuits and antennas. *IET Microwaves Antennas & Propagation, 5*(8), 909–920. https://doi.org/10.1049/iet-map.2010.0463

12. Liu, B., Hong, W., Tian, L., Zhu, H., Jiang, W., & Wu, K. (2006). Half mode substrate integrated waveguide HMSIW multi-way power divider. In: *Asia-Pacific Microwave Conference.* https://doi.org/10.1109/APMC.2006.4429562

13. Lee, D., An, Y., & Yook, J. G. (2012). J.: An eight-way radial switch based on SIW power divider. Journal of electromagnetic. *Engineering and Science, 12*, 216–222. https://doi.org/10.5515/JKIEES.2012.12.3.216

14. Song, K., Zhang, F., Chen, F., & Fan, Y. (2013). Wideband millimeter-wave four-way spatial power combiner based on multilayer SIW. *Journal of Electromagnetic Waves and Applications, 27*(13), 1715–1719. https://doi.org/10.1080/09205071.2013.823123

15. Parment, F., Ghiotto, A., Vuong, T., Duchamp, J., & Wu, K. (2015). Air-filled substrate integrated waveguide for low-loss and high power-handling millimeter- wave substrate integrated circuits. *IEEE Transactions of Microwave Theory and Techniques, 63*(4), 1–11. https://doi.org/10.1109/TMTT.2015.2408593

16. Zhuo, Y., Wang, H., Li, J., & Jin, H. (2015). A wideband four-way power divider/combiner based on substrate integrated waveguide and double layer finline. *IEICE Electronics Express, 12*(23), 1–6. https://doi.org/10.1587/elex.12.20150861

17. Jin, H., & Wen, G. (2008). A novel four-way Ka-band spatial power combiner based on HMSIW. *IEEE Microwave and Wireless Components Letters, 18*, 515–517. https://doi.org/10.1109/LMWC.2008.2001007

18. El-Akhdar, A., El-Hennawy, H., & El-Tager, A. (2016). A study on double via row configuration for SIW based structures. In: *International Conference on Microelectronics (ICM),* pp. 77–80. https://doi.org/10.1109/TMTT.2015.2408593

19. Song, K., Fan, Y., & Zhang, Y. (2009). Broadband radial waveguide power amplifier using a spatial power combining technique. *IET Microwaves, Antennas & Propagation, 3*(8), 1179–1185. https://doi.org/10.1049/iet-map.2008.0299

20. Fathy, A., Lee, S., & Kalokitis, D. (2006). A simplified design approach for radial power combiners. *IEEE Transactions on Microwave Theory and Techniques, 54*(1), 247–255. https://doi.org/10.1109/TMTT.2005.860302

Using Geometric Algebra for Formulating Electromagnetic Propagation Ray-Tracing Preprocessing

Ahmad H. Eid, Sherif M. Abuelenin, and Heba Y. Soliman

Abstract Ray tracing is an effective computational electromagnetic method that is used for studying electromagnetic wave propagation in complex scenarios. In this chapter, we present the formulation of the geometric processing of electromagnetic ray tracing using geometric algebra (GA). The presented formulation is more compact and uniform and geometrically significant compared to traditional formulations based on linear algebra. Simulation results show the correctness of the presented geometric processing formulations.

Keywords Computational electromagnetics · Geometric optics · Geometric algebra · Ray tracing

1 Introduction

Ray tracing (RT) is an effective computational electromagnetic modeling method that that can provide estimates of path loss, received signal strength, angle of arrival/ departure, and time delays of electromagnetic waves in complex propagation scenarios [1, 2]. RT algorithms solve the Maxwell's equations in high-frequency regime techniques, when the dimensions in the studied scenario are much larger than the wavelength of the electromagnetic wave. In RT procedures, electromagnetic waves are treated as rays that travel in straight lines. Rays are traced from the source to the receiving point by considering the couplings that the rays encounter. Geometric optics (GO) is used to trace direct, transmitted, and reflected rays, and the uniform theory of diffraction (UTD) [3] is used to take into account the effects of wave diffraction.

There are two classes of ray-tracing algorithms: the image method and the shooting-and-bouncing rays (SBR) method. The basic idea of SBR, which is also

A. H. Eid (✉) · S. M. Abuelenin · H. Y. Soliman
Department of Electrical Engineering, Faculty of Engineering, Port-Said University, Port-Fouad, Egypt
e-mail: ahmad.eid@eng.psu.edu.eg

© Springer Nature Switzerland AG 2020
M. H. Farouk, M. A. Hassanein (eds.), *Recent Advances in Engineering Mathematics and Physics*, https://doi.org/10.1007/978-3-030-39847-7_19

known as the brute-force method, is to trace a discrete number of rays launched from the source. Tracing stops after reaching a preset number of reflections or propagating a certain distance. The computation time in SBR depends on the spatial resolution in launching rays. The accuracy of RT is greatly reduced for rays traveling long distances from the transmitter [4, 5]. And problems such as double counting of rays may occur. In general, the method is approximate, and care must be taken in evaluating the received rays.

On the other hand, the deterministic image method recursively determines ray paths with multiple reflections by utilizing the image theory, i.e., images of the transmitting electromagnetic source are generated by reflecting it on all surfaces. Images serve as secondary sources. The total field at a receiver point is then determined, using the image method, by vector summing the direct field and the reflected field emanating from the image sources and arriving at the receiving point:

$$\vec{E}_t = \frac{E_0}{d_{\mathrm{LOS}}} e^{-jkd_{\mathrm{LOS}}} + \Gamma \frac{E_0}{d_{\mathrm{ref}}} e^{-jkd_{\mathrm{ref}}}$$

where k is the wavenumber and Γ is the reflection coefficient. In the case of multiple rays and multiple reflections, the second term becomes a summation of all rays considering the total propagation distance each has traveled and the total reflection coefficient it encountered (the product of all reflection coefficients of the surfaces it was reflected on).

The method is more accurate than SBR but may not be efficient when there are many reflection surfaces. The main computational bottleneck in an electromagnetic RT algorithm is the geometric processing stage, which can take up to 90% of the processing time in naïve implementations of ray-tracing algorithms [6, 7]. In the geometric processing (preprocessing), the RT algorithm recursively finds all valid ray paths from the transmitter to the receiver point [8], which requires initial solution of the environment. When the environment contains N reflecting surfaces, there are N first-order images (N rays arriving from the transmitter to the receiver that have been reflected only once), $N(N - 1)$ second-order images, $N(N - 1)^2$ third-order images, etc. In general, there is a total of M images when considering "n_r" reflections, where M is given by

$$M = \sum_{i=1}^{n_r} N(N - 1)^{n_r - 1}$$

Following, to generate the image sources by reflecting the primary EM source on all surfaces of the studied environment, valid ray paths must be found. Not every image can be "seen" everywhere. Images contribute to the total field only if the receiver is located in a point where the line connecting the image and the receiver intersects the last wall it was reflected on. To find a valid ray path from transmitter to receiver on a given sequence of surfaces, some geometric processing steps are typically performed repeatedly to determine if a ray path to the receiver point is obstructed by any surface of the scene and which surfaces can participate in ray coupling (i.e., reflection, transmission, diffraction, etc.). The main geometric processing steps include point reflection on a surface and testing and computing ray-surface intersection.

Such operations, as well as the transformations involved, are commonly formulated using linear algebra (vector and matrix algebra). Linear algebra has many limitations in the representation of geometric objects and transformations [9]. Clifford's geometric algebra (GA) is a much more powerful mathematical framework in that regard. GA is basically a comprehensive mathematical language that is well suited for the description of geometric transformations [9]. It unifies different mathematical systems such as complex numbers, vectors, and linear algebra. Therefore, GA provides a tool for compacting and simplifying the formulation in many cases [10]. The GA concept of "multivector" can be used to treat points, lines, planes, etc., in a single algebraic framework [11, 12]. For example, GA was used to combine the four Maxwell's equations into one compact formula that represents them as a single multivector [10, 13]. Other applications of GA are steadily appearing [10].

Compared with linear algebra, GA offers straightforward calculations with geometric objects and transformations, geometric intuitiveness, and compactness. The intuitive and compact descriptions based on GA lead to reduction in development time and robustness of the implementation. Therefore, GA formulations can lead to better algorithm development process.

In the rest of this chapter, we review the basics of GA and show how the main geometric processes of EM RT can be formulated in a compact way using GA, and then results and conclusions are presented.

2 GA-Based Geometric Processing for Ray Tracing

In this work, a simple type of GA, namely, the four-dimensional homogeneous (or projective) geometric algebra (4D HGA) [11], is used to represent basic geometric primitives and geometric processing steps in 2D and 3D RT. We focus on the following basic geometric steps: reflecting a source on a surface to determine the image, testing for intersections between rays and surfaces to determine if the ray path is blocked, and determining the intersection point (if existed) between a ray and a surface.

The homogeneous geometric algebra can be favorably applied to problems in Euclidean geometry because of the computational advantage that only one extra dimension is needed [14]. The 4D HGA is an algebraic extension of 3D Euclidean space with standard orthonormal basis vectors $\{e_x, e_y, e_z\}$. The extension is made initially by adding another basis vector e_w orthogonal to all three basis vectors e_x, e_y, e_z to obtain a 4D homogeneous vector space [15]. Following, the geometric product (GP) [11] is used to construct a set of 16 basis blades $\{1, e_x, e_y, e_z, e_w, e_{xy}, e_{xz}, e_{yz}, e_{xw},$ $e_{yw}, e_{zw}, e_{xyz}, e_{xyw}, e_{xzw}, e_{yzw}, e_{xyzw}\}$ from the basis vectors $\{e_x, e_y, e_z, e_w\}$. Blades are a special kind of multivectors used here to represent lines and planes constructed from points belonging to them, as shown in Table 1. For example, the basis blade e_{xyw} is defined using the GP as $e_{xyw} \equiv e_x e_y e_w$. The GP is an associative, noncommutative, and bilinear product between vectors. A general 4D HGA multivector is a linear combination of the 16 basis blades.

Some helpful algebraic operations are defined on arbitrary HGA multivectors A, B including the outer product (OP) $A \wedge B$, the left contraction product (LCP) $A \rfloor B$, and the blade inverse operator B^{-1} [11]. The OP is a bilinear product of multivectors used in this work to construct blades from vectors. Each nonzero blade B in HGA has a unique inverse B^{-1} with respect to the GP defined as $B^{-1} \equiv (1/B^2)B$. B^{-1} is essentially a scaled version of B because $B^2 = BB$ is always a scalar for any blade in GA. The LCP is another bilinear product of multivectors used in this work to compute signed distances (SDs) between geometric primitives as described in Table 2. A signed distance numerically encodes the relative orientation of geometric objects in space in addition to the actual distance between the objects.

SDs enable the use of compact expressions for geometric processing steps. In this work, when testing and finding the intersection of two line segments $\overline{p_1 p_2}$ and $\overline{q_1 q_2}$ in 2D, four point-line SDs are utilized: $a_i = d\left(p_i, \overleftrightarrow{q_1 q_2}\right)$ and $b_j = d\left(q_j, \overleftrightarrow{p_1 p_2}\right)$ where $i, j \in \{1, 2\}$. The intersection occurs if and only if $\text{sign}(a_1) = -\text{sign}(a_2)$ and $\text{sign}(b_1) = -\text{sign}(b_2)$. The intersection point r, if exists, is $r = tp_2 + (1 - t)p_1$ where $t = a_1/(a_1 - a_2)$. Similarly, when testing and finding the intersection of line segment $\overline{p_1 p_2}$ and triangle $\Delta q_1 q_2 q_3$ in 3D, two point-plane SDs are utilized: $a_i = d\left(p_i, \overleftrightarrow{q_1 q_2 q_3}\right)$ where $i \in \{1, 2\}$ in addition to three line-line SDs: $b_1 = d\left(\overleftrightarrow{p_1 p_2}, \overleftrightarrow{q_2 q_3}\right)$, $b_2 = d\left(\overleftrightarrow{p_1 p_2}, \overleftrightarrow{q_3 q_1}\right)$, and $b_3 = d\left(\overleftrightarrow{p_1 p_2}, \overleftrightarrow{q_1 q_2}\right)$. In this case, the intersection occurs if and only if $\text{sign}(a_1) = -\text{sign}(a_2)$ and $\text{sign}(b_1) = \text{sign}(b_2) = \text{sign}(b_3)$. As in the 2D case, the intersection point r, if exists, is $r = tp_2 + (1 - t)p_1$ where $t = a_1/(a_1 - a_2)$.

By inspecting Tables 1 and 2, the compactness and uniformity of HGA algebraic formulations in 2D/3D compared to VA formulations are apparent. Such geometric-algebraic representational features of GA enable the expression of geometric ideas more concisely. This enables the possibility of designing new classes of efficient geometric processing algorithms for electromagnetic ray tracing. This goal is mainly achieved by freeing the algorithm designers from the limitations of the algebraic

Table 1 HGA multivector representations for 2D/3D vectors, points, lines, and planes [15]

Euclidean primitive	HGA multivector representation
2D direction vector, $u = (u_x, u_y)$	$U = u_x e_x + u_y e_y$
3D direction vector, $u = (u_x, u_y, u_z)$	$U = u_x e_x + u_y e_y + u_z e_z$
2D point, $q = (q_x, q_y)$	$Q = q_x e_x + q_y e_y + e_w$
3D point, $q = (q_x, q_y, q_z)$	$Q = q_x e_x + q_y e_y + q_z e_z + e_w$
2D/3D line l passing through points q_1 and q_2	$L = Q_1 \wedge Q_2$
3D plane p passing through points $q_1, q_2,$ and q_3	$P = Q_1 \wedge Q_2 \wedge Q_3$

Table 2 Basic geometric processing steps used in ray tracing and corresponding HGA [15] and vector algebra (VA) [16] expressions

Geometric processing step	HGA expression	VA expression
(2D) Find reflection b of point a on line $\overrightarrow{q_1q_2}$ passing through points q_1, q_2	$B = Q_1 + V(A - Q_1)V^{-1}$ where $V = Q_1 - Q_2$	$b = 2\left(q_1 + \frac{(a-q_1)\cdot v}{v\cdot v}v\right) - a$ where $v = q_2 - q_1$
(3D) Find reflection b of point a on plane $\overrightarrow{q_1q_2q_3}$ passing through points q_1, q_2, q_3	$B = Q_1 + V(A - Q_1)V^{-1}$ where $V = (Q_1 - Q_2) \wedge (Q_2 - Q_3)$	$b = a - 2\frac{n\cdot(a-q_1)}{n\cdot n}n$ where $n = (q_1 - q_2) \times (q_2 - q_3)$
(2D) Compute signed distance $SD\left(a,\ \overrightarrow{q_1q_2}\right)$ from point a to the line $\overrightarrow{q_1q_2}$ passing through points q_1, q_2	$\mathrm{SD} = (A \wedge Q_1 \wedge Q_2)\rfloor e_{xyw}^{-1}$	$\mathrm{SD} = \frac{1}{\|q_2-q_1\|}\left(\begin{vmatrix} a_x & a_y \\ q_{2x} - q_{1x} & q_{2y} - q_{1y} \end{vmatrix} - \begin{vmatrix} q_{1x} & q_{1y} \\ q_{2x} & q_{2y} \end{vmatrix}\right)$ where $a = (a_x, a_y)$, $q_1 = (q_{1x}, q_{1y})$, $q_2 = (q_{2x}, q_{2y})$
(3D) Compute signed distance $SD\left(a,\ \overrightarrow{q_1q_2q_3}\right)$ from point a to the plane $\overrightarrow{q_1q_2q_3}$ passing through points q_1, q_2, q_3	$\mathrm{SD} = (A \wedge Q_1 \wedge Q_2 \wedge Q_3)\rfloor e_{xyzw}^{-1}$	$\mathrm{SD} = \frac{1}{\|n\|}n \cdot (a - q_1)$ where $n = (q_1 - q_2) \times (q_2 - q_3)$
(3D) Compute signed distance $SD\left(\overrightarrow{p_1p_2},\ \overrightarrow{q_1q_2}\right)$ from line $\overrightarrow{p_1p_2}$ to line $\overrightarrow{q_1q_2}$	$\mathrm{SD} = (P_1 \wedge P_2 \wedge Q_1 \wedge Q_2)\rfloor e_{xyzw}^{-1}$	$\mathrm{SD} = \frac{1}{\|n\|}n \cdot (p_1 - q_1)$ where $n = (p_1 - p_2) \times (q_1 - q_2)$

language of vectors and their operations. The alternative use of the more powerful GA formulations and operations is key in this process.

3 Results

To test the correctness of the presented formulations, a simple indoor propagation models were created. Figures 1 and 2 visualize the rays between source and destination points in simple 2D and 3D scenarios. The rays show that the visibility and reflections of rays are computed correctly.

The compactness of the GA implementation is expected to lead to more efficient coding and therefore better computational performance. To test the effectiveness of the described formulation, we implemented the basic operations (i.e., determining the source image by reflecting the source point on a surface and testing for ray and surface intersection to determine whether or not the ray is blocked or is to be reflected) using both the GA and VA formulations. The performance improvement is estimated by running both algorithms on the same machine, for a large number of operations, and comparing the algorithm execution time for both cases. The performance metric we used is the speedup factor, which is defined as the ratio between the execution time of the GA and VA algorithms: T_{GA}/T_{VA}. For the reflection and ray/intersection testing operations, we generated a large number of random points, rays, and surfaces; the average speedup factor was found to be around 130%.

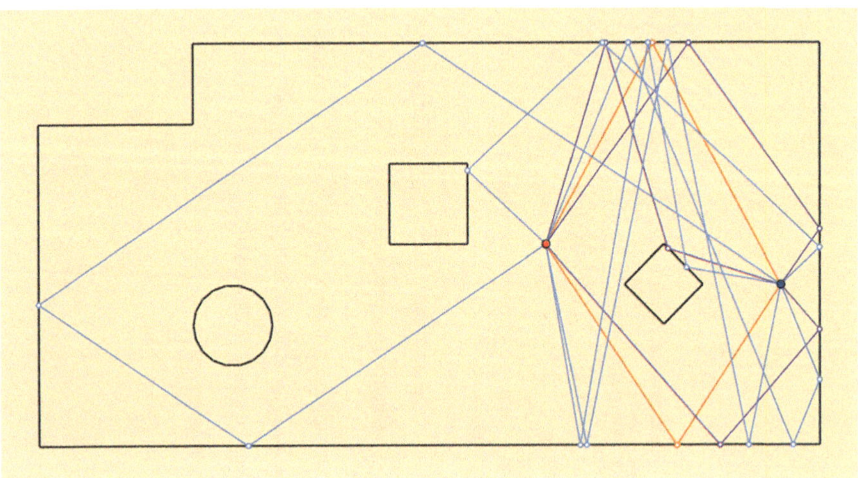

Fig. 1 Visualization of EM rays between a source and destination in simple 2D scenario using the developed GT-RT formulation, considering up to three reflections

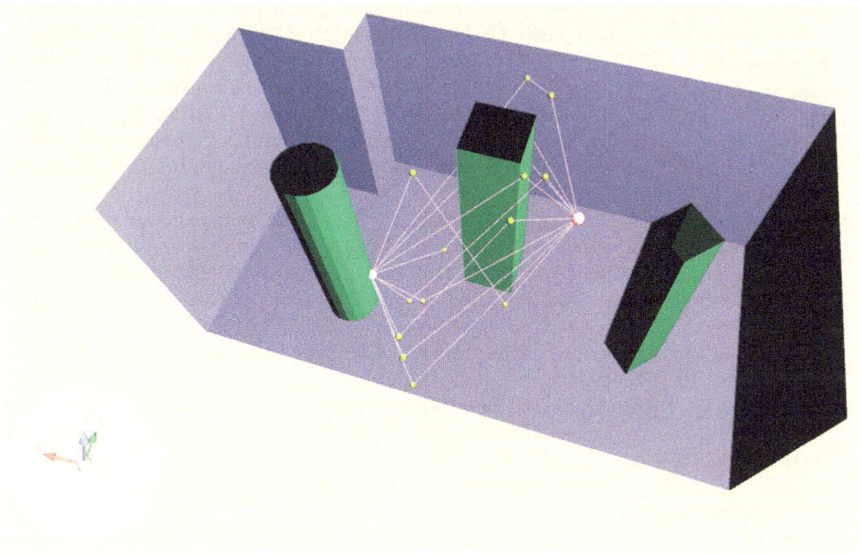

Fig. 2 Simple 3D indoor ray-tracing scenario with maximum of two reflections per ray path. The red and blue spheres indicate transmitter and receiver positions, respectively, while the smaller green spheres are the reflection points in the computed ray paths shown here as lines (the ceiling and one of the walls are made invisible, but rays that are reflected on them are shown)

4   Conclusion

GA is a unifying mathematical language that can easily represent different geometric objects and handle their transformations. Therefore, GA is very suitable for building RT applications. We demonstrated how using GA to perform the main geometric processing steps in electromagnetic propagation can lead to more compact formulation. In addition, the GA formulation is considerably more uniform in 2D and 3D cases. This leads to the possibility of devising new classes of geometric processing algorithms for electromagnetic ray tracing that are unified for both 2D and 3D cases without losing the efficiency of the derived implementations. The algorithm implementation details and further analysis shall be reported later.

Acknowledgments This work was supported by the ITIDA-ITAC program, Egypt (grant CFP-133).

References

1. Yun, Z., & Iskander, M. F. (2018). Radio propagation modeling and simulation using ray tracing. In *The world of applied electromagnetics* (pp. 275–299). Cham: Springer International Publishing.

2. Yun, Z., & Iskander, M. F. (2015). Ray tracing for radio propagation Modeling: Principles and applications. *IEEE Access, 3*, 1089–1100. https://doi.org/10.1109/ACCESS.2015.2453991

3. Kouyoumjian, R. G., & Pathak, P. H. (1974). A uniform geometrical theory of diffraction for an edge in a perfectly conducting surface. *Proceedings of the IEEE, 62*, 1448–1461.

4. He, D., Ai, B., Guan, K., Wang, L., Zhong, Z., & Kürner, T. (2019). The design and applications of high-performance ray-tracing simulation platform for 5G and beyond wireless communications: A tutorial. *IEEE Communications Surveys and Tutorials, 21*, 10–27. https://doi.org/10.1109/COMST.2018.2865724

5. Agelet, F. A., Formella, A., María, J., Rábanos, H., De Vicente, F. I., & Fontán, F. P. (2000). Efficient ray-tracing acceleration techniques for radio propagation modeling. *IEEE Transactions on Vehicular Technology, 49*, 2089–2104. https://doi.org/10.1109/25.901880

6. Cátedra, M. F., Pérez, J., Saez De Adana, F., & Gutierrez, O. (1998). Efficient ray-tracing techniques for three-dimensional analyses of propagation in Mobile communications: Application to picocell and microcell scenarios. *IEEE Antennas and Propagation Magazine, 40*, 15–27. https://doi.org/10.1109/74.683539

7. Iskander, M. F., & Yun, Z. (2002). Propagation prediction models for wireless communication systems. *IEEE Transactions on Microwave Theory and Techniques, 50*, 662–673. https://doi.org/10.1109/22.989951

8. Hussain, S. (2017). *Efficient ray-tracing algorithms for radio wave propagation in urban environments*. Diss: Dublin City University

9. Lopes, W. B., & Lopes, C. G. (2019). Geometric-algebra adaptive filters. *IEEE Transactions on Signal Processing, 67*, 3649–3662. https://doi.org/10.1109/TSP.2019.2916028

10. Chappell, J. M., Drake, S. P., Seidel, C. L., Gunn, L. J., Iqbal, A., Allison, A., & Abbott, D. (2014). Geometric algebra for electrical and electronic engineers. *Proceedings of the IEEE, 102*, 1340–1363. https://doi.org/10.1109/JPROC.2014.2339299

11. Doran, C., & Lasenby, A. (2003). *Geometric algebra for physicists*. Cambridge: Cambridge University Press.

12. Dorst, L., & Mann, S. (2002). Geometric algebra: A computational framework for geometrical applications part 1. *IEEE Computer Graphics and Applications, 22*, 24–31. https://doi.org/10.1109/MCG.2002.999785

13. Arthur, J. W. (2011). *Understanding geometric algebra for electromagnetic theory*. Hoboken: Wiley.

14. Hitzer, E., Nitta, T., & Kuroe, Y. (2013). Applications of Clifford's geometric algebra. *Adv. Appl. Clifford Algebr., 23*, 377–404. https://doi.org/10.1007/s00006-013-0378-4

15. Kanatani, K. (2015). *Understanding geometric algebra: Hamilton, Grassmann, and Clifford for computer vision and graphics*. CRC Press.

16. Goldman, R. (2009). *An integrated introduction to computer graphics and geometric modeling*. Boca Raton: CRC Press.

Part V
Information Technology

Robust GPS Anti-jamming Technique Based on Fast Orthogonal Search

Mohamed Tamazin and Aboelmagd Noureldin

Abstract Despite the major significance of signal processing methods used nowadays, GPS receivers still face substantial challenges, such as jamming, which remain a dominant source of ranging error. The presence of GPS jamming influences the acquisition and tracking modules inside the receiver leading to biased measurements or loss of lock of the GPS satellite signal. Consequently, GPS receivers cannot provide reliable position, velocity, and time solutions. The main aim of this research is to introduce a robust GPS anti-jamming technique based on fast orthogonal search (FOS). A SPIRENT GSS6700 GPS simulator is utilized to examine the performance of the proposed methods under deep jamming scenarios. The robustness of the proposed method is compared to the performance of the NovAtel Propak V2 commercial receiver under simulated jamming scenarios. The solution availability of the proposed method is increased by 72% compared to the commercial GPS receiver.

Keywords GPS · Jamming · Fast orthogonal search

1 Introduction

The global positioning system (GPS) technology plays a vital role in positioning and navigation applications due to the increasing demands for a positioning system that can provide continuous, accurate, and reliable position. However, this may not always be possible due to different environmental effects on GPS signals such as multipath and interference [1]. Therefore, efforts need to be exerted to make the

M. Tamazin (✉)
Electronics and Communications Engineering Department, Arab Academy for Science, Technology and Maritime Transport, Alexandria, Egypt
e-mail: Tamazin@aast.edu

A. Noureldin
Electrical and Computer Engineering Department, Royal Military College of Canada, Kingston, ON, Canada

© Springer Nature Switzerland AG 2020
M. H. Farouk, M. A. Hassanein (eds.), *Recent Advances in Engineering Mathematics and Physics*, https://doi.org/10.1007/978-3-030-39847-7_20

civilian part of the system more accurate, reliable, and available for challenging navigation environments.

Since the GPS signals are weak, they are vulnerable to RF interference; interference can be intentional or unintentional interference. Any effect that causes degradation in carrier-to-noise density ratio (C/N_0) can be considered as interference. Interference can result in poor geometry and large position errors or system unavailability [2]. The unintentional interference consists of any unwanted disturbance within a useful frequency band, including distortions of phase and amplitude (e.g., atmospheric effects, multipath), and noise. Intentional (in-band) jamming is the emission of RF energy of sufficient power and characteristics to prevent receivers in the target area from tracking the GPS signals.

The jamming interference can result in degraded navigation accuracy or complete loss of tracking the GPS receiver. The effect of RF interference on code correlation and loop filtering is to reduce the C/N_0 for all the incoming signals [3]. The receiver can lose lock if the effective C/N_0 is reduced under the tracking threshold.

The objective of this paper is to develop a robust GPS anti-jamming technique based on fast orthogonal search (FOS) algorithm. In this research, a SPIRENT GSS6700 simulator is utilized to provide a controlled environment to examine the performance of the proposed methods under certain circumstances such as deep jamming scenarios.

2 Proposed Method

In this research, the FOS algorithm [4–8] is used to model the distorted correlation function to estimate TOA of the GPS signal accurately. The motivation behind the development of the proposed technique based on FOS was to find such an algorithm that improves the code-delay estimation inside the tracking loop of a GPS receiver, especially under navigation environments. The main idea is utilizing the high-resolution estimation capabilities of FOS to mitigate the jamming signals [9]. The general architecture of the proposed technique inside the tracking loop is shown in Fig. 1.

As shown in Fig. 1, after the essential front-end processing and after the carrier has been wiped off, the received signal is passed to a bank of correlators. The numerically controlled oscillator (NCO) and PRN generator block produce a group of early and late versions of replica codes. In the case of the classical delay locked loop (DLL), the received signal is correlated with each replica in the bank of correlators. Several code tracking techniques (called discriminator in Fig. 1) use the output of the correlator bank values as input to generate the estimated GPS signal delay as output [10]. A loop filter then provides a smooth output of the discriminator.

The proposed algorithm that is highlighted in red in Fig. 1 uses the corresponding output of the bank of correlators as input to the FOS algorithm to model the measured correlation function (i.e., distorted correlation function due to jamming effects) and estimates the delay parameters along with the GPS signal [11].

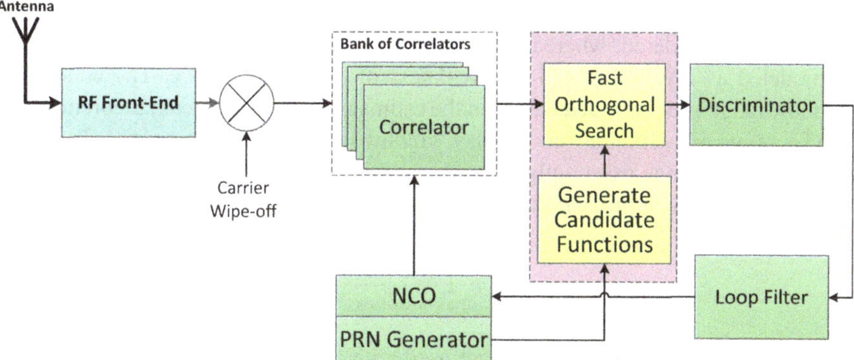

Fig. 1 The general architecture of FOS-based jamming mitigation technique inside the tracking loop

The FOS algorithm utilizes an arbitrary set of non-orthogonal candidate functions $P_m(n)$ and finds a functional expansion of a measured correlation function $\widehat{\mathscr{R}}(n)$ in order to minimize the mean squared error (MSE) between the measured correlation function and the functional expansion.

The functional expansion of the input $\widehat{\mathscr{R}}(n)$ in terms of the arbitrary candidate functions $P_m(n)$ is given by

$$\widehat{\mathscr{R}}(n) = \sum_{m=0}^{M} a_m P_m(n) + e(n) \tag{1}$$

where a_m is the set of weights of the functional expansion, $P_m(n)$ are the model terms selected from the set of candidate functions, and $e(n)$ is the modeling error.

These model terms are a set of GPS L1 ideal correlation functions $\mathscr{R}_{\text{ideal}}(n)$ for the middle/prompt correlator of a certain code-delay window range with several delays, phases, and amplitudes thereof:

$$P_m(\alpha, \tau, \varphi) = \alpha_m \, \mathscr{R}_{\text{ideal}}(\tau_m) e^{j\varphi_m} \tag{2}$$

where α is the relative amplitude of lth jamming signal, τ is the relative delay, and φ is the relative phase with respect to GPS signal. These candidate functions are generated offline and saved in a lookup table in memory. The set of candidate function parameters can be defined as follows:

$$0 \leq \alpha \leq 1$$
$$0 \leq \tau \leq 2T_c \tag{3}$$
$$0 \leq \varphi \leq 2\pi$$

where T_c is GPS C/A code chip duration, since the focus here is to search for those components with sub-chip level delays. Thus, we consider the search region to be

± 2 chip around the correlation peak. Herein, only one signal from one of the satellites is considered, whereas the contribution of the signals from other satellites is modeled as AWGN due to the weakness of their interference. The code-delay window range is determined based on the estimated correlation peak; theoretically, it can be anywhere within the code-delay window range of $\pm \tau_w$ chips [12]. The code-delay window range essentially depends on the number of correlators (i.e., M) and the spacing between the correlators (i.e., Δ) according to

$$\tau_w = \pm \frac{(M-1)}{2} \Delta \tag{4}$$

For example, if 201 correlators are used with a correlator spacing of 0.02 chips, then the resulting code-delay window range will be ± 2 chips with respect to the prompt correlator.

By choosing non-orthogonal candidate functions, there is no unique solution for Eq. (1). However, FOS may model the measured correlation function with fewer model terms than an orthogonal functional expansion [6].

FOS begins by creating a functional expansion using orthogonal basis functions such that

$$\widehat{\mathcal{R}}(n) = \sum_{m=0}^{M} g_m w_m(n) + e(n) \tag{5}$$

where $w_m(n)$ is a set of orthogonal functions derived from the candidate functions $P_m(n)$, g_m is the weight, and $e(n)$ is an error term. The orthogonal functions $w_m(n)$ are derived from the candidate functions $P_m(n)$ using the Gram-Schmidt (GS) orthogonalization algorithm. The orthogonal functions $w_m(n)$ are implicitly defined by the GS coefficients α_{mr} and do not need to be computed point by point [13].

The GS coefficients α_{mr} and the orthogonal weights g_m can be found recursively using the equations [6]:

$$w_0 = p_0(n) \tag{6}$$

$$D(m,0) = \overline{p_m(n)p_0(n)} \tag{7}$$

$$D(m,r) = \overline{p_m(n)p_r(n)} - \sum_{i=0}^{r-1} \alpha_{ri} D(m,i) \tag{8}$$

$$\alpha_{mr} = \frac{\overline{p_m(n)w_r(n)}}{\overline{w_r^2(n)}} = \frac{D(m,r)}{D(r,r)} \tag{9}$$

$$C(0) = \overline{\widehat{\mathscr{R}}(n)p_0(n)} \tag{10}$$

$$C(m) = \overline{\widehat{\mathscr{R}}(n)p_m(n)} - \sum_{r=0}^{m-1} \alpha_{mr}C(r) \tag{11}$$

$$g_m = \frac{C(m)}{D(m,m)} \tag{12}$$

In its last stage, FOS calculates the weights of the original functional expansion a_m (Eq. (1)), from the weights of the orthogonal series expansion g_m and GS coefficients α_{mr} [13]. The value of a_m can be found recursively using

$$a_m = \sum_{i=m}^{M} g_i v_i \tag{13}$$

where $v_m = 1$ and

$$v_i = -\sum_{r=m}^{i-1} \alpha_{ir}v_r, \quad i = m+1, m+2, \ldots, M \tag{14}$$

FOS requires the calculation of the correlation between the candidate functions and the calculation of the correlation between the input and the candidate functions. The correction between the measured correlation function and the candidate function $\overline{\widehat{\mathscr{R}}(n)p_m(n)}$ is typically calculated point by point once at the start of the algorithm and then stored for later quick retrieval.

The MSE of the orthogonal function expansion has been shown to be

$$\overline{\varepsilon^2(n)} = \overline{\widehat{\mathscr{R}}^2(n)} - \sum_{m=0}^{M} g_m^2 \overline{w_m^2(n)} \tag{15}$$

It then follows that the MSE reduction given by the mth candidate function is given by

$$Q_m = g_m^2 \, \overline{w_m^2(n)} = g_m^2 \, D(m,m) \tag{16}$$

The candidate with the greatest value for Q is selected as the model term, but optionally its addition to the model may be subject to its Q value exceeding a threshold level [13]. The residual MSE after the addition of each term can be computed by

$$\text{MSE}_m = \text{MSE}_{m-1} - Q_m \tag{17}$$

The search algorithm may be stopped when an acceptably small residual MSE has been achieved (i.e., a ratio of the MSE over the mean squared value of the measured correlation function [14] or an acceptably small percentage of the variance of the time series being modeled). The search may also stop when a certain number of terms have been fitted.

3 Experimental Work and Results

The SPIRENT GSS8000 hardware simulator was used to generate the jamming scenario for GPS signals. The simulator outputs the RF GPS signal in addition to adjustable additive jamming signals. The starting point was chosen at a point in Kingston, Ontario, Canada, at latitude 44° 13.726', longitude −76° 27.948', and height 100 m. The RF signals are collected through a NovAtel front end using a sampling frequency of $f_s = 10$ MHz and quantified in 4 bits. The front end rotates the received frequencies to the baseband and saves the sampled data for post-processing. The modified NavINST's software receiver that runs under a MATLAB™ platform is used to process the raw GPS data. The software is capable of performing GPS signal acquisition and tracking using the proposed methods and different tracking algorithms. In this section, the performance of the developed software receiver is compared to the performance of the NovAtel Propak V2 commercial receiver under the simulated jamming scenarios.

There are 10 GPS satellites available above a 5-degree elevation mask at the initial location, as shown in Fig. 2. The number of satellites acquired by the GPS software receiver is shown in Fig. 3.

Three jamming scenarios are considered at both static and dynamic receiver modes. The trajectory simulated used in this scenario is shown in Fig. 4. The signal specifications of these scenarios are given in Table 1. In the first scenario, the received signal is corrupted by a swept continuous wave (CW) interference for 1-min duration when the receiver is static. The frequency range and swept time of the swept-like interference are 1574.42–1576.42 MHz and 16 s, respectively. Two different jamming types were added to GPS signals when the receiver was in dynamic mode. The first one is swept CW with the same specifications mentioned

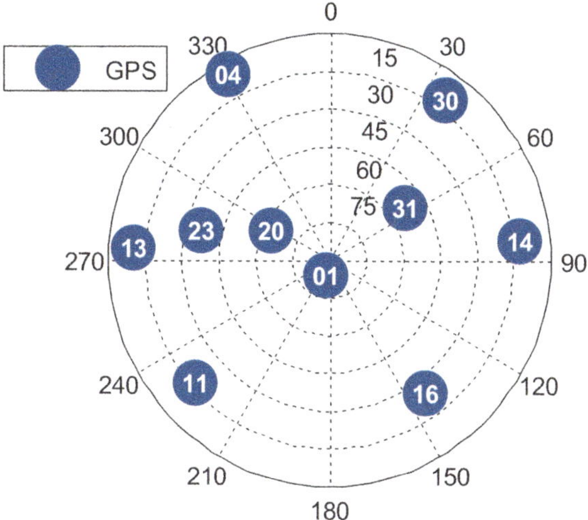

Fig. 2 Skyplot of the GPS satellites in view during the experiment—jamming scenario

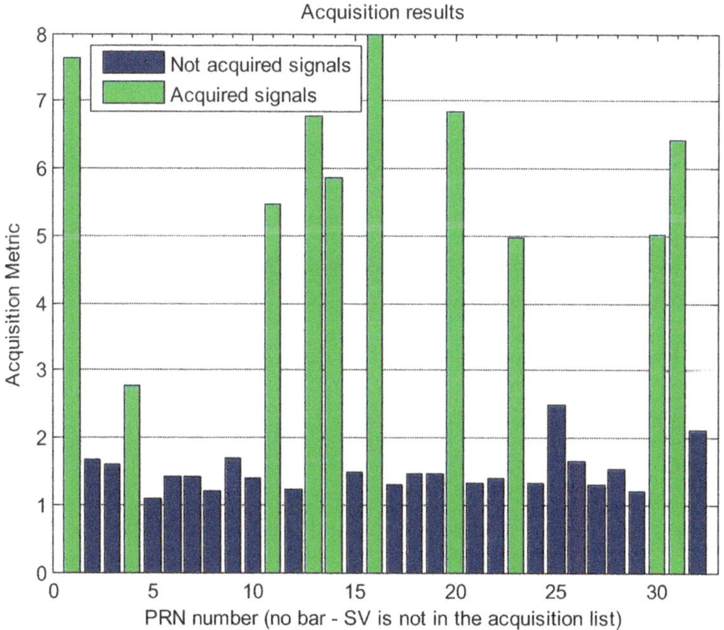

Fig. 3 Acquisition results of the GPS software receiver—jamming scenario

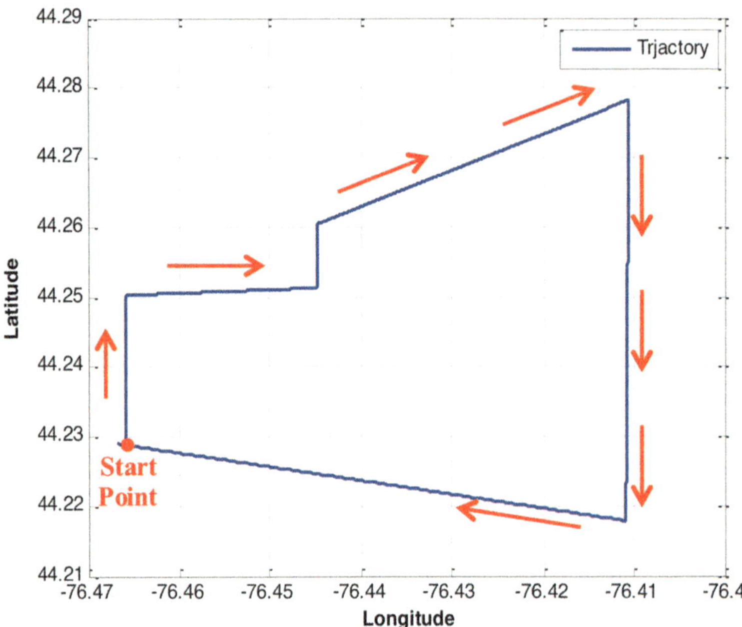

Fig. 4 Simulated trajectory

Table 1 Interference signal specifications

Scenario	Receiver mode	Interference type	Frequency (MHz)	Power (dBm)	Duration (min)
1	Static	Swept continuous wave	1574.42–1576.42	−80	1
2	Dynamic	Swept continuous wave	1574.42–1576.42	−80	1
3	Dynamic	Interfere noise	$f_{center} = 1575.42$ (BW = 1 MHz)	−80	2

before, for a duration of 1 min. The second type is noise interference at center frequency $f_{center} = 1575.42$ MHz and bandwidth 1 MHz for a duration of 2 min.

To analyze the jamming interference effect, the unjammed C/N_0 (dB-Hz) has to be first calculated. C/N_0 at baseband can be calculated as [2].

$$\frac{C}{N_0} = P_r + G_d - 10\log\left(kT_0\right) - N_f - L \tag{18}$$

where P_r is the received GPS signal power (dBW), G_d is the antenna gain (dBic), k is the Boltzmann's constant, T_0 is the thermal noise reference temperature (K), N_f is the noise figure of the receiver including antenna and cable losses (dB), and L is the implementation loss.

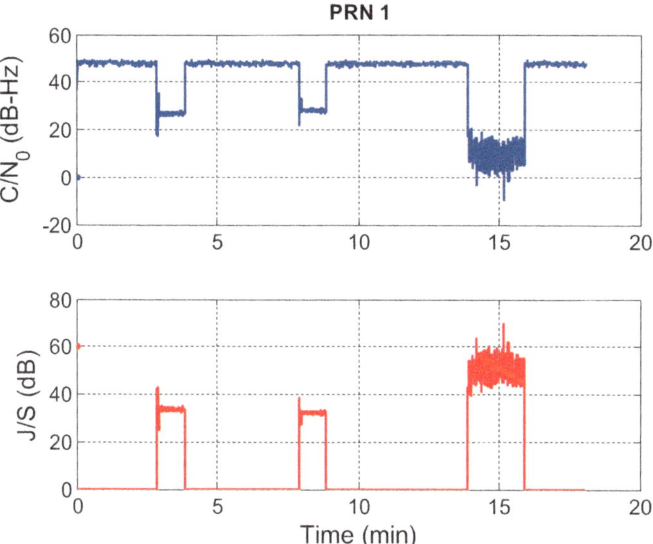

Fig. 5 Effect of J/S on C/N$_0$ level changes for PRN 1

The level to which the unjammed C/N_0 is reduced by RF interference is called the equivalent C/N_0 or $(C/N_0)_{eq}$, which can be calculated as [1].

$$(C/N_0)_{eq} = -10 \log \left[10^{-\frac{C/N}{10}} + \frac{10^{\frac{J/S}{10}}}{QR_c} \right] \tag{19}$$

where R_c is the PRN code chipping rate, Q is a dimensionless spread spectrum processing gain adjustment factor (different for each interference type), and $J/_S$ is the jammer-to-signal power ratio (dB). Equation (21) can be rearranged to calculate $J/_S$ as

$$\frac{J}{S} = 10 \log \left[QR_c \left(10^{\overline{(C/N)}} - 10^{\overline{C/N}} \right) \right] \tag{20}$$

Figure 5 demonstrates the interference effects on the signal quality, particularly the signal C/N_0 where N_0 refers to the Gaussian noise plus interference. C/N_0 is estimated during FLL-assisted-PLL tracking and computed from the correlator outputs using a narrowband/wideband power estimator [15]. Lower elevation satellites lose lock at the simulated interference levels as noticed for PRNs 4, 11, 13, 14, and 30. Figure 6 shows the estimated Doppler shift of PRN 1. The changes in C/N_0 due to the increasing interference levels affect the discriminators' output noise level and hence the estimated Doppler noise. These effects are clearly visible in Fig. 6. On the

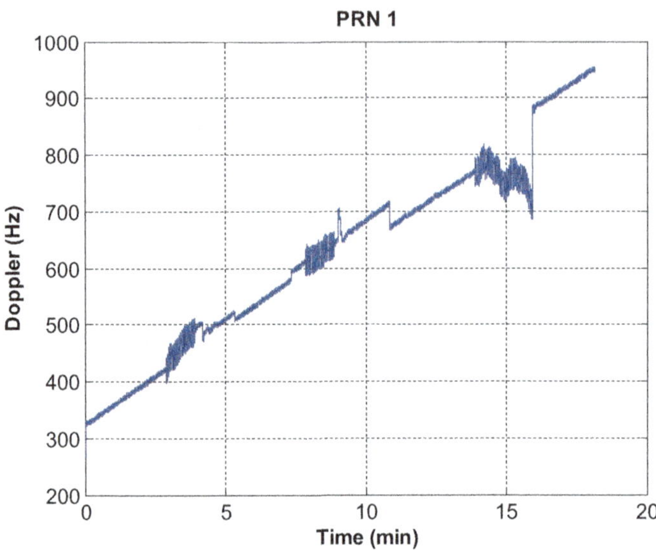

Fig. 6 Estimated Doppler shift of PRN 1 under the jamming scenario

contrary, these changes are almost absorbed by the proposed methods; thus, the C/N_0 changes have a minimum effect on its estimated Doppler noise.

The position solution was calculated using least-squares estimation technique. A single-point epoch-by-epoch least-squares solution was computed at a 1-Hz rate. If there were insufficient satellites in view, at a given epoch, or the solution failed to converge in 10 iterations, no solution would be computed.

Figure 7 shows the trajectories obtained using the NovAtel Propak V2 commercial receiver (shown in red) and using the software receiver in which the proposed methods are implemented (shown in green). It is observed that throughout the simulated trajectory, the commercial receiver was unable to fix position solutions during all three jamming events. Moreover, after the jamming was turned off, the commercial receiver was still not able to fix a position which means it was not able to reestablish lock.

The advantages of the proposed methods are clearly demonstrated such that after the jamming was turned off, the proposed method provided an accurate code delay and Doppler shift, which in turn enables the tracking loops to relock onto the satellite signals. Therefore, the solution availability is increased by 72% compared to the commercial receiver. It is noticed that the commercial receiver reestablished lock and fixed a position solution towards the end of trajectory.

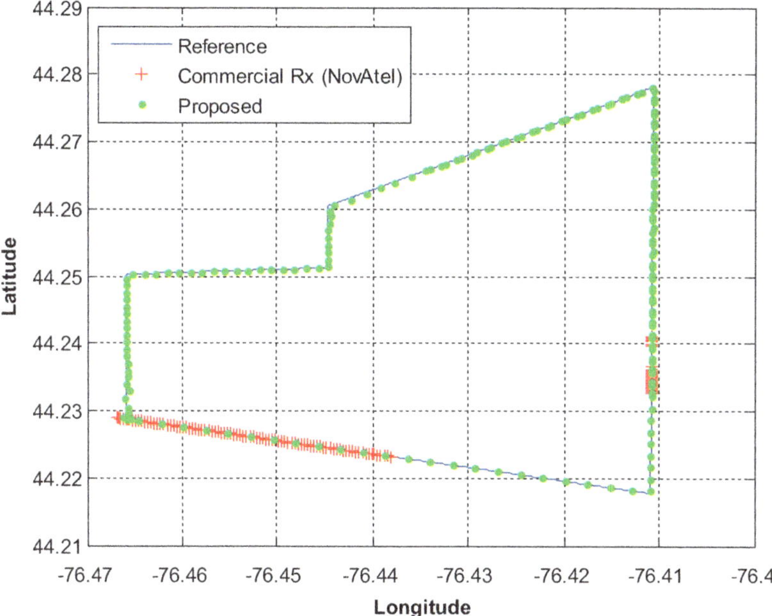

Fig. 7 Test trajectory (reference, commercial receiver, proposed)—jamming scenario

4 Conclusion and Future Work

A robust GPS anti-jamming technique has been introduced in this paper inside the tracking loop of a GPS receiver to mitigate jamming effects. The new algorithm is based on FOS, which is used to model the distorted correlation function and get better TOA estimation GPS signal. The performance of the proposed method has been tested using several realistic simulation scenarios using SPIRENT GSS6700 simulator. Three jamming scenarios are considered at both static and dynamic receiver modes. The experimental results showed that the proposed method increased the position solution availability by 72% compared to position provided by the NovAtel Propak V2 commercial receiver.

References

1. Kaplan, E. D., & Hegarty, C. J. (2017). *Understanding GPS/GNSS: Principles and applications*. Norwood: Artech House.
2. Kamel, A. M. M. (2011). *Context aware high dynamics GNSS-INS for interference mitigation* (Vol. 73).
3. Soubielle, J., Fijalkow, I., Duvaut, P., & Bibaut, A. (2002). GPS positioning in a multipath environment. *IEEE Transactions on Signal Processing, 50*, 141–150.

4. Adeney, K., & Korenberg, M. (1994). Fast orthogonal search for array processing and spectrum estimation. In *Vision, Image and Signal Processing, IEE Proceedings—IET* (pp. 13–18).
5. Adeney, K. M., & Korenberg, M. J. (1992). Fast orthogonal search for direction finding. *Electronics Letters, 28*, 2268–2269.
6. Korenberg, M. J. (1989). Fast orthogonal algorithms for nonlinear system identification and time-series analysis. In *Advanced methods of physiological system modeling* (pp. 165–177). Springer.
7. Korenberg, M. J. (1988). Identifying nonlinear difference equation and functional expansion representations: The fast orthogonal algorithm. *Annals of Biomedical Engineering, 16*, 123–142.
8. Tamazin, M., Noureldin, A., & Korenberg, M. J. (2012). Acquisition of weak GNSS signals using fast orthogonal search. In *Proceedings of the ION GNSS 2012*; September 17–21, 2012; Nashville, TN, USA (pp. 1399–1405).
9. Moussa, M. M. E., Osman, A., Tamazin, M., Korenberg, M., & Noureldin, A. (2016). Enhanced GPS narrowband jamming detection using high-resolution spectral estimation. *GPS Solutions, 21*, 475–485.
10. Ma, X., Cui, Y., & Stojmenovic, I. (2012). Energy efficiency on location based applications in mobile cloud computing: A survey. *Procedia Computer Science, 10*, 577–584.
11. Tamazin, M. (2015). *High resolution signal processing techniques for enhancing GPS receiver performance*. Queen's University, Electrical & Computer Engineering.
12. Bhuiyan, M. Z. H., & Lohan, E. S. (2010). Advanced multipath mitigation techniques for satellite-based positioning applications. *International Journal of Navigation and Observation, 2010*, 1–15.
13. Osman, A., Nourledin, A., El-Sheimy, N., Theriault, J., & Campbell, S. (2009). Improved target detection and bearing estimation utilizing fast orthogonal search for real-time spectral analysis. *Measurement Science and Technology, 20*, 065201.
14. Korenberg, M. J., & Adeney, K. M. (1998). Iterative fast orthogonal search for modeling by a sum of exponentials or sinusoids. *Annals of Biomedical Engineering, 26*, 315–327.
15. Parkinson, B., & Spilker, J. (1996). *Global positioning system: Theory and applications I*. Washington: American Institute of Aeronautics and Astronautics.

New Algorithm Based on S-Transform to Increase Defect Resolution Within Ultrasonic Images

Ahmed Benyahia, Ahmad Osman, Abdessalem Benammar, and Abderrezak Guessoum

Abstract Recent years have seen a notable advance in the quality of produced industrial ultrasonic data. This is due to two main factors. On the one hand, advances on hardware level permitted to design and trigger ultrasound sensing arrays and matrices that led to new acquisition strategies such as phased array method and full matrix capture technique. On the other hand, the development of algorithms and software components to reconstruct and process the measured data allowed a major improvement of the signal and image quality. Within this aspect, modern signal processing algorithms improved the defect resolution and thus their detection in ultrasound data. Mostly, methods based on time–frequency analysis are used. The measure of the improvement resulting from the signal processing methodology can be confirmed, for instance, by evaluating A-scans containing defects near the front and the back wall of inspected specimens. In this work, we describe a novel algorithm for processing one-, two- or three-dimensional ultrasonic data, in order to increase their defect resolution. The algorithm is demonstrated using simulation phantom as well as on a real specimen both including defects at different depths. The proposed enhancement method is based on the Stockwell transform and normalized

A. Benyahia (✉)
Algorithms/Signal- und Data Processing, Fraunhofer Institute for Nondestructive Testing IZFP, Saarbrücken, Germany

University of Blida 1, Blida, Algeria

Research Center in Industrial Technologies (CRTI), Algiers, Algeria
e-mail: ahmed.benyahia@izfp-extern.fraunhofer.de

A. Osman
Algorithms/Signal- und Data Processing, Fraunhofer Institute for Nondestructive Testing IZFP, Saarbrücken, Germany

University of Applied Sciences, htw saar, Saarbrücken, Germany

A. Benammar
Research Center in Industrial Technologies (CRTI), Algiers, Algeria

A. Guessoum
University of Blida 1, Blida, Algeria

© Springer Nature Switzerland AG 2020
M. H. Farouk, M. A. Hassanein (eds.), *Recent Advances in Engineering Mathematics and Physics*, https://doi.org/10.1007/978-3-030-39847-7_21

"""

Hilbert envelope. Proposed method can effectively improve the quality of the ultrasound data.

Keywords Ultrasound · B-scans · Defect enhancement · Stockwell transform · Hilbert envelope · Thresholding

1 Introduction

Nondestructive testing (NDT) with ultrasound (US) waves represents a well-established method to detect all types of defects within industrial specimens. An accurate interpretation of measured ultrasonic data is mandatory in order to extract the information regarding potential defects and successfully decide about the quality of the material under control [1]. The application of advanced signal processing methods can ease the interpretation task and help automate it by replacing the human expert with an artificial intelligence (AI)-based classifier. This is one major goal for the NDT of the future or NDT 4.0. Various researchers proposed algorithms for signal enhancement and defect detection in US data [7]. Results in [2] show that the studied signal processing methods (Hilbert transform, cross-correlation, and wavelet transform), for nondestructive testing of material, can efficiently characterize and determine the size and the location of defects in the material. Modified artificial immune system (AIS) algorithm was applied in [3]; as a result, the average value of location and length prediction errors was reported to be 0.1968%. In [4], a technique using wavelet transform for improving the SNR of ultrasonic signals is presented. Experimental results show that the used technique is effective for improving the ultrasonic signal and removing the white noise. Threshold modified S-transform (TMST) was proposed in [5, 17]; precisions of 1.8% for the thickness measurements and less than 3.7% for the closely positioned reflectors were obtained. High-frequency P- and S-wave algorithms were used in [6] and gave a 20% increase in the objective function chosen to quantify the delamination detection capability of the proposed sensor geometry optimization. Threshold modified S-transform and Shannon energy (TMSSE) was applied in [7]. Reported results showed that the detection of delamination depends on the position of defect: for defect close to the front face, TMSSE lost its accuracy in localization and the error was about 13%; for defect away from the front face of the sample, TMSSE detects defects in their proper positions and the error did not exceed 3%. In [7], Stockwell transform (ST) [8], as a famous technique in time–frequency analysis, proved to be a robust method and gave very promising results.

S-transform is an invertible time–frequency spectral localization technique that combines elements of wavelet transforms and short-time Fourier transform. The S-transform uses an analysis window in which width is decreasing with increasing frequency providing a frequency-dependent resolution. This transform is equivalent to phase-corrected continuous wavelet transform. The S-transform provides multiresolution analysis while retaining the absolute phase of each frequency [9].

The authors in [5, 7] proved the effectiveness of the S-transform in increasing the resolution and defect (or flaw) detection in B-scans. In this chapter, we propose a novel algorithm based on ST which is applied on the individual A-scans of B-scan images. The algorithm can be applied even on raw data considered for generating C-scans, on D-scans, and on 3D ultrasonic measurements. The rest of this chapter is organized as follows: in Sect. 2, required tools in the proposed algorithm are presented. B-scan image modeling and proposed algorithm are described in Sect. 3. Section 4 evaluates the performance of proposed method using simulated and experimental signals, and results are discussed. Finally, conclusions are given in Sect. 5.

2 Required Tools

2.1 S-Transform

Introduced by Stockwell et al. [10] and used to obtain the time–frequency representation of a time domain signal, the continuous version of S-transform $S(\tau, f)$ of a time signal $h(t)$ is defined as

$$S(\tau, f) = \int_{-\infty}^{\infty} h(t) \frac{|f|}{\sqrt{2\pi}} e^{-(\tau-t)^2 f^2/2} e^{-i2\pi ft} \, dt \tag{1}$$

$S(\tau, f_0)$ is a one-dimensional function of time for a constant frequency f_0, which shows how the amplitude and phase for this frequency f_0 changes over time. If the time series $h(t)$ is multiplied point by point with a Gaussian function $g(t)$, then the resulting spectrum is

$$H(f) = \int_{-\infty}^{\infty} h(t)g(t)e^{-2\pi ft} \, dt \tag{2}$$

The generalized $g(t)$ Gaussian function is defined as

$$g(t) = \frac{1}{\sigma\sqrt{2\pi}} e^{-t^2/2\sigma^2} \tag{3}$$

where σ is the dilatation parameter.

The choice of a Gaussian function is in the following reasons [11]:

- Symmetry in time and frequency domain (Fourier transform of a Gaussian is Gaussian),

- There is only one fundamental lobe, and there are no secondary lobes in a Gaussian function.
- Uniquely minimizes the quadratic time–frequency moment about a time–frequency point.

The Gaussian function will have a translation parameter τ and dilatation parameter σ:

$$S(\tau, f, \sigma) = \int_{-\infty}^{\infty} h(t) \frac{1}{\sigma\sqrt{2\pi}} e^{-(t-\tau)^2/2\sigma^2} e^{-2\pi f t} \, dt \tag{4}$$

This function presents a special case of the multiresolution Fourier transform due to the reason of the consideration of three independent variables τ, f, and σ, simplified by adding the constraint restricting the dilatation coefficient σ to be inverse of the frequency f (or proportional to the period T):

$$\sigma(f) = \frac{1}{|f|} \tag{5}$$

The discrete S-transform of the signal $h[kT]$ is given by

$$S\left[jT, \frac{n}{NT}\right] = \sum_{m=0}^{N-1} H\left[\frac{m+n}{NT}\right] e^{-2\pi^2 m^2/n^2} e^{i2\pi mj/N} \tag{6}$$

where $H[n/NT]$ H is the Fourier transform of $h[kT]$, where the integers j, m, $n = [0, 1, \ldots, N-1]$.

2.2 Otsu Thresholding Algorithm

The aim of applying a threshold on the image is to separate pixels belonging to the foreground from these who belong to the background. Suppose that the intensity of a gray level of an image can be expressed in L gray levels. The number of points with gray level $i \in [1, 2, \ldots, L]$ is denoted as x_i, and the entire number of points can be expressed as $X = x_1 + x_2 + \ldots x_L$. The occurrence of the gray level i in the histogram of the image is regarded as a probability [12]:

$$p(i) = \frac{x_i}{X}, \quad x_i \geq 0, \quad \sum_{i=1}^{L} x_i = 1 \tag{7}$$

In the presence of front, back, and potential defect(s) echoes, the image pixels are to be divided into two parts—C0 (representing the foreground) and C1 (background part)—by applying a threshold t on the singular pixel values. The class C0 represents pixels within levels $[1, 2, \ldots, t]$, and C1 denotes pixels within levels $[t + 1, \ldots, L]$. The occurrence probabilities ω_0 and average μ_0 of class C0 can be expressed as in Eqs. (8) and (10). The computation of occurrence probabilities ω_1 and average μ_1 of class C1 is obvious and provided in Eqs. (9) and (11):

$$\omega_0 = \omega(t) = \sum_{i=1}^{t} p(i) \tag{8}$$

$$\omega_1 = 1 - \omega(t) = \sum_{i=t+1}^{L} p(i) \tag{9}$$

$$\mu_0 = \sum_{i=1}^{t} \frac{i.p(i)}{\omega_0} = \frac{1}{\omega(t)} \sum_{i=1}^{t} i.p(i) \tag{10}$$

$$\mu_1 = \sum_{i=t+1}^{L} \frac{i.p(i)}{\omega_1} = \frac{1}{1 - \omega(t)} \sum_{i=t+1}^{L} i.p(i) \tag{11}$$

The total mean of the image can be written as

$$\mu_T = \sum_{i=1}^{L} i.p(i) \tag{12}$$

and can be expressed as

$$\mu_T = \omega_0 \mu_0 + \omega_1 \mu_1 \tag{13}$$

The between-class variance σ_B^2 of the two classes C0 and C1 is given by

$$\sigma_B^2 = \omega_0 (\mu_0 - \mu_T)^2 + \omega_1 (\mu_1 - \mu_T)^2 \tag{14}$$

The separability degree η of the classes, in the discrimination analysis, is

$$\eta = \max_{1 \le t \le L} \sigma_B^2 \tag{15}$$

Finally, optimization is done by maximizing σ_B^2 to find the optimal threshold t^* known as Otsu threshold:

$$t^* = \operatorname*{argmax}_{1 \le t \le L} \sigma_B^2 \tag{16}$$

2.3 Hilbert Envelope

For a real-valued function $x(t)$, its Hilbert transform $y(t)$ is defined as [13]

$$y(t) = H[x(t)] = \frac{1}{\pi} \int_{-\infty}^{\infty} \frac{x(\tau)}{t - \tau} d\tau \tag{17}$$

$y(t)$ can be further written as

$$y(t) = x(t) * \frac{1}{\pi t} \tag{18}$$

where "$*$" indicates convolution operator.

Hilbert transform is often interpreted as a $90°$ phase shifter $H[x(t)] = y(t) = -x(t)$. Moreover, from the given signal $x(t)$, a complex analytic signal $z[x(t)]$ can be expressed as

$$z[x(t)] = x(t) + jH[x(t)] = |z(t)| e^{j\theta(t)} \tag{19}$$

where $|z(t)|$ is the instantaneous amplitude of $x(t)$, $|z(t)|$ is named the envelope of signal by Hilbert spectrum, $\theta(t)$ is the instantaneous phase,

$$a(t) = |z(t)| = \left| x^2(t) + y^2(t) \right|^{1/2} \tag{20}$$

$$\theta(t) = \arctan \frac{y(t)}{x(t)} \tag{21}$$

and instantaneous frequency is $f(t)$,

$$f(t) = \frac{1}{2\pi} \frac{d\theta(t)}{dt} \tag{22}$$

$a(t)$ governs the intensity non-stationarity (or the temporal variation of amplitude), whereas $f(t)$ dominates the frequency non-stationarity (or the temporal variation of frequency content) of the signal.

3 Methodology

3.1 *Mathematical Modeling of the B-scan Signal*

The backscattered ultrasonic A-scan signal $s(t)$ composed of a single echo reflected by a flat surface can be modeled as follows:

$$s(t) = \beta e^{-\alpha(t-\tau)^2} \cos\left(2\pi f_c(t - \tau) + \Phi\right) \tag{23}$$

where β is the amplitude, τ is the arrival time, f_c is the central frequency, α is the bandwidth, and Φ is the phase. An A-scan signal containing several echoes M can be modeled as [14]

$$x(t) = y(t) + n(t) = \sum_{m=1}^{M} \beta_m e^{-\alpha_m(t-\tau_m)^2} * \cos\left(2\pi f c_m(1 - \tau_m) + \Phi_m\right) + n(t) \tag{24}$$

In this equation, $x(t)$ represents a noise-embedded A-scan signal, $y(t)$ is the model-based noise-free signal, and $n(t)$ is the added white Gaussian noise.

Ultrasonic B-scan of several stacked A-scans can be modeled as [15]

$$x(p,t) = \sum_{m=1}^{M} \beta_m(p) e^{-\alpha_m(t-\tau_m(p))^2} * \cos\left(2\pi f c_m(t - \tau_m(p)) + \Phi_m\right) + n(t) \tag{25}$$

This model was used in this work to generate simulated A-scans which were stacked into a B-scan image.

It is worth to mention that this model is not completely realistic as ultrasound images suffer from a dominant multiplicative noise (speckle) [16]; the effect of additive noise is insignificant compared to speckle effect. However, this is out of the scope of this study and will be addressed in a future work.

3.2 *Proposed Algorithm*

The B-scan image is a 2D view of ultrasound data (Fig. 1). Where the scan position is presented in the horizontal axis and the time in the vertical axis, every row in a B-scan matrix represents a single A-scan acquired from different points along a line. The B-scan data is the input of the algorithm. Once the algorithm starts, the first A-scan of the B-scan image is selected (Fig. 2); then, time–frequency domain representation of the selected A-scan using S-transform is done as presented in Eq. (6). Figure 3a shows the time–frequency representation of selected A-scan using S-transform. In the next step, a mask window is applied to the time–frequency representation of the selected A-scan to remove noises. Finding the maximal energy

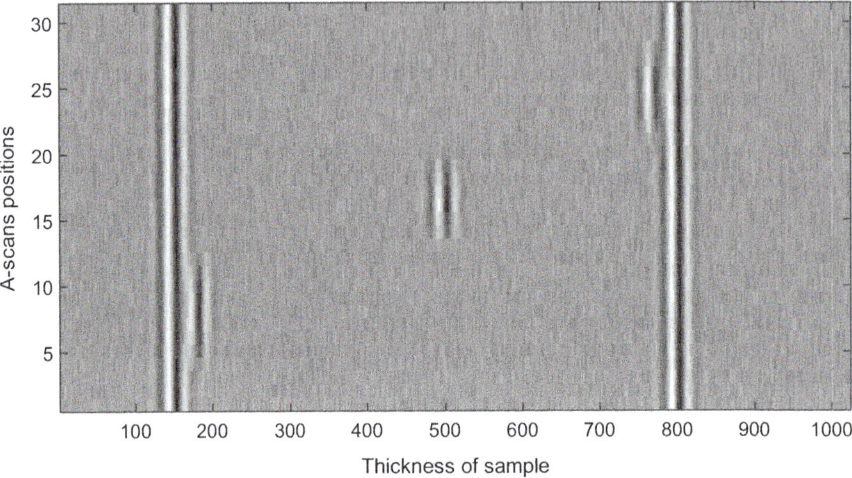

Fig. 1 Simulated B-scan with three defects

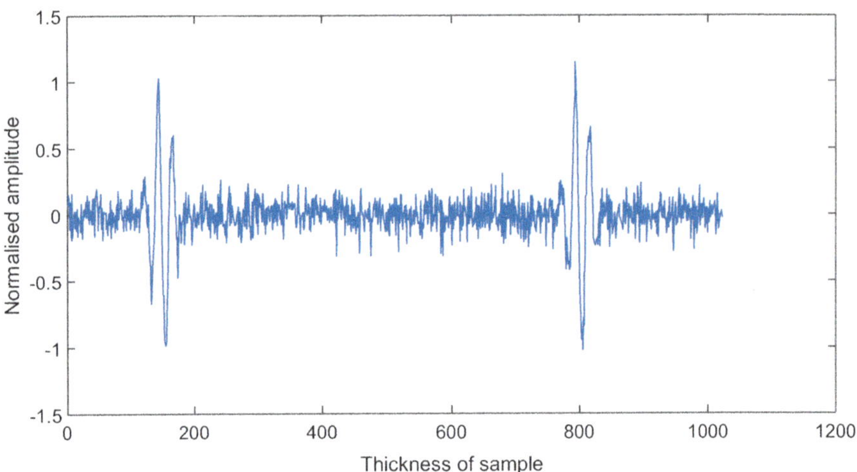

Fig. 2 Exemplary A-scan of the B-scan image

is done by multiplying the time–frequency representation by the mask window that is unity near and at max energy positions in the matrix and zeros elsewhere. Figure 3b shows the mask output of the time–frequency representation. Basic noise echoes of the individual A-scan signals remain in the resultant ST matrix obtained after masking. A global Otsu thresholding technique is applied to remove the remaining noise. The time–frequency representation of signal after masking and thresholding is shown in Fig. 3c. The denoised A-scan signal is achieved using inverse discrete S-transform of ST matrix after masking and thresholding. Note that

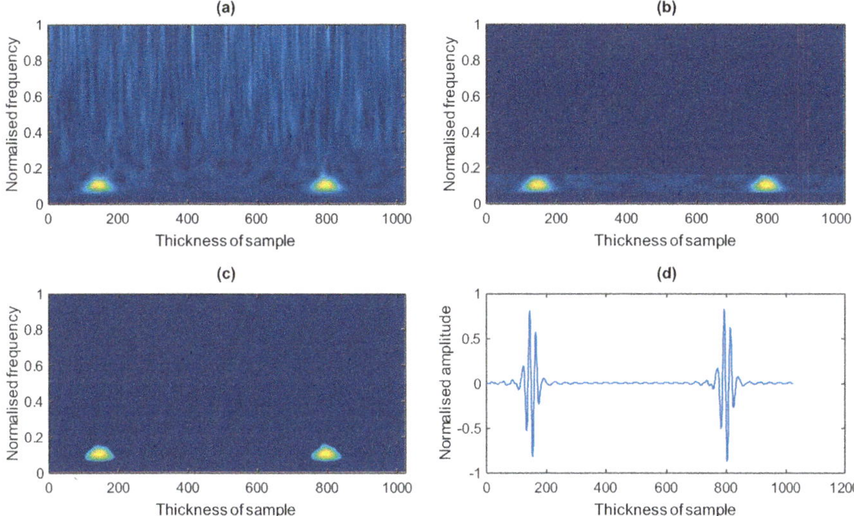

Fig. 3 (**a**) S-transform of the selected A-scan using max energy masking matrix output (**b**), time–frequency representation after masking and thresholding (**c**), and inverse S-transform (**d**)

the denoised signal preserves the important time–frequency domain features after masking and thresholding. The inverse S-transform equation is given as

$$h(kt) = \frac{1}{N} \sum_{n=0}^{N-1} \left\{ \sum_{j=0}^{N-1} S\left[\frac{n}{NT}, jT\right] \right\} e^{j2\pi nk/N} \tag{26}$$

where $h(kt)$ is the reconstructed signal from the denoised time–frequency representation. Figure 3d represents the inverse S-transform. Finally, envelope calculation using Hilbert transform (Fig. 4) is done as a last step. This algorithm is applied on each of the A-scans of the image. The resulting B-scan is shown in Fig. 5. The proposed enhancement procedure is illustrated in the block diagram in Fig. 6.

4 Results and Discussion

4.1 Simulated Model Analysis

The simulated B-scan data is composed of 31 A-scans of length 1024 point. Each A-scan is drowned in noise with signal-to-noise ratio of 5 dB. The simulated data contain three defects in different positions (close to front face, in the middle, and close to the back wall), as seen in Fig. 1. The proposed algorithm in Sect. 3 is applied to enhance the simulated B-scan ultrasonic data.

                                                          A. Benyahia et al.

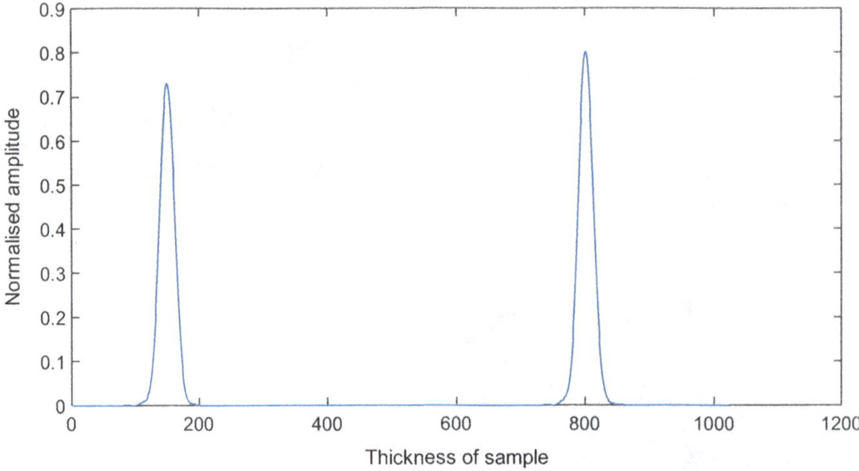

Fig. 4 Hilbert envelope after A-scan denoising

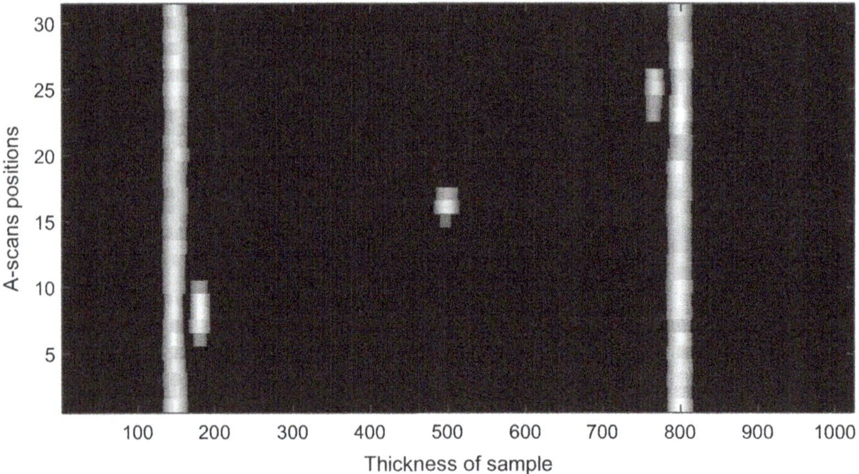

Fig. 5 Enhanced simulated B-scan using the proposed algorithm

The performances of the proposed method is evaluated on the A-scans using the signal-to-noise ratio (SNR) and root mean square error (RMSE) metrics.

The SNR can be represented as follows:

$$\text{SNR} = \frac{\sum_{t=0}^{L-1} h'(t)^2}{\sum_{t=0}^{L-1} h(t)^2} \tag{27}$$

The RMSE is defined as follows:

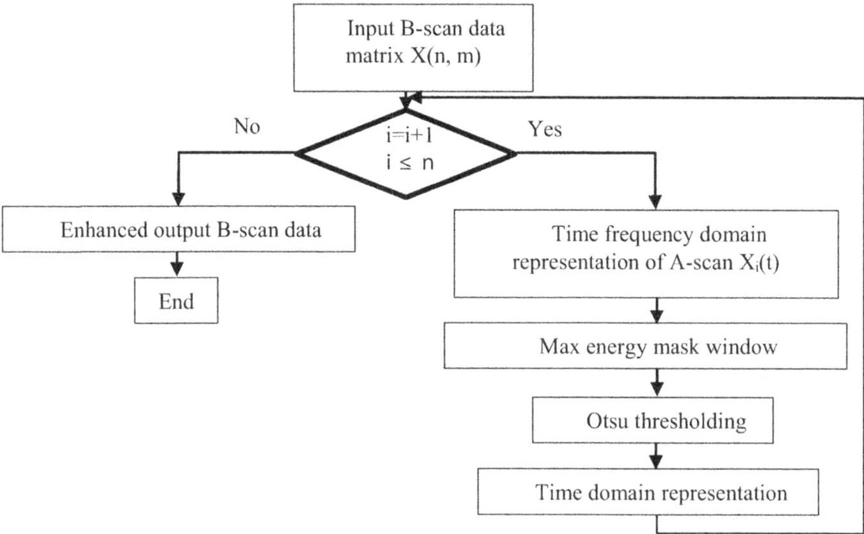

Fig. 6 Block diagram of the proposed algorithm

$$\text{RMSE} = \sqrt{\frac{\sum_{t=0}^{L-1} h(t)^2 - h'(t)^2}{L}} \tag{28}$$

where $h'(t)$ is the output reconstructed signal and $h(t)$ is the input noisy signal of L length.

Figure 7 shows the average RMSE obtained on all the A-scans using the proposed algorithm with Otsu thresholding methods which was compared to the iterative, local thresholding methods. This was done by simply replacing the Otsu thresholding step in Fig. 6 by the two other methods, respectively, and the result was obtained by TMST algorithm in [17] (Fig. 9). The diagram of Fig. 7 clearly indicated that the proposed method with Otsu thresholding gives lower RMSE than the iterative, local thresholding and TMST algorithm. The SNR results for the different thresholding method and TMST algorithm are depicted in the bar diagram in Fig. 8. The proposed method with Otsu thresholding shows better performance compared to other methods.

Indeed, Table 1 summarizes the comparison of the RMSE and SNR results obtained by using different denoising methods for a randomly selected A-scan (number 17). The conclusion is however valid for all the other A-scans. The proposed algorithm with Otsu shows SNR value 4.34 dB and RMSE 0.1013, whereas the TMST algorithm shows 3.6036 dB SNR and 0.1338 RMSE. The proposed method with Otsu provides the smallest RMSE and the highest SNR, thus demonstrating its capability to effectively enhance the A-scans (i.e., high SNR) with less loss of important foreground information (i.e., low RMSE).

A. Benyahia et al.

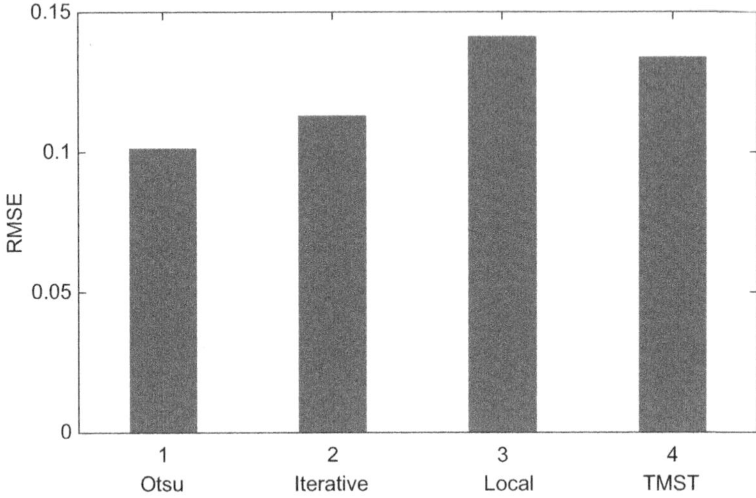

Fig. 7 Comparison of the RMSE obtained by using the proposed algorithm with Otsu, iterative, local thresholding, and TMST algorithm, respectively

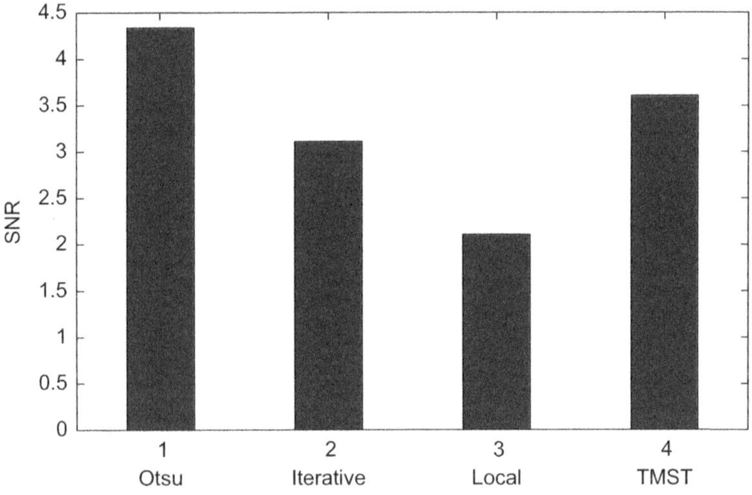

Fig. 8 Comparison of the SNR obtained by using the proposed algorithm with Otsu, iterative, local thresholding, and TMST algorithm, respectively

4.2 Experimental Data Analysis

For the experimental study of the algorithm, a carbon-fiber-reinforced polymer (CFRP) composite material with delamination defect was considered. The specimen was scanned in phased array mode using an OMNISCAN MX acquisition

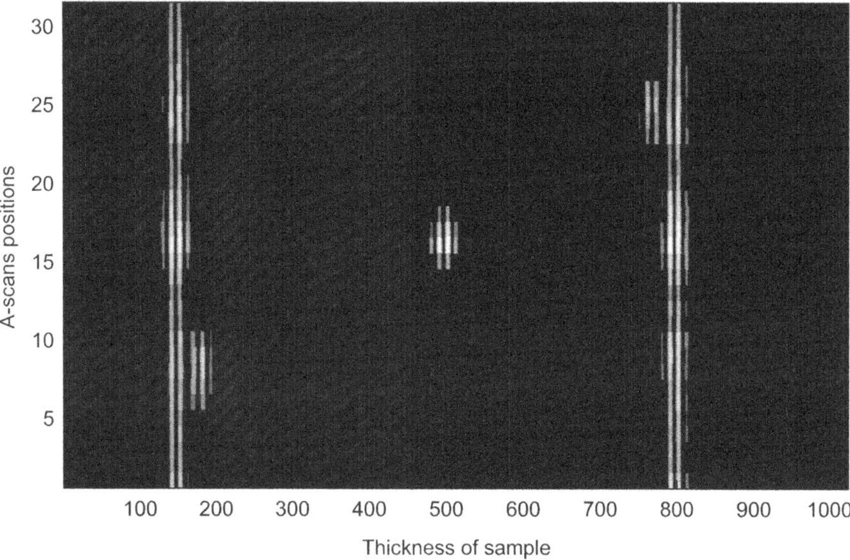

Fig. 9 Enhanced simulated B-scan using TMST algorithm [17]

Table 1 Performance results for proposed algorithm with different thresholding methods and TMST algorithm

	Otsu		Iterative		Local		TMST	
A-scan	SNR	RMSE	SNR	RMSE	SNR	RMSE	SNR	RMSE
17	4.34	0.1013	3.11	0.1130	2.1088	0.1413	3.6036	0.1338

system (illustrated in Fig. 10). The used phased array transducer is composed of 32 linear piezoelectric elements, and its central frequency is 5 MHz. Longitudinal waves are used, and the sound velocity in the material under test is 2830 m/s.

Figure 11 shows that the proposed algorithm in this work provides clear improvement of the B-scan image quality, by reducing the noise effects and enhancing the defects, thus giving a better detection of delamination defects (Fig. 11c) compared to the original B-scan (Fig. 11a).

5 Conclusion

A novel algorithm based on S-transform for B-scan image enhancement is proposed in this chapter. The proposed technique is applied on noisy image with defects near the front wall, in the middle, and near the back wall of the simulated B-scan. Comparison between the proposed algorithm with Otsu, iterative, local thresholding

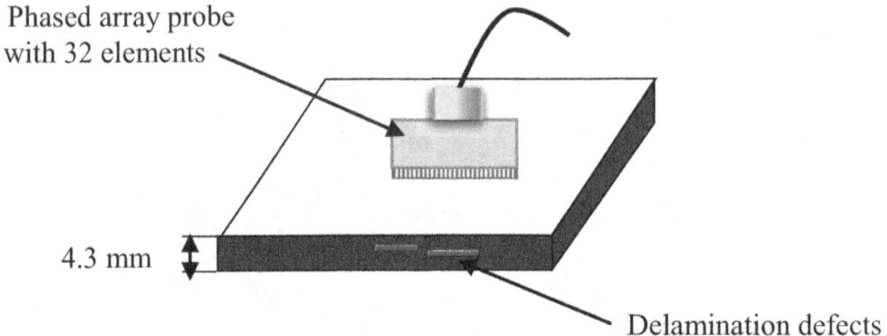

Fig. 10 Illustrative figure of CFRP sample scanned with phased array ultrasonic transducer

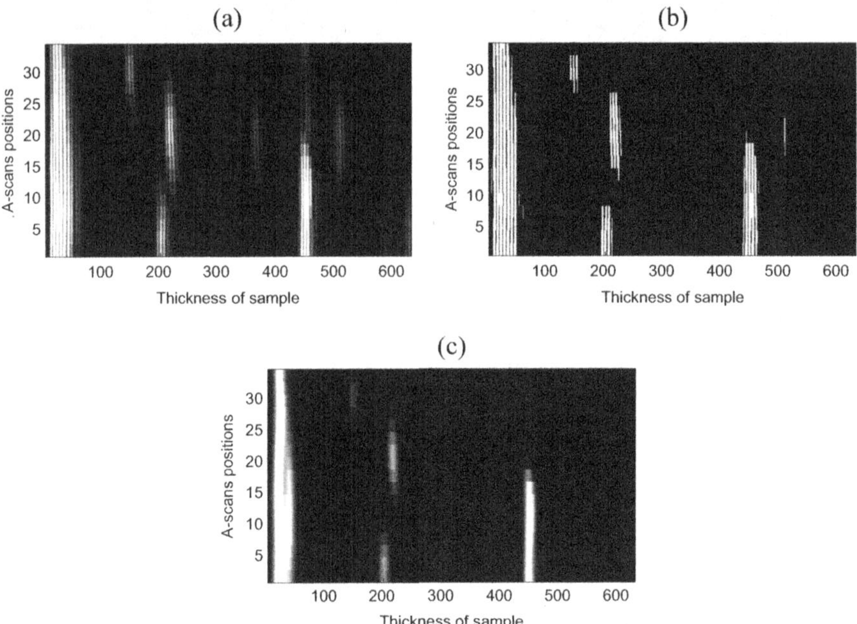

Fig. 11 (**a**) Experimental B-scan obtained from CFRP composite material, (**b**) enhanced experimental B-scan using TMST algorithm [17], and (**c**) enhanced experimental B-scan using the proposed algorithm

methods, and TMST algorithm on simulated data showed that the combination with Otsu threshold gives better SNR than iterative thresholding method, local thresholding, and TMST algorithm. Similarly, the performances of the proposed algorithm with Otsu threshold show also lower RMSE compared to other techniques. The enhancement of the defect resolution and thus the improved defect detectability are clearly visible on experimental data as seen in Fig. 9 using the proposed method in comparison to the TMST algorithm on experimental data from CFRP composite material with delamination defect.

References

1. Liao, X., Wang, Q., & Yan, T. (2013). Data-processing for ultrasonic phased array of austenitic stainless steel based on wavelet transform. In *Lecture notes in electrical engineering* (Vol. 237).
2. Bouden, T., Dib, S., Aissaous, K., & Grimes, M. (2009). Signal processing methods for materials defects detection. In *2009 IEEE international ultrasonics symposium*, Rome (pp. 1–4).
3. Mohebbi, B., Abbasidoust, F., & Mettefagh, M. (2012). Delamination detection in CFRP composite beam using modified AIS algorithm. In: *IEEE*.
4. Song, S.-P., & Que, P.-W. (2006). Wavelet based noise suppression technique and its application to ultrasonic flaw detection. *Ultrasonics, 44*(2), 188–193.
5. Benammar, A., Drai, R., & Guessoum, A. (2014). Ultrasonic flaw detection using threshold modified S-transform. *Ultrasonics, 54*, 676–683.
6. Pasquali, M., & Lacarbonara, W. (2015). Delamination detection in composite laminates using high-frequency P- and S-waves—Part I: Theory and analysis. *Composite Structures, 134*, 1095–1108.
7. Benammar, A., Kechida, A., & Drai, R. (2017). Signal quality improvement using a new TMSSE algorithm: application in delamination detection in composite materials. *Journal of Nondestructive Evaluation, 36*, 16.
8. Stockwell, R. G., Mansinha, L., & Lowe, R. P. (1996). Localisation of the complex spectrum: the S transform. *IEEE Transactions on Signal Processing, 44*, 998–1001.
9. Pinnegar, R., & Mansinha, L. (2003). The S-transform with windows of arbitrary and varying shape. *Geophysics, 68*. https://doi.org/10.1190/1.1543223
10. Stockwell, R. G. (2007). A basis for efficient representation of the S-transform. *Digital Signal Processing, 17*(1), 371–393.
11. Pei, S. C., & Wang, P. W. (2010). Discrete inverse S-transform with least square error in time-frequency filters. *IEEE Transactions on Signal Processing, 58*(7), 3557–3568.
12. Otsu, N. (1979). A threshold selection method from gray-level histograms. *IEEE Transactions on Systems, Man, and Cybernetics, 9*(1), 62–66.
13. Guanlei, X., Xiaotong, W., & Xiaogang, X. (2009). Generalized Hilbert transform and its properties in 2D LCT domain. *Signal Processing, 89*(7), 1395–1402.
14. Kirchhof, J., Krieg, F., Römer, F., Ihlow, A., Osman, A., & Del Galdo, G. (2017). Sparse signal recovery for ultrasonic detection and reconstruction of shadowed flaws. In *2017 IEEE International Conference on Acoustics, Speech and Signal Processing (ICASSP)*, New Orleans, LA (pp. 816–820).
15. Demirli, R., & Saniie, J. (2001). Model-based estimation of ultrasonic echoes. Part I: analysis and algorithms. *IEEE Transactions on Ultrasonics, Ferroelectrics, and Frequency Control, 48*(3), 787–802.
16. Osman, A., & Kaftandjian, V. (2017). Characterization of speckle noise in three dimensional ultrasound data of material components. *AIMS Materials Science, 4*(4), 920–938.
17. Abdessalem, B., Redouane, D., Ahmed, K., Lyamine, D., & Farid, C. (2015). Enhancement of phased array ultrasonic signal in composite materials using TMST algorithm. *Physics Procedia, 70*, 488–491.

Towards a Learning Style and Knowledge Level-Based Adaptive Personalized Platform for an Effective and Advanced Learning for School Students

Wafaa S. Sayed, Mostafa Gamal, Moemen Abdelrazek, and Samah El-Tantawy

Abstract This chapter proposes an artificial intelligence (AI)-based adaptive personalized platform for an effective and advanced learning. Most e-learning platforms target adult and lifelong learners. Yet, the educational process for younger people can be much enhanced through e-learning and artificial intelligence (AI) support to suit each learner's pace and learning style. In addition, it complements the role of classroom teacher in providing one-to-one tutoring for each learner, which is matched to his/her capabilities, preferences, and needs. Based on the mathematical definition of the problem, it is found that reinforcement learning (RL) is the most suitable AI technique for the proposed adaptive personalized e-learning system for school students. A literature review of the related research works is provided focusing on personalized e-learning systems for school students on the one hand and utilizing RL in this problem on the other hand. Learning styles, visual, aural, read/write, and kinesthetic (VARK), and Bloom's taxonomy are considered in the proposed system design. A website is designed based on Moodle learning management system (LMS) as the e-learning platform. An artificial intelligence module (AIM) responsible for adaptation is developed using multitask deep Q-learning. The module is implemented and trained using an ε-greedy policy. Its performance is evaluated using the running mean of the reward function, the total taxonomy loss, and the VARK loss. The performance metrics validate the convergence of the RL algorithm.

Keywords Adaptive personalized e-learning · Artificial intelligence · Bloom's taxonomy · Learning styles · Multitask deep Q-learning · Primary school education · Reinforcement learning · ε-Greedy policy

W. S. Sayed (✉) · M. Gamal · M. Abdelrazek · S. El-Tantawy
Faculty of Engineering, Cairo University, Giza, Egypt
e-mail: wafaasayed@eng1.cu.edu.eg; mostafa.gamal95@alumni.eng.cu.edu.eg; moemen.abdelrazek96@alumni.eng.cu.edu.eg; samah.el.tantawy@cu.edu.eg

1   Introduction

Education quality is a main factor of knowledge level [1], prosperity [2], and economic growth. Yet, a major drawback of conventional educational systems is focusing on a unified quality of education while ignoring differences among students and capabilities among teachers.

The theory of mastery learning [3, 4] considers different learning rates and individual student needs. It proposes that each student must achieve true mastery of a topic before moving on to the next, more advanced one, even if this takes longer time than his/her colleagues.

One-to-one learning is the optimal solution that can provide each student with a personalized learning experience that maximizes his/her performance [3–6]. Although this solution is practically intractable at school classrooms, it can be applied using e-learning systems that are adaptive and personalized.

The development in information and communication technology (ICT) was accompanied by a breakthrough in e-learning, as well as making it personalized and adaptive [7, 8].

A famous classification of learning objectives is Bloom's taxonomy [9] and its revised form [10]. The taxonomy covers the learning objectives in cognitive (knowledge-based), affective (emotion-based), and psychomotor or sensory (action-based) domains. The cognitive domain levels of objectives—remember, understand, apply, analyze, evaluate, and create—are frequently used to structure curriculum learning objectives, assessments, and activities.

Meanwhile, learning style models were widely researched and extensively used in personalization. Learning style is defined as "A description of the attitudes and behaviors which determine an individual's preferred way of learning" or "The composite of characteristic cognitive, affective, and physiological factors that serve as relatively stable indicators of how a learner perceives, interacts with, and responds to the learning environment" [11]. Several learning style models have been proposed for people of college age and older; yet, some of them provide equivalent questionnaires for kids. Neil Fleming's VARK model expanded upon earlier notions of sensory modalities such as the VAK model and was launched in 1987 with the four sensory modalities: visual, auditory, read/write, and kinesthetic [11]. VARK website provides a questionnaire for younger people [1], where the resulting four sensory modalities are widely adopted in children psychology [12].

Children need more support in identifying learning styles and knowledge level. In addition, their characteristics are subject to dynamic changes. Consequently, it is more beneficial to dynamically model children behaviors through AI as targeted in our system. In light of the above challenges in the current education system, the objective of this chapter is to propose a framework of an artificial intelligence-based adaptive personalized platform for an effective and advanced learning.

Section 2 introduces the mathematical model for our problem, where reinforcement learning (RL) is found to be the most suitable approach to solve it and satisfy the objectives of the proposed platform. Section 3 reviews the definitions of the main modules of an adaptive personalized e-learning system and focuses on the systems developed specifically for school students on the one hand and utilizing RL approach on the other hand and summarizes their contributions and limitations. Section 4 presents the novel e-learning platform, which is designed based on Moodle learning management system (LMS). Section 5 proposes the platform architecture and the customized implementation of RL-based artificial intelligence module (AIM). Section 6 presents the experimental work and simulation results evaluating the performance of the proposed RL-based AIM. Finally, Sect. 7 summarizes the main contributions of the chapter and suggests future work directions.

2 Objective and Mathematical Problem Definition

The chapter aims at developing an artificial intelligence module (AIM) that provides the specified functionalities. First, the mathematical model of the problem is introduced in order to accordingly investigate the appropriate method for the solution. The system can be formulated as a stochastic control problem, more specifically Markov decision process (MDP), where the state of environment is represented by the interaction of the student/learner with the e-learning platform. Such interactions are partially random and partially affected by the platform learning style presentation and exercise level adaptation. The optimal presentation and adaptation to the e-learning platform aims at enhancing the academic performance of the learner/ student. The problem of MDP is optimally solved using dynamic programming and practically solved by its approximation using reinforcement learning (RL) due to the curse of modeling and curse of dimensionality issues. Consequently, RL is the most appropriate setup for the proposed system.

RL is a form of machine learning where the system determines the ideal behavior by receiving reward feedback when its action is evaluated as being better than other choices. It represents a kind of mapping situations to actions in order to maximize the long-term reward. It differs from supervised and unsupervised machine learning approaches in the presence of learning agent. The learning agent senses the environment (despite its uncertainty), chooses the action that will maximize the rewarding function, and updates the state accordingly. Learners' behaviors, interactions, and learning styles are uncertain and have stochastic features especially at early age where the personality traits and character are still being developed.

An RL-based framework for personalized learning systems consists of the state-action-reward triplet. Generally, such a framework can consider, for example, the state as the static/dynamic data obtained from the user, the action as suggesting a recommended learning material/object, and the reward as the satisfaction, interactivity level, and long-term student's performance enhancement.

3   Adaptive Personalized E-Learning Literature

The state-of-the-art research subdivides an adaptive educational system into three main modules: the domain module, the learner module, and the adaptation module. The domain (content) model is a representation of the course content being offered. The learner module models learners' differences in background knowledge, skills, abilities, demographic, sociocultural and affective variables, learning styles, and interactions with the education system. The adaptation module is responsible for conveying the contents of the domain module to the learners in an adaptive way. It provides an adaptive presentation of the adaptive learning content.

3.1   School Students' E-Learning Systems

Young students need assistance as they are not mature enough yet to control and adapt their learning experience by themselves. However, the majority of the studies of adaptive personalized e-learning focus on adult and lifelong learners.

One of the limitations of the relatively older state-of-the-art research on adaptive personalized e-learning systems for young school students is the dependence on an initial static learner model [13–15], which is not updated using interactions. They either depend on a pretest of knowledge to create a student profile [14] or use basic information collected in a preliminary way [15] to ultimately detect the same result of the static model, not enhance it. More recent studies employ dynamic modeling; however, they do not mention how the system is initiated without a static model [16].

Some studies addressed developing a learning style model only [17], while others have extended the learner model to include more learner characteristics. In some studies employing dynamic learner modeling, learning styles are automatically detected, and an accuracy test of the model was applied through comparing the results to those of the static model [18]. Other studies have initiated the model using a questionnaire and tuned it afterward based on dynamic behaviors [19]. Most importantly, several studies only performed learners' data collection and sometimes classification without implementing an adaptation module or providing adaptation rules [14, 16, 19–21]. For the rest of the studies, they do not cover all elements of the learning process or branches of adaptation of content presentation and exercise navigation.

Our work aims at bridging the gap between these studies by developing a dynamic learner model, which detects and corrects the learner model in terms of learning style and cognitive level. In addition, it provides the corresponding adaptation strategies to the content presentation and exercise navigation.

3.2 RL-Based Adaptive Personalized E-learning Systems

There is a variety of research works and commercial products targeted toward college level, adult, and lifelong learners; yet, there is a lack in conducting similar experimental work for school students or research on how to modify the systems to suit them although they need it more. Machine learning and soft computing techniques utilized in adaptive personalized e-learning systems include association rule mining, support vector machine, artificial neural networks, Bayesian networks, clustering, fuzzy logic, decision trees, evolutionary algorithms, heuristic algorithms, RL, and hybrids of two or more techniques [22, 23]. In this section, we focus on systems which are based on the AI approach that best suits our problem, which is RL as discussed in Sect. 2.

Although RL is proven to be an effective approach for an adaptive and personalized e-learning system [24–32], objectives of the studies focused on either learning style detection or academic level improvement, while there is a lack of studies that consider both dimensions. In addition, most of the papers did not clearly describe their RL models so there is a lack of benchmark and simulation environment that can help the research community to have real evaluation of their models and validate their results.

4 The Proposed E-learning Platform

The basic e-learning platform, which represents the domain module, is built and presented in this section. It represents the front end of the proposed system, which includes all user interface components: course display, exercise interactions, viewing results, etc. This e-learning platform provides users with different forms of the main components of the educational experience, which fit their styles and levels. Through the platform, a student can access his/her enrolled courses, take exercises, and view the reports and results of his/her progress. The platform-building process included four main steps. First, the host server and domain were provided by the "Bluehost." Server machine is dedicated to the scripts that handle the communication between the RL implementation and website domain. The domain is dedicated to our proposed system, where all user interactions can take place via an official website dedicated to the platform. Second, the website was designed based on Moodle learning management system (LMS), as shown in Fig. 1. Moodle provides several advantages such as open source, free registration, tutorial availability, downloadable user reports, course reports, plug-ins, and free cloud deployment. Third, the Egyptian Ministry of Education curriculum is used as the material, where lessons are recreated in different styles and exercises representing three scaffolding difficulties for each of the six Bloom's taxonomy categories are selected for the platform adaptive usability. An example from primary two mathematics curriculum is

Fig. 1 The proposed platform's website interface

shown in Fig. 2. Finally, the web developed interface encompasses information about students' interactions and the corresponding adaptation.

5 The Proposed Adaptive Personalized System Architecture

Figure 3 shows the high-level architecture of the proposed system and how the components of RL algorithm are mapped to the domain, learner, and adaptation modules.

5.1 Student Simulator

There is an established historical need and importance of having a simulator before applying data from real students to any tutoring system [33]. Our problem, which we aim at solving using the proposed system, is a model-based RL problem. This is opposed to model-free RL problems in which learning occurs on the fly through

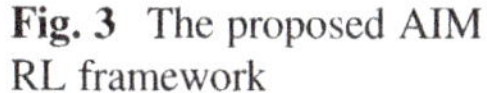

Fig. 2 Primary 2 mathematics curriculum example on the platform's website

Fig. 3 The proposed AIM RL framework

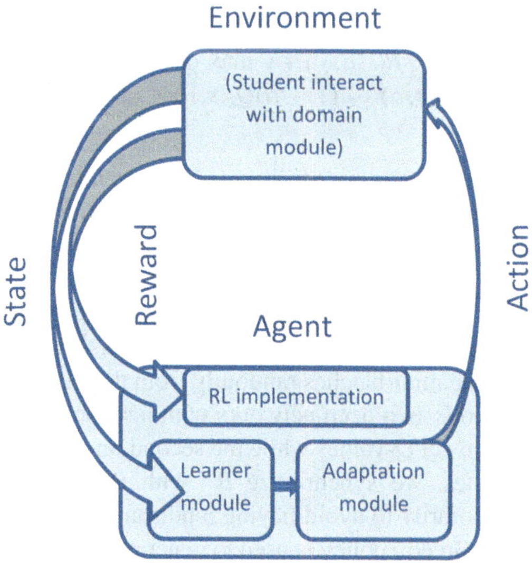

interactions from real students with the system. It is extremely difficult, impractical, and computationally inefficient to learn using real students in the first stage, and, hence, we need to simulate their behavior.

5.2 *Deep* Q-*learning RL Algorithm*

The action is either generated directly using a policy-based RL algorithm or according to a greedy policy based on a value-based RL algorithm. In value-based RL and given the state, the possible actions are investigated to reach a high action value based on the predicted value function to go through. Hence, it is called greedy policy and is mainly used when we have a finite set of actions. On the other hand, policy-based RL does not predict a value function. It recommends the action directly based on a predefined score function. It is more suitable when we have a large number or almost infinite number of actions, which are also continuous. Policy-based RL behaves better than value-based after convergence; yet, it is harder to reach convergence or takes longer time due to several factors such as the choice of the score function.

Q-learning is a value-based RL algorithm, which learns the state-action value function $Q(s, a)$. The Q-learning RL algorithm is given as follows:

Start with $Q_0(s, a)$ for all s, a.
Get initial state s
For $k = 1, 2, \ldots$ till convergence
    Sample action a, get next state s'
    if s' is terminal:
    target $= R(s, a, s')$
    Sample new initial state s'
    else:
    target $= R(s, a, s') + \gamma \max Q_k(s', a')$
    $Q_{k+1}(s, a) \leftarrow (1 - \alpha)Q_k(s, a) + \alpha[\text{target}]$
    $s \leftarrow s'$

Previously in Q-learning, RL systems used to build a memory table $Q[s, a]$ to store Q-values for all possible combinations of s and a. Nowadays, because of the infinite number of states, deep neural network is used as a function approximator. To ensure correct learning, we need the updates to be independent and identically distributed. Two solutions were proposed: experience replay and target network. In experience replay, we store millions of transition tuples in a buffer. Then, we sample mini-batches randomly from the buffer to avoid correlated updates. In target network, two deep networks with parameters θ^- and θ are created. We use the first one to get Q-values while the second one is getting updated. After a fixed number of updates, we synchronize θ^- with θ. The purpose is fixing the Q-value target temporarily to avoid having a nonstationary target.

ε-Greedy policy is used to generate actions as follows. Select the action that gives the maximum Q-value with probability $(1 - \varepsilon)$ or select a random action with probability ε. The probability ε decreases through learning as the number of iterations increases, which is known as annealing. A proper loss function should be selected to guarantee convergence of the deep Q-learning algorithm. Vanilla DQN loss function is given by

$$L_i(\theta_i) = \mathbb{E}_{s,a,s',r} \ _D\left(r + \gamma Q\left(s', \arg\max_{a'} Q(s',a';\theta);\theta_i^-\right) - Q(s,a;\theta_i)\right)^2, \quad (1)$$

which depends on the reward and predicted current and next states.

Huber loss function reduces dramatic changes which hurts DQN convergence and is given by

$$L_\delta(a) = \begin{cases} \dfrac{1}{2}a^2 & \text{for } |a| \leq \delta, \\ \delta\left(|a| - \dfrac{1}{2}\delta\right), & \text{otherwise.} \end{cases} \quad (2)$$

where a is the previous loss function. Dramatic change reduction is achieved through squaring the low-valued components only, while absolute value is obtained for larger components.

Learning several tasks given the same extracted features from the state requires a multitask deep Q-learning architecture [34]. Our neural network architecture results in multiple decisions to cover both VARK and Bloom's taxonomy dimensions. Decisions are represented as action heads in the last layer. A Q-learning Huber loss function is constructed for each branch. Then, the complete loss function is computed by summing the loss functions of each branch. When an agent takes a certain action in the environment, it replies back with a reward.

6 Experimental Work and Simulation Results

6.1 Experimental Setup

We implemented our deep network using PyTorch 0.4 framework under Python 3.6. We trained it using NVIDIA GTX1070TI graphics card for one million episodes. Each episode starts with a simulated student selected randomly. Then, it is terminated in two ways: either by reaching the final state in which the student masters the lesson or the depth of unrolling the episodic tree has reached one hundred states. We used a batch size of 4096 state-action-reward-next state quadruples, a learning rate of $7e^{-4}$, an experience replay memory of one million quadruple, and a discount factor of 0.999.

We applied an ε-greedy policy that starts with a high ε of 0.9 to encourage state exploration. Then, it decays exponentially as the number of training iterations increases until it reaches 0.005 which exploits the best action taken by our agent.

6.2   Results

To evaluate the performance of our agent, we used the running mean of the reward function as our main metric. We also used the total taxonomy loss as well as the VARK loss as our side metrics that give us intuition about the convergence of the RL algorithm. Tensorboard provided by Google was used to record the learning curves. To better visualize the results, a smoothing factor of 0.9 was used.

Fig. 4 Performance metrics of the RL agent

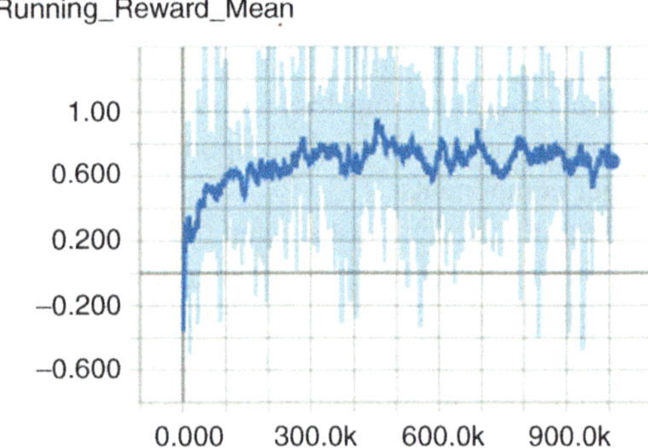

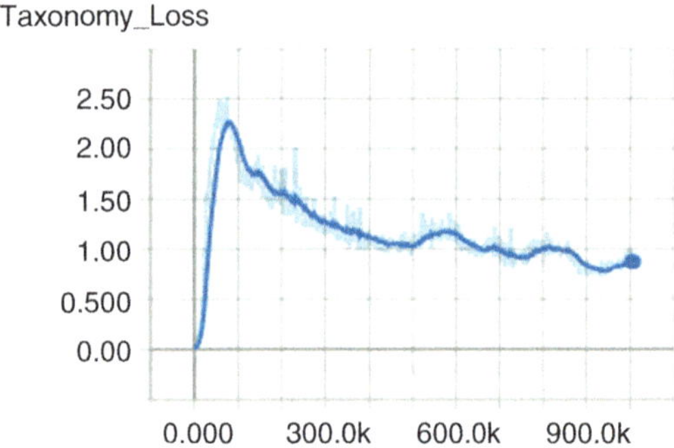

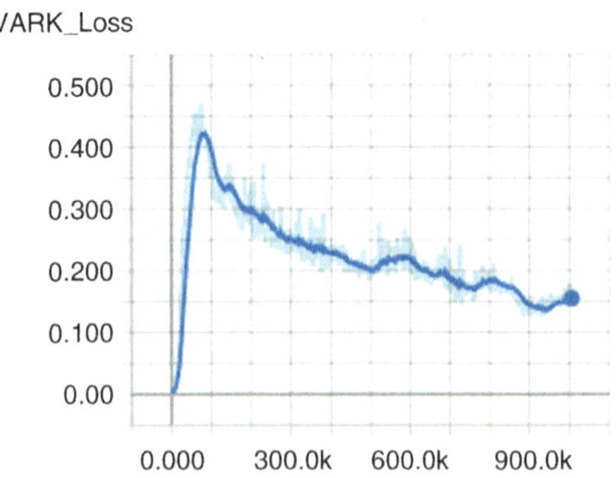

As shown in the losses curves in Fig. 4, the losses increase at the beginning of the training process due to high exploration. In addition, in the running mean reward, there are fluctuations after the termination of an episode and the start of a new one.

7 Conclusions and Future Work Directions

This chapter proposed an adaptive personalized platform for an effective and advanced learning. The mathematical model of the problem was defined where RL technique is demonstrated to be the most suitable methodology to achieve the objective. This chapter provided a literature review of the studies focusing on personalized and adaptive e-learning systems for school students on the one hand and the studies implementing RL techniques on the other hand. We built the basic educational platform based on Moodle as a dedicated website providing learning material for the students and keeping track of their interactions. A students' simulator model was designed to provide an analogy of the interaction data from real students. The simulator code was validated across the theoretical model. A deep Q-learning RL algorithm was implemented to provide the required adaptation of presentation and assessment according to the corresponding learning style and academic level. The multitask deep Q-learning network was implemented and trained using an ε-greedy policy. Its performance was evaluated using the running mean of the reward function, the total taxonomy loss, and the VARK loss. The performance metrics validate the convergence of the RL algorithm.

For future work, the back-end developed AIM will be integrated with the front-end reward-based educational platform, checking different forms of adaptation. Consequently, this will enable us to finalize the proposed adaptive personalized platform that provides the specified functionalities and test its performance and effectiveness on real students.

Acknowledgments This work was supported by the National Developmental Initiatives Grants from Academy of Scientific Research and Technology (ASRT) in Egypt. The authors would also like to acknowledge the efforts done by an engineer from Valeo, Ahmed Noeman, and an undergraduate student, Basant Elsayed.

References

1. GlobalKnowledge. (cited 2018). Retrieved from https://www.globalknowledge.com/
2. LegatumInstitute. (2018). *The Legatum Prosperity Index*™.
3. Bloom, B. S. (1968). Learning for mastery. Instruction and curriculum. Regional Education Laboratory for the Carolinas and Virginia, topical papers and reprints, number 1. *Evaluation Comment, 1*(2), n2.
4. Bloom, B. S. (1984). The 2 sigma problem: The search for methods of group instruction as effective as one-to-one tutoring. *Educational Researcher, 13*(6), 4–16.

5. Bloom, B. S. (1964). *Stability and change in human characteristics*. New York: Wiley.
6. Guskey, T. R. (2007). Closing achievement gaps: Revisiting Benjamin S. Bloom's "learning for mastery". *Journal of Advanced Academics, 19*(1), 8–31.
7. Kay, J., et al. (2013). MOOCs: So many learners, so much potential. *IEEE Intelligent Systems, 28*(3), 70–77.
8. Vandewaetere, M., Desmet, P., & Clarebout, G. (2011). The contribution of learner characteristics in the development of computer-based adaptive learning environments. *Computers in Human Behavior, 27*(1), 118–130.
9. Bloom Benjamin, S., & Krathwohl, D. R. (1956). *Taxonomy of educational objectives: The classification of educational goals, by a committee of college and university examiners. Handbook I: Cognitive domain*. New York: Longmans, Green.
10. Anderson, L. W., et al. (2001). *A taxonomy for learning, teaching, and assessing: A revision of Bloom's taxonomy of educational objectives, abridged edition*. White Plains, NY: Longman.
11. Truong, H. M. (2016). Integrating learning styles and adaptive e-learning system: Current developments, problems and opportunities. *Computers in Human Behavior, 55*, 1185–1193.
12. Mahdjoubi, L., & Akplotsyi, R. (2012). The impact of sensory learning modalities on children's sensitivity to sensory cues in the perception of their school environment. *Journal of Environmental Psychology, 32*(3), 208–215.
13. Dolenc, K., & Aberšek, B. (2015). TECH8 intelligent and adaptive e-learning system: Integration into technology and science classrooms in lower secondary schools. *Computers & Education, 82*, 354–365.
14. Wang, T.-H. (2014). Developing an assessment-centered e-learning system for improving student learning effectiveness. *Computers & Education, 73*, 189–203.
15. Chen, C.-H. (2014). An adaptive scaffolding e-learning system for middle school students' physics learning. *Australasian Journal of Educational Technology, 30*(3), 342–355.
16. Prabha, S. L., & Shanavas, A. M. (2014). Educational data mining applications. *Operations Research and Applications: An International Journal (ORAJ), 1*(1), 1–6.
17. Jaiswal, A. K., Singh, N., & Ahuja, N. J. (2017). Learning styles based adaptive intelligent tutoring systems: Document analysis of articles published between 2001. and 2016. *International Journal of Cognitive Research in Science Engineering and Education:(IJCRSEE), 5*(2), 83–98.
18. Özyurt, Ö., & Özyurt, H. (2015). Learning style based individualized adaptive e-learning environments: Content analysis of the articles published from 2005 to 2014. *Computers in Human Behavior, 52*, 349–358.
19. Mosharraf, M. (2016). Tuning primary learning style for children with secondary behavioral patterns. *Interdisciplinary Journal of e-Skills and Lifelong Learning, 12*, 19–32.
20. Tashtoush, Y. M., et al. (2017). Adaptive e-learning web-based English tutor using data mining techniques and Jackson's learning styles. In *2017 8th International Conference on Information and Communication Systems (ICICS)*. IEEE.
21. Hsieh, S.-W., et al. (2011). Effects of teaching and learning styles on students' reflection levels for ubiquitous learning. *Computers & Education, 57*(1), 1194–1201.
22. Kardan, A. A., Aziz, M., & Shahpasand, M. (2015). Adaptive systems: A content analysis on technical side for e-learning environments. *Artificial Intelligence Review, 44*(3), 365–391.
23. Tsortanidou, X., Karagiannidis, C., & Koumpis, A. (2017). Adaptive educational hypermedia systems based on learning styles: The case of adaptation rules. *International Journal of Emerging Technologies in Learning (iJET), 12*(05), 150–168.
24. Balasubramanian, V., & Anouncia, S.M. (2016), Learning style detection based on cognitive skills to support adaptive learning environment—a reinforcement approach. *Ain Shams Engineering Journal, 9*(4), 895–907.
25. Beck, J., Woolf, B. P., & Beal, C. R. (2000). ADVISOR: A machine learning architecture for intelligent tutor construction. *AAAI/IAAI, 2000*(552–557), 1–2.
26. Iglesias, A., et al. (2009). Reinforcement learning of pedagogical policies in adaptive and intelligent educational systems. *Knowledge-Based Systems, 22*(4), 266–270.

27. Iglesias, A., et al. (2009). Learning teaching strategies in an adaptive and intelligent educational system through reinforcement learning. *Applied Intelligence, 31*(1), 89–106.
28. Martin, K. N., & Arroyo, I. (2004). AgentX: Using reinforcement learning to improve the effectiveness of intelligent tutoring systems. In *International Conference on Intelligent Tutoring Systems*. Springer.
29. Tetreault, J. R., & Litman, D. J. (2006). Using reinforcement learning to build a better model of dialogue state. In *11th Conference of the European Chapter of the Association for Computational Linguistics*.
30. Lin, H.-T., Lee, P.-M., & Hsiao, T.-C. (2015). Online pedagogical tutorial tactics optimization using genetic-based reinforcement learning. *The Scientific World Journal, 2015*, 1.
31. Shawky, D., & Badawi, A. (2018). A reinforcement learning-based adaptive learning system. In *International Conference on Advanced Machine Learning Technologies and Applications*. Springer.
32. Shawky, D., & Badawi, A. (2019). Towards a personalized learning experience using reinforcement learning. In *Machine Learning Paradigms: Theory and Application* (pp. 169–187). Springer.
33. Vanlehn, K., Ohlsson, S., & Nason, R. (1994). Applications of simulated students: An exploration. *Journal of Artificial Intelligence in Education, 5*, 135–135.
34. Caruana, R. (1997). Multitask learning. *Machine Learning, 28*(1), 41–75.

A Context-Aware Motion Mode Recognition System Using Embedded Inertial Sensors in Portable Smart Devices

Omar Sheishaa, Mohamed Tamazin, and Iman Morsi

Abstract The expeditious market transformation to smart portable devices has created an opportunity to support activity recognition using the embedded sensors of these devices. Over the last decade, many activity recognition approaches have been proposed for various activities in different settings. The motion mode recognition or transition in modes of the device is needed in many technological domains. This approach detects a variety of motion modes for a human using a portable device. The approach includes many aspects: usability, mounting and data acquisition, sensors used, signal processing, methods employed, features extracted, and classification techniques. This chapter sums up with a comparison of the performance of several motion mode recognition techniques. In this research, multiple behaviors were distinguished using embedded inertial sensors in portable smart devices. In our experiments, we selected four types of human activity, which are walking, standing, sitting, and running. A combination of one of the embedded mobile sensors and machine learning techniques have been proposed in order to do this kind of classification. The proposed system relies on accelerometer data to classify user activities. The results show that using SVM classifier showed better accuracy for detection compared to the outcomes of the other classifiers like KNN and ensemble classifiers. For future work, classification of other human activities like cycling, driving, and swimming will be investigated.

Keywords Motion mode recognition · Machine learning · Smartphone accelerometer data

O. Sheishaa (✉) · M. Tamazin · I. Morsi
Electronics and Communications Engineering Department, Arab Academy for Science, Technology and Maritime Transport, Alexandria, Egypt

© Springer Nature Switzerland AG 2020

M. H. Farouk, M. A. Hassanein (eds.), *Recent Advances in Engineering Mathematics and Physics*, https://doi.org/10.1007/978-3-030-39847-7_23

1 Introduction

The need to recognize the mode of motion for a person, i.e., knowing whether the user is walking, standing, or sitting, is required in several fields, including childcare and sports services using inertial sensors with low costs. These sensors are already embedded in the portable smart devices like a smart mobile phone. Accordingly, this provides flexibility to perform tests and observe changes irrelevant to the surrounding environment [1].

In literature [2, 3], there are several techniques for motion mode recognition as integrating vision and inertial sensors. The technique based on computer vision approach is widely used in several applications like motion mode recognition, assistance system for driving, and surveillance [4]. Since smartphones are essential for daily life, Ericsson Consumer Lab reported that smartphone users increased from 36% in 2013 to 70% by 2020 [5, 6]. Smartphones nowadays come with several embedded sensors such as accelerometers and gyroscopes. The applications concerned with monitoring activity have been developed to detect body movements during daily living [7, 8]. Human activity recognition using embedded phone sensors through the approach of the proposition of the convolutional neural network by exploiting the characteristics inherited of the activities achieved accuracy beyond 90% [9].

Not only accelerometer sensor is used for activity recognition on smartphone but also the gyroscope sensor. Based on two-stage approach, initially, the first-level continuous hidden Markov model has been used for the activity recognition according to the collected data from the accelerometer and gyroscope sensors of the smartphone, while for fine classification a second-level continuous hidden Markov model was used [10].

Both accelerometer and gyroscope embedded in the smartphone were used, where feature selection was proposed for the selection of a subset of discriminant features and construction for online recognition of the activity, which yield at the end less consumption of the smartphone power [11].

The importance of the health care of the elderly and people with special needs is increasing every day. Accordingly, smartphone inertial sensors and wireless body sensors were used for performing this task along with a variety of classifiers like Naive Bayes, random forest, decision tree, and random committee [12].

Further approaches are developed for detection of human and human-human interaction for enhanced classification using multi-class support vector machine (SVM) [13].

The need to detect multiple motion modes is tremendously increasing that leads to the detection of praying positions [14]. This approach depends mainly on using the readings of accelerometer sensor, and then classifiers like Naive Bayes, J48, and KNN have been used which showed good accuracy for bowing and prostration positions. Not only Android-based smartphones are involved in human activity recognition, but BlackBerry phones were also used [15] through using accelerometer sensor along with the GPS, video camera, and a timer for identification between static activities and movement-related activities.

The Android-based smartphone has been chosen in this research since the Android operating system is an open source.

Moreover, Android-based smartphone is dominant in the smartphone market. The development of activity recognition applications using smartphones is growing tremendously since it has several advantages such as comfortable device portability without the need for additional equipment for mounting the sensors and comfort to the user due to the unobtrusive sensing. This is in contrast with other existed activity recognition approaches, like sensors at the surroundings, which could be located in specific locations like in a room or in a building to detect the presence or motion of a user [16, 17], or those who use dedicated hardware devices such as in [18] or sensor body networks [19]. Although the use of many sensors could improve the output performance of a recognition algorithm, it is not practical to expect that the public will use them in their daily life activities because of the difficulty and the time required to wear them. On the other hand, a drawback of the smartphone-based approach is the limitation of both energy and services as they are shared with other applications. This chapter introduces a robust motion mode recognition system based on the accelerometer embedded sensor in a portable smartphone.

2 Terminology

In this chapter, the following terms shall be used:

- *Motion mode* refers to the way the user who is holding a portable device moves from one location to another. The term motion mode shall be interchangeably used with the mode of transit when referring to some motion modes associated with transportation or transit methods (such as train, plane, or elevator).

  - Examples of motion mode include sitting, walking, driving, cycling, standing in an elevator, flying in a plane, etc.

- *Motion mode recognition* (MMR) refers to the recognition or detection of the motion mode.
- *Activity recognition* (AR) refers to the recognition or detection of the motion mode as well as other activities which are not related to motion (such as writing or reading). It usually has applications such as monitoring patients and smart homes.

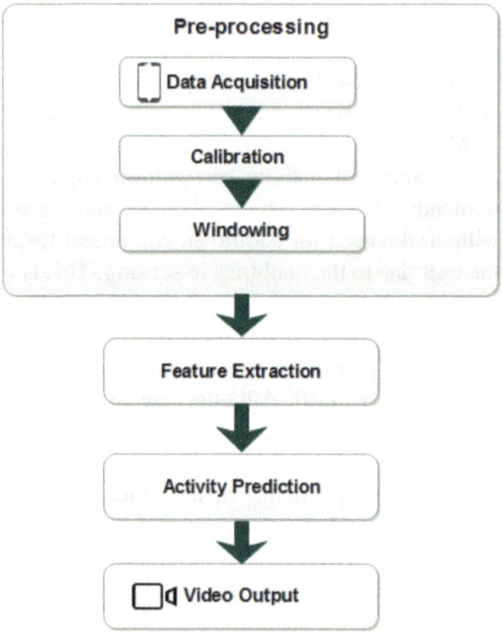

Fig. 1 Block diagram of the proposed motion recognition system

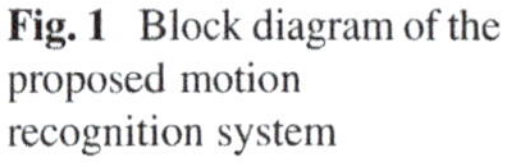

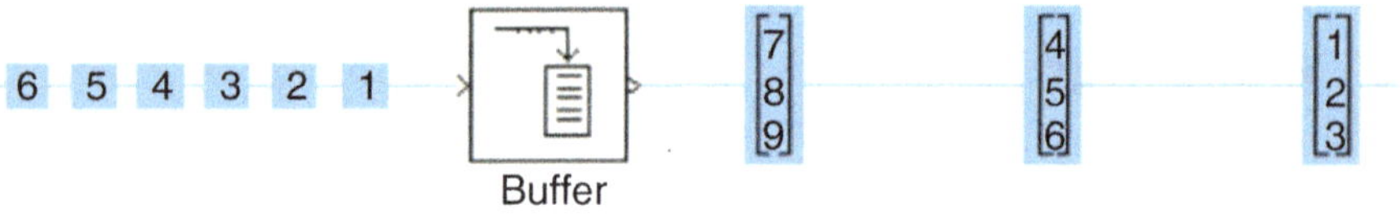

Fig. 2 Frame output at the lower rate

3   Methodology

The proposed method used in the research is shown in Fig. 1. The initial stage is acquiring data where raw signals obtain raw acceleration data from accelerometer sensor of the portable device. It is observed that the data collected from the sensors located at the phones are often affected by noise, which may affect the whole process. Initially, the preprocessing takes place through using a second-order Butterworth low-pass filter with cutoff frequency of 4 Hz. The following stage is the data calibration that is used to calibrate the acquired values from the accelerometer sensor, which is done through a static calibration technique.

The third stage is windowing. An input sample rate of 10 Hz is sent to each buffer block, which then outputs a buffered frame that includes 32 samples per 2 s, converting scalar samples to a frame output at a lower rate as in Fig. 2. After collecting 20 samples, each buffer block joins the 20 samples with 12 samples from the previous frame and passes the total 32 samples to the block of feature extraction. The windowing stage is followed by feature extraction where the features or properties are extracted from the output of variables in the preprocessing stage, which mainly consists of five main features. Then, the fifth stage is the activity prediction, where it loads the model that has been trained using any of the classifiers that will be discussed in the following Sects. 3.1, 3.2, and 3.3.

The last stage is the video output, which uses a multiport switch block to choose the corresponding user activity image data to display on the portable device; also, the detected activity ID along with the timestamp is being sent using ThingSpeak for better result collection.

3.1 Support Vector Machines (SVMs)

It is a discriminative classifier formally defined by a separating hyperplane. In other words, given labeled training data (supervised learning), the algorithm outputs an optimal hyperplane which categorizes new examples. In two-dimensional space, this hyperplane is a line dividing a plane into two parts wherein each class lay in either side as in Fig. 3. SVMs are a form of supervised machine learning. These learning systems use linear functions in multidimensional feature space to learn trained data based on statistical learning methods. SVMs are very powerful and can outperform most other systems in a wide variety of applications [20]. Included in the SVM framework is the ability to use various forms of kernels. The kernel of an SVM describes the basic principles on which the target function is created.

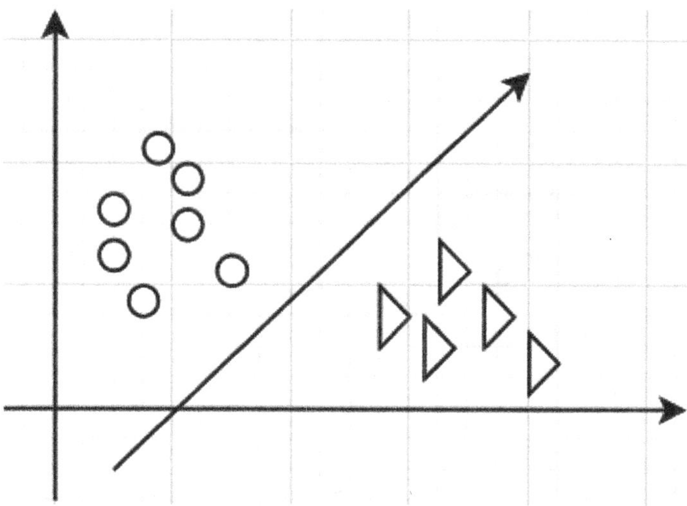

Fig. 3 The separation of the plane represented in an inclined line reveals two classes

3.2 K-Nearest Neighbors

K-nearest neighbor complexity is considered to be a simple one as it saves all scenarios and decides based on how close the new case is. In other words, it categorizes data based on how its neighbors are classified. It has often showed excellent results and greatly enhanced the state of the art [21, 22].

As an overview for KNN classification algorithm, we take as an example two classes A and B where class A is for the stars, while class B is for triangles, as shown in Fig. 4. The idea here is that the yellow shape needs to be classified into one of the defined classes. As shown in Fig. 4, a circle was drawn that contains a known shape and an unknown one. Accordingly, the unknown shape is classified to be a star, since it has only one neighbor, which is the star shape.

To say how it works in details, initially, known categories or classes are defined, and then an unknown instance or figure is added which does not belong to any of the predefined classes. The instance will be categorized based on the nearest neighbor where in our case, $K = 1$. However, if the case is that $K = 7$, then the nearest seven neighbors will be put into consideration where several neighbors will be classified according to their classes. Every class will have votes according to the neighbors around the unknown instance. Then the instance will belong to the class that has the highest votes.

In order to know who are the nearest neighbors, KNN classifier calculates the distance between the unknown instance and the other classified instances through calculating the Euclidean distance. As in Eq. (1), the distance will be calculated [23] according to the coordinates of the instance and those of its neighbors. Then, the values of the distances are sorted, and according to value of K, we know how many neighbors will be put into consideration. Furthermore, the Euclidean distance is being calculated between the unknown instance and other neighbors, while the votes of the neighbors that have least distance will be put into consideration:

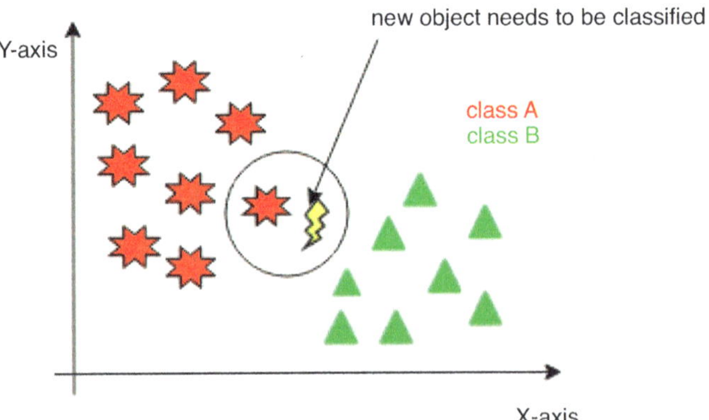

Fig. 4 How the instance is classified based on a single nearest neighbor

$$\sqrt{(A - X)^2 + (B - Y)^2} \tag{1}$$

where (X, Y) are the coordinates for the unknown instance, while (A, B), (C, D), etc., are the coordinates for the other known instances.

3.3 Ensemble Learning

In machine learning, ensemble learning is using multiple learning algorithms for the same task in order to have enhanced predictions compared to the outcome of individual learning models. In other words, we have an object that needs to be classified concerning a training dataset. The query object will be classified according to different learning algorithms where the outcome of each learning algorithm is recorded as a vote, and the final prediction will be decided according to the highest votes. This way of learning could be used in medical fields like predicting the type of the tumor in cancer as it is much better to have the result from many learning ways other than from a single one. One of the advanced ensemble techniques is boosting if an instance is wrongly predicted by the first learning method and the rest of the single methods. Accordingly, boosting will correct the errors of the previous model. Initially, a subset is built from the primary dataset, where all objects are equally weighted. Then a model is created which will be used for making predictions for the rest of the dataset as in Fig. 5.

The first model M1 consists of eight objects which are divided into four objects as plus sign, and the other four objects are as star sign where each of the objects has equal weight. The classification of this model had two plus signs wrongly classified as stars.

Fig. 5 The initial model M1 created

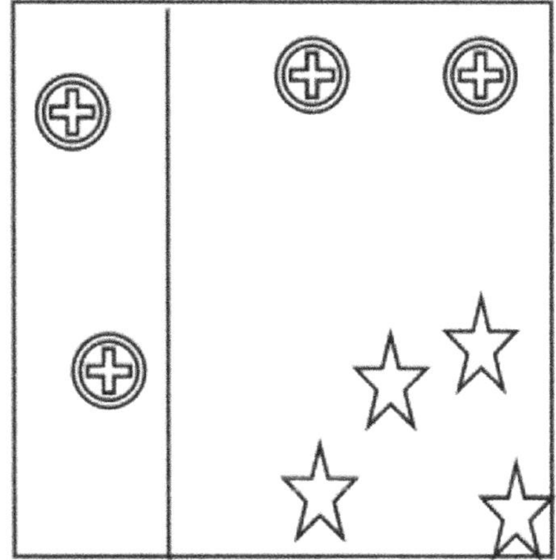

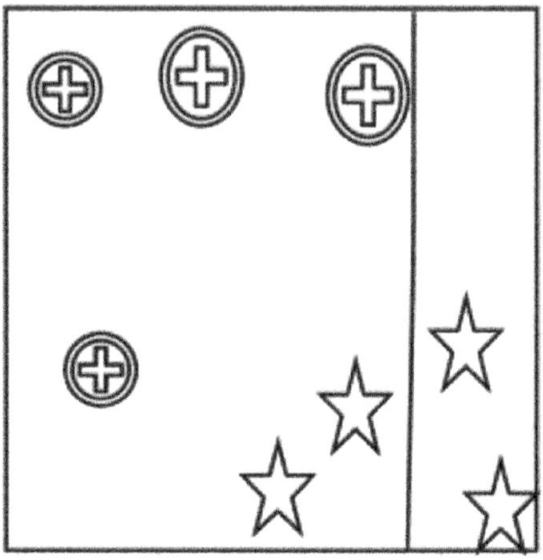

Fig. 6 Model M2 shows that the two misclassified plus signs are given higher weight

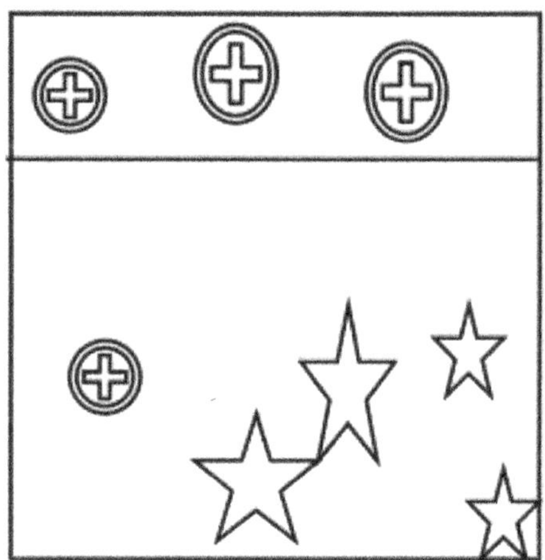

Fig. 7 Two stars that have higher weight are correctly classified in this model

Regarding the errors, the instances that are wrongly classified are given higher weight. As from the previous figure, two plus signs were wrongly categorized. Consequently, a new model is built as in Fig. 6, which will work on correcting the misclassification of the previous model.

The second model M2 shown in Fig. 6 has the same number of objects which will improve the previous classification since two plus signs were misclassified as stars so that the current model tries more to classify the wrong instances correctly. This model generates a vertical separator line which correctly classifies the previous wrongly classified instances.

In model M2, two objects are misclassified as plus. Accordingly, these two objects will have higher weight, and the next model M3 will try to classify them correctly. Consequently, the third model M3 will build a horizontal line that will correctly classify the previous wrongly classified objects, as shown in Fig. 7.

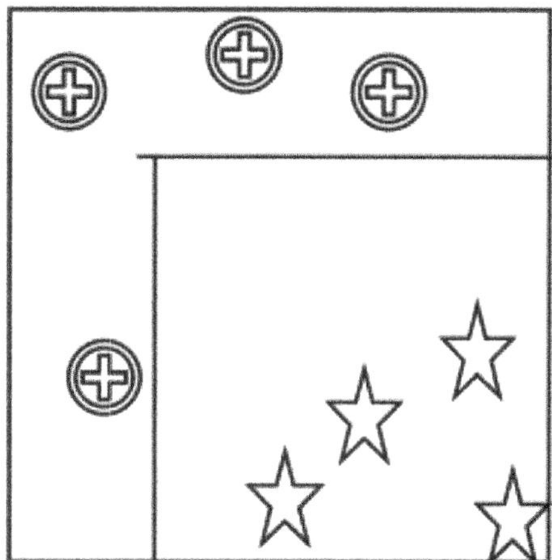

Fig. 8 Active learner has been created from the previous weak models

Recursively, several models are built where each of them works on correcting the misclassification of the previous ones. In the end, a model is built from the weighted mean of the previous models as in Fig. 8 which combines the three previous models M1 + M2 + M3, and this is considered to be much stronger than other individual models.

4   Experimental Work

In this research, experiments were conducted for different activities. These activities are sitting, standing, walking, and running. A commercially available armband was used to secure a smartphone at the upper part of the arm. The experiments done were while the smartphone was either upside down in the pocket or the armband mounted the same as in the pocket. During the experiments, three similar Samsung Galaxy S3 devices were used having a processor of 1.4 GHz quad-core and 1 GB RAM, where each of these devices had a different trained classifier. Each experiment lasted for 12 min, where 3 min was dedicated for performing each of the activities like sitting, standing, walking, and running. The entire activity recognition process pipeline is shown in Fig. 9.

4.1   Dataset

The dataset used consisted of 24,075 instances for different scenarios like sitting, standing, walking, running, and dancing. Each instance has 60 features extracted from the previously collected data by the smartphone accelerometer sensors. Regarding the device position, it was placed in the pocket as capsized where the screen was

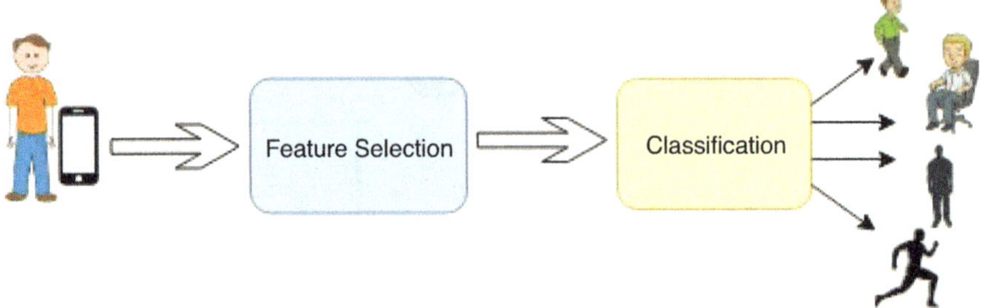

Fig. 9 Motion recognition process

facing the person. The dataset was partitioned to a ratio of 9 to 1 for training and testing, respectively.

4.2 Accelerometer Sensor

An accelerometer is a type of inertial sensor that can measure acceleration along its sensitive axis. The typical operation principle of accelerometers is based on a mechanical sensing element that comprises a proof mass attached to a mechanical suspension system, for a reference frame. Three common types of accelerometers are available, namely, piezoelectric, piezoresistive, and capacitive accelerometers [24]. Piezoresistive and capacitive accelerometers can provide dual acceleration components and have higher stability. Thus, these types of accelerometers are suitable for measuring the motion status in the human gait [25, 26]. Meanwhile, the used sensor is piezoelectric as it is SMB380.

5 Results

In this chapter, three classifiers have been used (ensemble, SVM, KNN), which were trained using the same dataset. The experiments were done while the devices were mounted at the arm and in the pocket for persons other than those involved in the training stage, as shown in screenshots in Fig. 10. After that, we plotted the outcome of our experiments as shown in Figs. 11, 12, 13, 14, 15, and 16 where each graph represents all the activities involved over the time interval of 12 min where the experiments started with standing position followed by sitting and then walking and ended with running. Each one of these classifiers has been deployed on a different Android phone. A duration of 3 min was set for testing every mode of motion (sitting, standing, walking, running), as shown in Fig. 9 and as plotted in Fig. 10.

We have analyzed the collected data from the experiments made using the three classifiers (ensemble, SVM, KNN) along with the mounting position of the portable

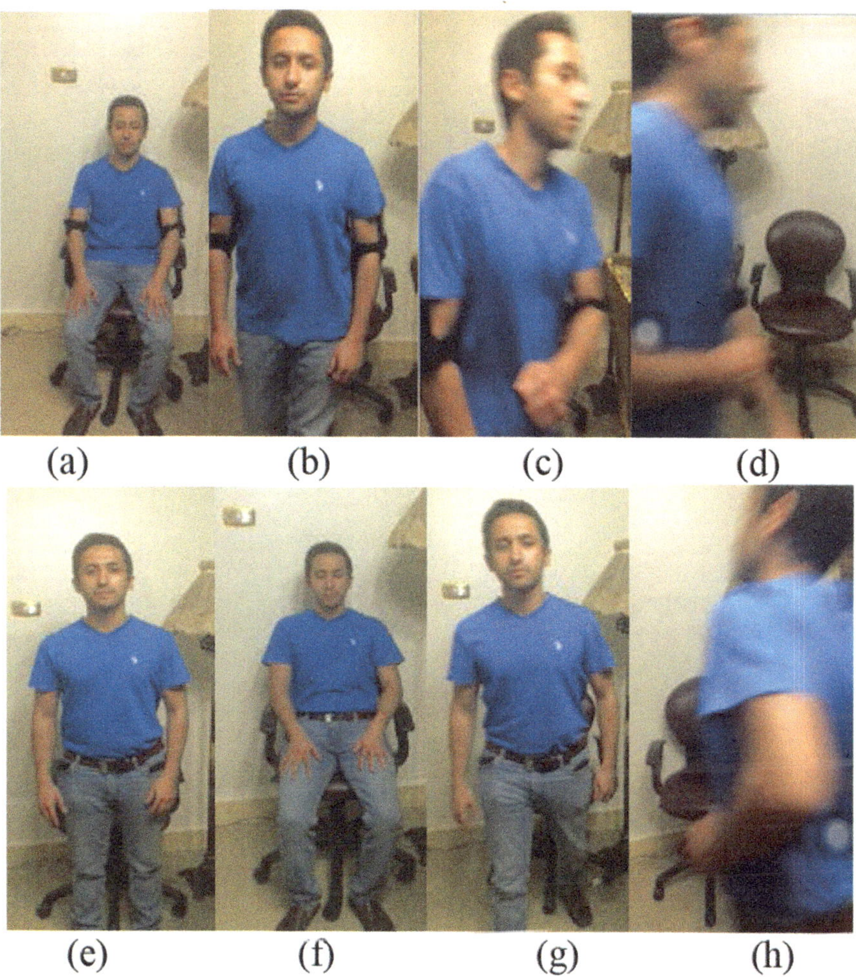

Fig. 10 Four activities took place while the devices were mounted at the arm and in the pocket

devices. Then, the accuracy of every classifier has been calculated as shown in Tables 1 and 2 according to the number of successful detections compared to the failed ones during the time interval of every position which was 3 min.

6   Conclusion

Accordingly, we went through using SVM classifier as from accuracy perspective SVM is preferred as it is usually robust to noise data and robust to overfitting and is known to be one of the best classifiers in terms of generalization. Also, SVM is well

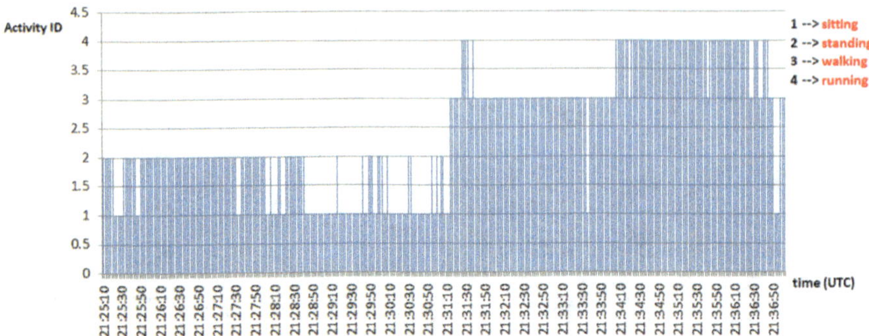

Fig. 11 Recognition accuracy using KNN while the devices were mounted at the arm

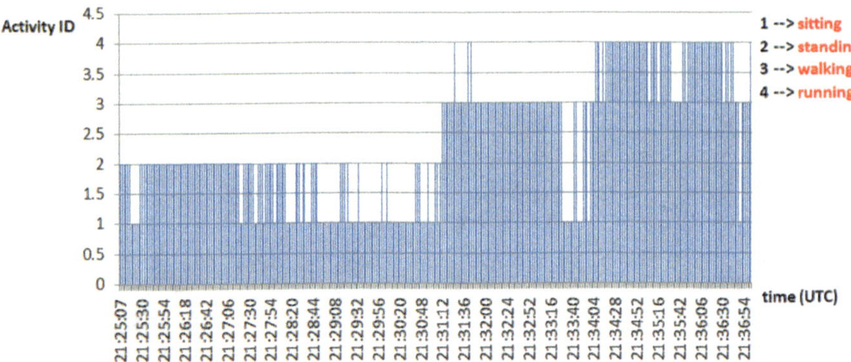

Fig. 12 Recognition accuracy using ensemble while the devices were mounted at the arm

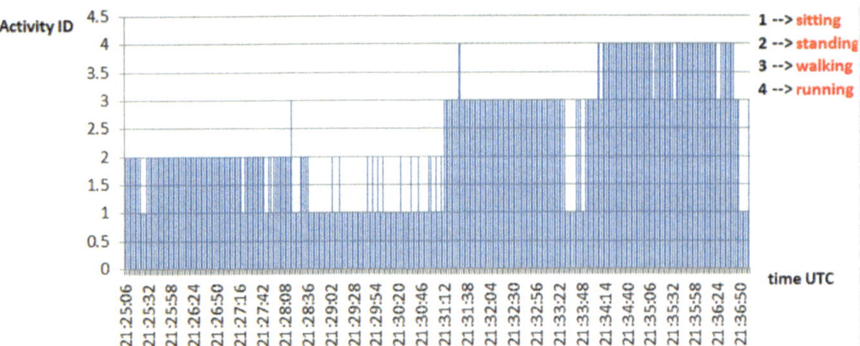

Fig. 13 Recognition accuracy using SVM while the devices were mounted at the arm

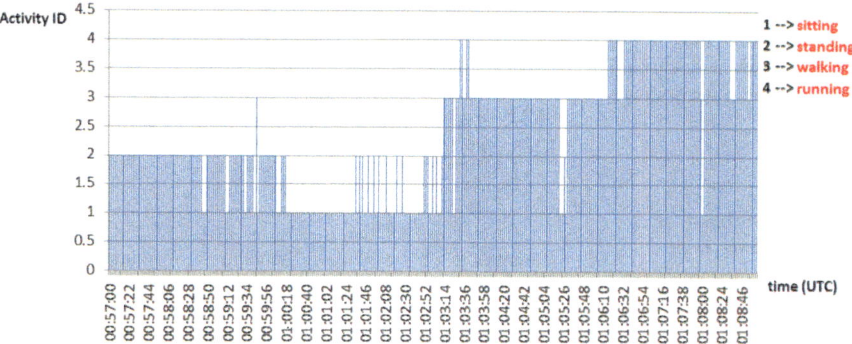

Fig. 14 Recognition accuracy using KNN while the devices were in the pocket

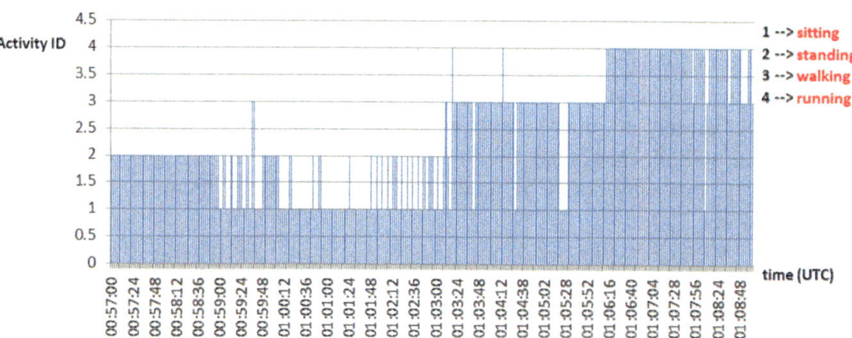

Fig. 15 Recognition accuracy using ensemble while the devices were in the pocket

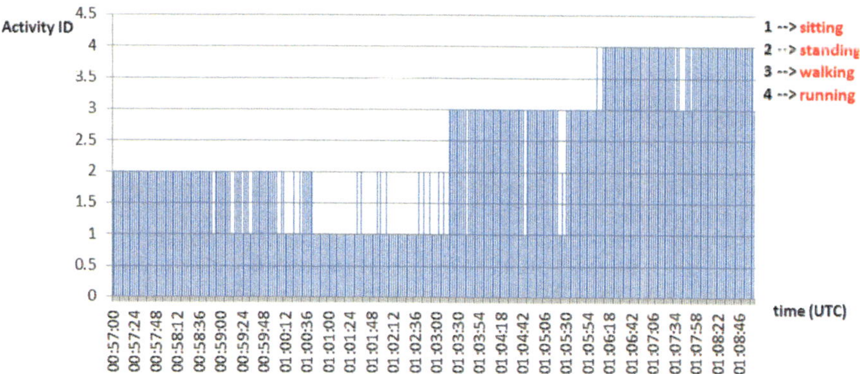

Fig. 16 Recognition accuracy using SVM while the devices were in the pocket

Table 1 Recognition accuracy of the three classifiers when the devices were placed in the pocket

Classifier	Mode			
	Sitting	Standing	Walking	Running
Ensemble (%)	71.42	84.61	83.14	81.31
SVM (%)	78.02	93.40	86.20	89.87
KNN (%)	74.72	91.20	83.51	82.41

Table 2 Recognition accuracy of the three classifiers when the devices were placed at the arm

Classifier	Mode			
	Sitting	Standing	Walking	Running
Ensemble (%)	70.45	87.5	79.54	73.33
SVM (%)	72.59	92.04	83.33	82.22
KNN (%)	71.11	89.53	86.66	79.12

known in motion mode recognition literature and extensively used in many types of research [27, 28]. All these advantages interpreted the improvement of detection compared to that detected using other classifiers. In addition, the accuracy achieved was through using a single sensor which was the accelerometer, unlike previous approaches that rely on other sensors which will require more computational calculations that will be restricted by the limited resources of the portable device. Accordingly, this way could be applied to the majority of them, along with the older versions.

Ethical Approval All procedures performed in studies involving human participants were in accordance with the ethical standards of the institutional and/or national research committee and with the 1964 Helsinki declaration and its later amendments or comparable ethical standards. Informed consent was obtained from all individual participants included in the study.

In addition, all the human figures provided are for me "Omar Sheishaa" while performing the experiments.

References

1. Frank, K., Nadales, M., & Robertson, P. (2010). Reliable real-time recognition of motion related human activities using MEMS inertial sensors. In *Proceedings of 23rd international technical meeting satellite division of the institute navigation (ION GNSS)* (pp. 2919–2932).
2. Bussmann, J. B. J., Martens, W. L. J., Tulen, J. H. M., Schasfoort, F. C., van den Berg-Emons, H. J., & Stam, H. J. (2001). Measuring daily behavior using ambulatory accelerometer: The activity monitor. *Behavior Research Methods, Instruments, & Computers, 33*(3), 349–356.
3. Chen, L., Hoey, J., Nugent, C. D., Cook, D. J., & Yu, Z. (2012). Sensor-based activity recognition. *IEEE Transactions on Systems, Man, and Cybernetics Part C: Applications and Reviews, 42*(6), 790–808.
4. Xu, X., Tang, J., Zhang, X., Liu, X., Zhang, H., & Qiu, Y. (2013). Exploring techniques for vision-based human activity recognition: Methods, systems, and evaluation. *Sensors, 13*(2), 1635–1650.
5. Ericson Mobility Report on the Pulse of the Networked Society. Retrieved from http://www.ericsson.com/res/docs/2013/ericssonmobility-report-june-2013.pdf

6. Ericsson Mobility Report: 70 percent of the world's population using smartphones by 2020. Ericsson.com, 03-June-2015. [Online]. Retrieved from www.ericsson.com/newsl1925907

7. Anguita, D., Ghio, A., Oneto, L., Parra, X., & Reyes-Ortiz, L. (2012). Human activity recognition on smartphones using a multiclass hardware-friendly support vector machine. In *Ambient assisted living and home care* (pp. 216–223). Berlin: Springer.

8. Khan, A. M., Lee, Y.-K., Lee, S. Y., & Kim, I.-S. (2010). Human activity recognition via an accelerometer-enabled-smartphone using kernel discriminant analysis. In *5th international conference on future information technology*, May 2010 (pp. 1–6).

9. Ronao, C. A., & Cho, S.-B. (2016). Human activity recognition with smartphone sensors using deep learning neural networks. *Expert Systems with Applications, 59*, 235–244. https://doi.org/10.1016/j.eswa.04.032.

10. Ronao, C. A., & Cho, S.-B. (2014). Human activity recognition using smartphone sensors with two-stage continuous hidden Markov models. In *10th International Conference on Natural Computation* (ICNC). https://doi.org/10.1109/icnc.6975918

11. Wang, A., Chen, G., Yang, J., Zhao, S., & Chang, C.-Y. (2016). A comparative study on human activity recognition using inertial sensors in a smartphone. *IEEE Sensors Journal, 16*(11), 4566–4578. https://doi.org/10.1109/jsen.2545708.

12. Chetty, G., & White, M. (2016). Body sensor networks for human activity recognition. In *3rd International Conference on Signal Processing and Integrated Networks (SPIN)*. https://doi.org/10.1109/spin.7566779.

13. Kumar, S. S., & John, M. (2016). Human activity recognition using optical flow based feature set. In *IEEE International Carnahan Conference on Security Technology (ICCST)* https://doi.org/10.1109/ccst.7815694.

14. Eskaf, K., Aly, W. M., & Aly, A. (2016). Aggregated activity recognition using smart devices. In *3rd International Conference on Soft Computing & Machine Intelligence (ISCMI)*. https://doi.org/10.1109/iscmi.2016.52.

15. Hui Hsien, W., Lemaire, E. D., & Baddour, N. (2011). Change-of-state determination to recognize mobility activities using a BlackBerry smartphone. In *2011 Annual International Conference of the IEEE Engineering in Medicine and Biology Society*. https://doi.org/10.1109/iembs.2011.6091299.

16. Tunca, C., Alemdar, H., Ertan, H., Incel, O. D., & Ersoy, C. (2014). Multimodal wireless sensor network-based ambient assisted living in real homes with multiple residents. *Sensors, 14*(6), 9692–9719.

17. Wan, J., O'Grady, M. J., & O'Hare, G. M. P. (2015). Dynamic sensor event segmentation for real-time activity recognition in a smart home context. *Personal and Ubiquitous Computing, 19*(2), 287–301.

18. Rodríguez-Molinero, A., Pérez-Martínez, D., Samà, A., Sanz, P., Calopa, M., Gálvez, C., Pérez López, C., Romagosa, J., & Catalá, A. (2007). Detection of gait parameters, bradykinesia and falls in patients with Parkinson's disease by using a unique triaxial accelerometer. In *World Parkinson Congress*, Glasgow.

19. Mannini, A., & Sabatini, A. M. (2010). Machine learning methods for classifying human physical activity from on-body accelerometer. *Sensors, 10*, 1154–1175.

20. Anguita, D., Ghio, A., Oneto, L., Parra, X., & Reyes-Ortiz, J. L. (2012). Human activity recognition on smartphones using a multiclass hardware-friendly support vector machine. In *International Workshop of Ambient Assisted Living (IWAAL 2012)*, Vitoria-Gasteiz, Spain.

21. Hamamoto, Y., Uchimura, S., & Tomita, S. (1997). A bootstrap technique for nearest neighbor classifier design. *IEEE Transactions on Pattern Analysis and Machine Intelligence, 19*(1), 73–79.

22. Alpaydin, E. (1997). Voting over multiple condensed nearest neighbours. *Artificial Intelligence Review, 11*, 115–132.

23. Li, F., & Klette, R. (2011). In *Euclidean shortest paths*. London: Springer.

24. Westbrook, M. H. (1994). Accelerometers. In M. H. Westbrook & J. D. Turner (Eds.), *Automotive sensors* (pp. 150–174). Bristol: Institute of Physics Publishers.

25. Bouten, C. V. C., Koekkoek, K. T. M., Verduin, M., Kodde, R., & Janssen, J. D. (1997). A triaxial accelerometer and portable data processing unit for the assessment of daily physical activity. *IEEE Transactions on Biomedical Engineering, 44*, 136–147.
26. Anguita, D., Ghio, A., Oneto, L., Parra, X., & Reyes-Ortiz, J. L. (2012). Human activity recognition on smartphones using a multiclass hardware-friendly support vector machine. In *4th International workshop of ambient assisted living, IWAAL,* Vitoria-Gasteiz, Spain. *Proceedings. Lecture Notes in Computer Science* (pp. 216–223).
27. Cheng, W.-C., & Jhan, D.-M. (2013). Triaxial accelerometer-based fall detection method using a self-constructing cascade-AdaBoost-SVM classifier. *IEEE Journal of Biomedical and Health Informatics, 17*(2), 411–419.
28. He, Y., & Li, Y. (2013). Physical activity recognition utilizing the built-in kinematic sensors of a smartphone. *International Journal of Distributed Sensor Networks, 9*(4), 481580.

Color Restoration Survey and an Overdetermined System for Color Retrieval from Faded Images

Devin Haslam, Soad Ibrahim, and Ayman Elmesalami

Abstract This chapter presents a survey of the most widely used color restoration techniques from faded images. The purpose of the survey is to explain the reasons for color fading and emphasize how the color restoration techniques have evolved in recent years. The survey covers the color bleaching models, single-scale and multiscale retinex, gray world, max white, machine learning, and underwater color correction approaches. Image colorization, inpainting, and color constancy aspects are discussed in the machine learning portion of this survey. Numerous color restoration approaches are reviewed under the uncategorized techniques. An over-determined system for retrieving the degraded color components from old faded images is presented as part of this survey. The system reviews three different approaches which combine the traditional max white, gray world, and retinex white balancing algorithms. In each approach, one of the three algorithms is eliminated while the other two are combined by a set of second-order equations.

Keywords Color retrieval · Color restoration · Faded colors · Degraded colors · Faded images · Color restoration survey · Color retrieval survey

1 Introduction

Automatic color retrieval techniques have attracted the interest of many researchers and image specialists in the last few years. More specifically, major interest has been directed towards the protection of images, pictures, paintings, and documents of special historical values. Each of these art pieces describes a specific era of the human legacy. The problem is even more severe when the faded colors affect the readability of old documents, which signify the importance of the color retrieval techniques.

D. Haslam (✉) · S. Ibrahim · A. Elmesalami
Department of Computer Science, Old Dominion University, Norfolk, VA, USA
e-mail: Dhasl002@odu.edu; sfibrahi@odu.edu; aelmesal@odu.edu

© Springer Nature Switzerland AG 2020
M. H. Farouk, M. A. Hassanein (eds.), *Recent Advances in Engineering Mathematics and Physics*, https://doi.org/10.1007/978-3-030-39847-7_24

Although the exact reasons for color changes and fading are unknown, this has not stopped the development of many color retrieval techniques [1–8]. The bleaching effect and light-absorbing color material, which exist in all natural and artificial dies, are among the most significant reasons for color fading. The human observation for different colors is directly linked to such material, yet the exact characteristics for this link are imprecisely known. The ongoing research on such topic is attempting to solve the ambiguity and precisely describe the characteristics of the color fading phenomenon. Moreover, the sensitivity of the human visual system (HVS) to different colors varies based on some chemical ties as well as the absorption of light at different wavelengths. Some lights at specific wavelengths (i.e., ultraviolet) can break the chemical ties and consequently affect the quality of the colors (the bleaching effect). The first three algorithms (gray world, retinex, and max white) are pioneers for the color retrieval techniques in [1–3]. These three approaches are simple and easy to implement via a set of the first-order equations (linear color mapping). The three algorithms are built on intuitive uncomplicated assumptions to imitate the HVS in recognizing different colors. Furthermore, these algorithms are the founders of the three techniques in [6–8], which combine two color restoration approaches via a set of the second-order equations (nonlinear color mapping). The combined gray world and retinex (GWR) technique triggered the development of the other two algorithms, which study the effect of the max white (MW) algorithm on the gray world and retinex, respectively. Other approaches including the standard deviation-weighted gray world and multiscale retinex [4, 5] were later developed to enhance the performance; however, the two techniques achieved limited satisfactory results in the correction of the faded colors. Unfortunately, most of these techniques are image-dependent. In this chapter, we also summarize the last three techniques [6–8] in an overdetermined system for comparison purposes. The overdetermined system eliminates one of the three traditional algorithms (GW, retinex, or MW), while the other two are combined by a set of second-order equations. The quadratic mapping is used to overcome the limitations of the original linear mapping associated with the three founding algorithms. The system provides a complete set of the combined approaches for automatic color retrieval from faded images.

The rest of the chapter is organized as follows: Section 2 explains the reasons for color fading and degradation. Section 3 summarizes different methods for color retrieval from faded paintings. Section 4 explains the color bleaching models including the early models and the semiautomatic model of the principal component analysis (PCA). Section 5 provides a comparison between the single- and multiscale retinex-based color restoration approaches. Section 6 summarizes the standard deviation-weighted gray world (SDWGW) algorithm including the simple and combined approaches. Sections 7 and 8 explain the presented overdetermined color restoration system. Section 9 surveys the machine learning (ML) color retrieval approaches with special focus on the colorization, inpainting, and color constancy aspects. Section 11 focuses on the underwater color correction approaches. Section 12 reviews different unclassified color restoration techniques. The retrieval procedures in this section include historical art restorations such as the Terracotta

Warriors, color image mosaicking, dental shade selection, augmented reality color correction, color restoration methods, colorization, and steam turbine manufacturing applications.

2 Reasons for Color Fading

Many physical and chemical changes can seriously affect the color quality of old art pieces. However, the exact reasons for fading and color degradations are not precisely known yet. The slow saturation of active chemicals in old films, paints, and photographs is one of the main reasons for color degradation [9, 11, 30]. For example, the chemicals of films like color couples and stabilizers usually stay active for a long time. Moreover, such chemicals change with time and introduce a noticeable fading into the original colors of the captured films. The oxidation of shiny surfaces is another example of chemical changes that can badly affect the brightness and contrast of the original colors. Removal of this oxidation layer is very tedious and cannot guarantee the restoration of the original colors. The tremendous amount of smoke and air pollution, especially in big crowded cities, is also one of the major sources for such malicious oxidization gases. Moreover, the irresponsible practices of human can also introduce some dirt and stains to the historical art pieces. Some of these stains can be sturdy and hard to remove, which might require a reproduction of the lost color values. Some erosion in the very thin outer layer of the historical art pieces is also expected due to the numerous cautious touches of the art experts, especially with hard tools [35]. Losing tiny spots of this layer would also require a reproduction of the missing color values. On the other hand, the physical changes in colors are due to the unbalanced exposition of art pieces to lighting (i.e., natural and/or artificial) for a long time [31, 33]. Also, the partial increase in the temperature of these valuable art pieces can result in some irregular color fading, which is another example for the physical changes that affect the color quality. Both light and heat usually expedite the color-fading processes.

Therefore, the color and/or dye fading of the art pieces can be categorized into two different types: The first type is a linear or uniform bleaching of all colors, while the second type is irregular spots or stains randomly distributed over the whole paint. It is worth mentioning in this brief discussion that the colorimetric layers of any color space (i.e., RGB, CMY, HSI, CIELAB, etc.) are not equally affected by the same color-fading reasons. In fact, the fading amount in each channel varies based on the environmental factors and storage condition, which is the main assumption for many color restoration and white balancing techniques [9]. Moreover, the shorter the wavelength of light, the more aggressively it causes fading of dyes. According to the Illuminating Engineering Society of North America (IESNA), color temperature is "the absolute temperature of a blackbody radiator having a chromaticity equal to that of the light source" [32], which supports the previous assumption of different fading amount in each colorimetric layer based on the surrounding factors. Numerous manual, semiautomatic, and fully automatic techniques have been developed to restore the original color and look of the historical art pieces. Selecting the best technique requires knowledge of the reasons for color fading and degradations.

3   Color Restoration from Faded Paintings

The trial and error approaches are among the earliest and simplest approaches that were manually introduced to solve the color degradation problem of old paintings. In such techniques, a large number of the chemical cleaning materials are cautiously applied to small portions of the painting in order to check the effectiveness of the chemicals and also to decide on the most suitable substance to clean the entire art piece. Many major drawbacks were associated with this approach including the manual application of the chemicals, which is a time-consuming process. Also, the restoration of missing color components by chemical treatments is impossible as the bleaching of dyes is an irreversible process [9]. Moreover, this approach is not applicable to printed art pieces, which is a major limitation for this approach. Therefore, the automatic retrieval and reproduction of the missing colors were introduced using some of the digital image processing techniques to imitate the color restoration process. An estimation of the original colors based on the Volterra filters has been introduced by Cortelazzo et al. [10], where all the corrections are implemented in the RGB color space. In this technique, the original color information was extracted by utilizing sample painting regions before and after the chemical cleaning of the art piece. Implementation in the RGB color space is a major drawback of this approach.

Pappas et al. introduced five digital color restoration techniques, which can be used to simulate the original appearance of paintings [11]. The five techniques were implemented in the CIELAB color space, where the basic assumption is that several uniformly colored portions of the painting are chemically cleaned. The two images (before and after the cleaning) of the painting are digitized into small patches. This process provides a clean color image data set Cl_i and a corresponding digitized data set (oxidized) for the same regions before cleaning Oxi_i. The main goal of the five techniques is to find color transformation $Cl_i = f(Oxi_i)$ for the inverse of the degradation process from these sample data sets. The transformation function represents the color changes/fading. The authors of this approach proposed applying the transformation to the entire art piece for a complete color retrieval of the missing color components. A limited number of patches N were utilized to approximate this phenomenon in order to minimize the effort and limit the damage of the manual applications of the chemical substances. This number is image-dependent, and it usually varies based on the number of the colors in the damaged painting. Two vectors $\widehat{M}_{Cl}$ and $\widehat{M}_{Oxi}$ were calculated for each patch to represent the sample mean of the clean and oxidized regions. These vectors are extended for the three-dimensional color space CIELAB. The difference between the two vectors were calculated for each degraded region of the painting. An estimate $\widehat{Cl}_i = f(Oxi_i)$ of the clean color was also calculated. The five techniques used the mean square error (MSE) for the calculation of the transformation function, as follows:

The first approach, sample mean matching, is simple and easy to implement by classifying each pixel of the oxidized (degraded colors) region of the art piece to one of the color clusters. For each pixel in a faded region, the color vector of this pixel is classified to a corresponding clean color cluster based on the difference in color between the current processed pixel and the mean of each cluster. This difference is next compared with the difference between the mean and all pixels in the faded region. The overall difference is denoted by $\widehat{\Delta M}_i$. The authors estimated the original colors of the art pieces by adding the values of the regions before cleaning Oxi to the overall color difference $\widehat{\Delta M}_i$. Even though this technique is simple, it was proven effective especially for a large number of color samples.

In the second approach, linear approximation (LA), the approximation function was modified to include a coefficient matrix, which is calculated by the polynomial regression of the difference means in the faded region. The transformation (approximation) function is calculated as a function of the corrupted colors and approximation matrix.

The third approach, iterative closest point approximation (ICP), was introduced as an efficient technique for the registration of any two 3D data sets. The main assumptions of this approach included an estimation of a rotation matrix and a displacement vector to provide a new transformation function. Therefore, it introduces a rotation to the corrupted color components and combines the results with the displacement vector. This technique is computationally expensive especially for large data sets, where the required computation is a function of the squared number of the digitized patches, which is a major drawback of this approach. However, the fast implementation of the ICP was proven to significantly minimize the required time for calculations. This technique can also lead to local minima and therefore misleading color correction results, which is another major drawback of the technique.

The main assumption of the fourth approach, white point transformation (WPT), is looking for object variances based on the lighting conditions. Therefore, changing the lighting condition of a clean portion of an image usually provides the oxidized look of this image region. The difference between the two (clean and oxidized) is calculated based on the white points, and consequently the lighting source is characterized by its reference white value. This technique also determines a white point vector in order to estimate the clean sample as a function (nonlinear transformation) of the oxidized samples and the corresponding white values. Some function minimization routines were used to minimize the instantaneous error associated with this algorithm. This approach models the degradation process more precisely, which is a major and unique advantage of this algorithm. Moreover, this technique is mathematically inexpensive.

The fifth approach, RBF approximation, is based on the radial basis functions. These functions were originally used for universal approximations. In this technique, an arbitrary mapping is approximated based on some radial basis functions to separately perform the transformation on each of the included dimensions. The nonnormalized Gaussian function was utilized as an example for the calculations, where the parameters of the Gaussian functions are estimated for each one of the three color components in the CIELAB domain. In order to minimize the total squared error, Pappas et al. suggested performing the estimation by a steepest descent algorithm [11]. Some simplifications were suggested to limit the overall restoration performance of the algorithm.

The best techniques among the five approaches are the LA and WPT. Also, it is worth mentioning that the cautious attempts of adjusting the corrupted colors' values in these techniques are similar to the insertion process of hidden marks in all naturalness-preserving transform (NPT) data hiding techniques [45].

4   Color Bleaching Models

4.1   Early Models

Gschwind et al. presented the earliest techniques for automatic color restoration and corrections in [12–15]. In the first approach, a model was proposed for the artificial bleaching of different types of films [12]. The CMY color space was chosen for implementation, where the fading effect is assumed linear and uniformly distributed over the whole corrupted image. Such assumption was the main drawback of this technique and, consequently, the reason for the associated poor performance. Moreover, this approach ignored the random (spatial and/or time) characteristic of the color degradation process. Later, this technique was modified due to the numerous mathematical bleaching models which were developed [13, 14]. The models were implemented via several experiments to suit different types of films. The process involves two main stages, as follows: First, adequate information about the film type and age would facilitate the usage of an appropriate bleaching model. As a result, the color restoration process is reasonably simple and mathematically inexpensive. In the second stage, a user-guided color retrieval process is suggested by the authors when the film type and age are unknown. The color adjustments are predicted by some image experts (manual selection), which is a drawback for this technique. On the contrary, the most suitable matrix for the complete color retrieval is generated automatically, which is the main advantage of this approach. The effect of side absorption by dyes was also highlighted in this technique which generates an unexpected color degradation pattern unrelated to any known fading reasons. This approach was also tested with color movies as indicated in [15], where the color retrieval results were adequate for some specific films. Chambah et al. developed a similar color retrieval algorithm based on the same assumption of a linear bleaching [16]. This approach also assumes a manual selection for the reference color components.

4.2 Principal Component Analysis (PCA) Model: A Semiautomatic Approach

Chambah et al. introduced a color restoration approach based on the principal component analysis (PCA) [17]. In this approach, the PCA is utilized to represent the data. This approach involved some nonlinear color correlation and restoration for the hue layer. This was the first semiautomatic color adjustment approach, which compromised the manual and automatic color retrieval techniques. Despite the poor color correction results of this technique, it directed many researchers towards the automatic restoration of the faded and corrupted colors. This technique also brought some attention to the artificial intelligence (AI) tools and the color retrieval enhancements that can be achieved by the AI-based techniques (Yeung et al.) [41].

Another effort to perform color correction using the PCA and independent component analysis (ICA) was proposed by Shao et al. in [101]. The first step in this method is to analyze the correction property of the original image. After this analysis, if the global correction property is satisfied, a global PCA color correction is performed. Otherwise, if the correction property is not satisfied, the signal sources are divided according to dominant color extraction. Then the color correction relations can be created between the independent sources. Lastly, the ICA correction outcomes are combined to produce a corrected image. The results of this method show that the proposed process performs well compared to state-of-the-art methods.

Until 1986, most enhancements of color from multispectral images were based on "stretching" of three input images. Gillespie et al. recognizes that leaving the hue values unchanged while the color saturation is stretched presents many shortcomings [102]. To overcome these errors, a PCA decorrelation stretch and a stretch of the HSI color space is thoroughly discussed. While the PCA transformation is dependent on the scene, the HSI stretching method is shown to be invariant. After careful comparisons between the PCA decorrelation stretch, a linear stretch and the HSI stretch, it is found that the PCA method is the most effective at correcting the colors of an image.

5 Single- vs. Multiscale Retinex Techniques

5.1 Single-Scale Retinex

The retinex theory is one of the basic techniques that has been utilized to retrieve the faded color components from corrupted images. Many single-scale retinex-based algorithms (SSR) have been developed for this purpose, including the restoration approach presented by Jobson et al. [18]. In this approach, the SSR for an image is calculated as a function of the logarithmic difference of the input image and the convolution of this input with a small Gaussian surround function. This technique presented a practical implementation of the retinex. Moreover, the technique ignored the retinex model to simulate the human perception of lightness and color. Also, many things were investigated, including the best placement of the logarithmic function, functional form of the surround, size of the surround space constant, and the best treatment of the retinex outputs prior to final display. Therefore, the retinex theory was proven to be a powerful tool for combining dynamic range compression, color constancy, and lightness/color rendition. Jobson et al. also defined a new method of color restoration which corrects for the deficiency of extending the single-scale center/surround retinex to a multiscale version [19]. This approach fails to provide good color rendition for any image that does not fulfill the assumptions of the gray world. This technique provided good color restoration results at the cost of a modest dilution in color consistency, and it was successful to bridge the gap between color images and the human observation of scenes.

Chambah et al. proposed another retinex-based color restoration algorithm [43]. In this approach, modified versions of the retinex white patch (RWP) were introduced. The gray world (GW) algorithm was also utilized to eliminate the color cast of the illumination shift. All color enhancements were implemented in the in the CIELAB domain for simplicity.

Some other SSR-based color restoration approaches were presented in [20, 21]. In these techniques, an automatic color equalization (ACE) method was utilized to combine the RWP and gray world algorithms. This approach was inspired by the lightness and color constancy of the human visual system, where saturation enhancements of the faded frames are achieved via a new color balancing model based on the ACE algorithm. This technique was later modified to provide a new system for balancing the colors of multiple frames with different color degradation characteristics, as presented by Rizzi et al. [22]. Such algorithm is usually combined with another technique capable of splitting the movie into different shots and choosing the most representative frames from each shot. The same setting for all frames was imposed by the robustness requirements of the color correction method. Drawbacks of these techniques included some color cast in the output images and no evaluation of the overall performance.

Another technique based on SSR is introduced by Chiang et al. which uses a JND-based filter that is nonlinear [84]. This technique transforms an image from RGB space to HSV in order to enhance the S and V component of the image. This enhancement is only used under the white light assumption. After the nonlinear filter is applied, part of the log signal of illumination is subtracted from the image V. Following this step, histogram modeling is performed on V, and the S component is adjusted according to the V portion. Lastly, the RGB image is obtained from the enhanced V and S components. The results show that this method can outperform usual methods of histogram equalization. The following year, Chiang et al. improved the method by making some changes. In the new system, only the V component in the image is enhanced [85]. This is done by applying the JND-based nonlinear filter and then estimating the reflectance. Afterward, gamma correction is applied on the estimates and an enhanced RGB image is derived from the original H and S component as well as the corrected V component. This method provided better results than the algorithm created a year earlier.

5.2 *Multiscale Retinex*

The preliminaries of the multiscale retinex (MSR) were introduced by Jobson et al. as a weighted sum of several SSR outputs [18]. The authors of the MSR approach attempted to express a practical implementation of the retinex without validating an image with human observation tests. The decision of choosing the best number of scales is made based on a big number of factors and restrictions, where the most important restriction is to produce an output image with a graceful rendition without halo artifacts. The authors focused on the trade-off between rendition and dynamic range compression that is governed by the surround space constant. Various functional forms for the retinex surround were evaluated; however, the Gaussian form provides the best performance.

An approach combining the MSR and curvelet transform (CT) was presented by Starck et al. [5]. The CT represents the edges better than any other transformation including the wavelets. Therefore, the CT is suited for multiscale edge enhancement. After enhancement of edges in an image, edge detection and segmentation, among other techniques, can be performed more easily. When comparing this new approach to wavelet transformation methods and multiscale retinex, it is clear that CT outperforms all enhancement algorithms for noisy images. However, it is shown that on noiseless images, the proposed CT is not significantly better than wavelet enhancement.

The combined gray world and MSR is another form of the retinex-based color restoration techniques [23]. In this algorithm, the input image is preprocessed by the MSR to substitute for the effect of the neighboring pixels. The technique was implemented via two main stages. The first stage is based on the convolution part of the SSR calculations, where the input damaged image is preprocessed by a surrounding Gaussian function in order to imitate the retinex response to the neighboring pixels. On the other hand, only four scales were selected to satisfy the contradictory requirements of the MSR theory (fine details at the low scales and the color rendition at the large scales). In the second stage, the output of the MSR was introduced to the traditional combined GWR color restoration process. This approach was successful in restoring missing color values with a high quality and a clear distinction between all included image features. This technique provided a more superior spatial quality for the output image. The retinex-based techniques can provide much more for the automatic retrieval of the corrupted colors as indicated in [36–40].

Due to its computational complexity, much research has been done attempting to reduce computation. Wang et al. showed that by constructing average-value templates in advance, the computation can be simplified to arithmetic [86]. This method is practical due to its good results in very little time. A second MSR method that focused on low computational complexity is designed by An and Yu [87]. This approach first recognized that traditional retinex algorithms produce halo artifacts. Therefore, this novel creation restricted spatial contrast in regions of similar color. In addition, an adaptive Gaussian filter is used to greatly reduce computations. Another

MSR technique is proposed by Hanumantharaju et al. which modifies the color restoration process [88]. This system provides true color constancy and produces visually good results. In addition to its good results, the method is very computationally efficient.

While many MSR techniques may provide good results, countless are not automatic. Parthasarathy and Sankaran presented an automated MSR method that estimated the top and bottom clipping points by using the variance of the histogram and observing the frequency of each pixel in the image [89]. The results show that this automatic technique can provide visually good results. Similarly, Petro et al. attempted to create an automatic MSR method that is as similar as possible to the original retinex method [90]. Unlike other techniques, this algorithm requires that the user chooses if the changes should be applied to the colors directly, or just the luminance layer of the image. This idea produces results comparative to manual state-of-the-art algorithms. Parthasarathy and Sankaran attempted utilizing image fusion with the MSR to produce color-enhanced images [91]. After using a MSR technique to produce three separate output images, the authors attempted to fuse these three images to produce a more superior image. Several forms of wavelet transform were analyzed and shift invariant wavelet transform produced the best results. A novel MSR method with color restoration and an auto levels algorithm was proposed by Jiang et al. [92]. The authors noticed that many MSRCR algorithms produce good results, but the images could be further restored. While this novel method could produce very good state-of-the-art results, the computation was very complex. Therefore, the parallel version of this application was employed on GPUs.

Kim et al. recognized that while retinex performs well on low-contrast images, there is still a color tone change in these restored images [95]. In order to overcome this problem, the authors proposed a novel post-processing method to reduce the color shifting issue. After multiscale retinex is used on an image, the method projects each pixel from the new image on to the original image. This projection restores the color of the image better than any other single retinex method.

6 Standard Deviation-Weighted Gray World

6.1 Simple SDWGW

The standard deviation-weighted gray world (SDWGW) was introduced by Lam et al. as a new approach for white balancing and color adjustments in captured images [4]. In this technique, the input image is subdivided into a large number of blocks. The standard deviations and means were calculated for the input image on each block. These calculations were repeated for the three colorimetric layers (RGB), where the main focus was directed towards the enhancements of the gray world color restoration algorithm. The standard deviation-weighted average for each colorimetric layer is individually calculated. Next, the gain for each channel is calculated by adjusting the total standard deviations using Eq. (1). Poor color correction results were provided by this technique, which does not satisfy the image observers. Therefore, a more advanced technique was presented in an attempt to join the combined gray world and retinex approach with the SDWGW technique [23]. More colors were retrieved by this algorithm; however, the final results are image-dependent.

$$\text{Gain}_{(R,G,\text{or }B)} = \left(\sum_{k=(R,G,B)} \text{SDWA}_j \right) \Big/ \text{SDWA}_k \qquad (1)$$

6.2 *Combined SDWGW_CGWR*

Implementation of the SDWGW_CGWR color restoration technique involved three main stages [23]. A preprocessing for the input image is implemented in the first stage to increase the dynamic range representation of the faded color components. The output image of the first stage is subdivided into blocks. The choice of the best number of blocks is image-dependent and varies based on the total damage in the input faded image. In the second stage, the output of the subdivision is introduced to the SDWGW algorithm, where the variances and means of each channel are calculated for each block. The standard deviation-weighted averages are calculated for the three RGB channels. Finally, the gains for the three channels are calculated to compensate for the missing color values. In the third stage, the combined GWR is applied, where two quadratic equations for the red and blue colorimetric channels are used to perform the mapping. The green channel is kept without any changes in order to satisfy the main assumptions of the gray world. Also, the maximum and average requirements for the retinex theory and gray world techniques were satisfied. The output image of this stage includes all restored and corrected colors, and it is free from the pixelized looking which commonly happens in the results from the stand-alone SDWGW founding algorithm.

7 Principles of the Overdetermined System and Founding Algorithms

The gray world (GW), retinex, and max white (MW) algorithms are the basis and founding algorithms for many color retrieval techniques including the presented overdetermined system. All these techniques were developed to linearly correct and adjust the colors of the image pixels, which are drastically exposed to unbalanced illuminations (artificial or natural) for long periods of time. In this section, we simply provide the basic assumptions of each of these three algorithms, as follows:

7.1 Gray World Assumptions

The GW color restoration approach assumes that the red and blue channels of an RGB image are equally affected by any physical or chemical changes that occur into the target images [2]. However, the green channel is more resistive to such changes. Therefore, the green layer is kept without changes, which is a reasonable assumption to maximize the recoverable percentage of the lost color components. Also, in this technique the average intensities of the red, green, and blue colorimetric layers are initially assumed to be equal. Therefore, the color correction gain of the red channel is calculated using Eq. (2). A linear adjustment for the red channel is achieved using Eq. (3). Color correction for the blue channel is calculated analogously. Therefore, the output image of this color adjustment approach is decomposed from the three colorimetric layers $(\tilde{R}, G, \tilde{B})$, where the green channel is kept without any changes. Results from this technique were slightly poor and the output image includes a considerable color cast after restoration, which is a major drawback for this technique [35].

$$R_{\text{gain}} = \mu_G / \mu_R \tag{2}$$

$$\tilde{R}(i,j) = R_{\text{gain}} \times R(i,j) \tag{3}$$

7.2 Retinex Assumptions

The retinex theory is also similar to the GW algorithm with few differences as follows: This color restoration technique assumes that the perceived white color component is associated with the maximum cone signals of the human visual system (HVS) [1]. Similarly, all color degradations occur into the red and blue channels [2, 23, 34]. Therefore, the numerical values of the green layer are the references for the correction process. The mathematical solution of this approach involved an equalization process for the maximum values of the three RGB colorimetric layers. The gain for the faded red channel is calculated using Eq. (4). The final output image, which includes the corrected colors, of this technique is decomposed from the two adjusted layers (red and blue) combined with the original unaltered green channel. In conclusion, the red and blue channels of the faded images are linearly scaled so that the maximum values of the three RGB layers are equal and more specifically matching the value of the reference channel.

$$R_{\text{gain}} = \max\{I_G(x,y)\} / \max\{I_R(x,y)\} \tag{4}$$

7.3 Max White Assumptions

The MW algorithm is also a widely known color restoration technique, where the three colorimetric layers of an input image are assumed to be equally affected by any physical or chemical changes. Therefore, the green channel also fades, which is the major difference from the assumptions of the previous two techniques (GW and retinex). In this approach, the maximum values (brightest points) of an image corrected by this technique are represented by the white color, which are $(2^N - 1, 2^N - 1, 2^N - 1)$ for any image of N bits. The gain of any of the three channels is calculated using Eq. (5), and the three RGB layers are also linearly adjusted. The final output image (corrected colors) of this technique is decomposed of the three layers $(\tilde{R}, \tilde{G}, \tilde{B})$.

$$(R, B, \text{or } G)_{\text{gain}} = \left(2^N - 1\right) / \left((R, B, \text{or } G)_{\text{max}}\right) \tag{5}$$

All the GW, retinex, and MW white balancing algorithms are successful in restoring some of the lost and faded colors. However, none of the three techniques can completely restore all lost color components. Moreover, the three techniques provide different results for different faded and degraded images. Such difference was the real motive for developing the combined techniques to provide better results and overcome the limitations associated with each individual technique, especially the poor performance of the linear mapping.

8 The Overdetermined Color Restoration System

In this section, we introduce an overdetermined system for automatic color correction and retrieval from faded images. The system provides a complete set of the combined techniques, where two algorithms from the three founding techniques are combined together, as illustrated in Fig. 1. The corrections of the faded colors are calculated by a set of equations similar to the E. Lam method in [6]. A quadratic equation is used to perform the color corrections and adjustments in each colorimetric layer. This type of color mapping is proposed to solve some of the problems associated with the linear mapping of the traditional GW, retinex, and MW techniques.

The overdetermined system is developed to combine some advantages from any two of the participating algorithms, where the intensities of the green colorimetric layer are always assumed to be kept without any changes. Consequently, all color corrections and adjustments are only needed for the red and blue channels. Therefore, the color adjustments for any RGB input image of the size $m \times n$ are calculated as follows:

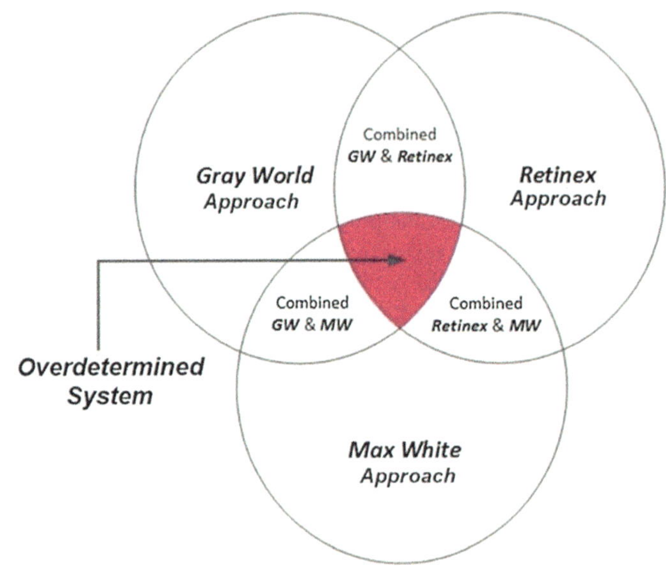

Fig. 1 Overdetermined color restoration system

$$R''(i,j) = e_r R^2(i,j) + f_r R(i,j) \tag{6}$$

$$B''(i,j) = e_b B^2(i,j) + f_b B(i,j) \tag{7}$$

where $e_r, f_r, e_b,$ and f_b are the mapping constants for the red and blue channels. The final output image is decomposed of the three colorimetric layers (R'', G, B''), where the green channel does not change. The four mapping constants can be utilized to fulfill the requirements of many hypothetical algorithms, which are also expected to produce different solutions with a wide range of the color correction results. Satisfactory color correction results vary based on the assumptions associated with the chosen set of algorithms. In this section, we limit our discussion to three combined algorithms, as follows:

8.1 Combined Gray World and Max White

Equations (6) and (7) are sufficient to perform the required nonlinear mapping for this color retrieval approach. Fulfilling the assumptions of the founding MW and GW algorithms requires using Eqs. (8) and (9) to adjust the values of the red channel. Adjusting the values of the blue channel is calculated similarly. Therefore, the four mapping constants (e_r, f_r, e_b, f_b) are calculated using Eqs. (10) and (11). The Gaussian elimination method or Cramer's rule can be used to solve the two equations [24]. The calculated values of the four constants are substituted into Eqs. (6) and (7) for the quadratic adjustments and retrieval of the faded colors.

$$e_\mathrm{r} \sum_{i=1}^{m} \sum_{j=1}^{n} R^2(i,j) + f_\mathrm{r} \sum_{i=1}^{m} \sum_{j=1}^{n} R(i,j) = \sum_{i=1}^{m} \sum_{j=1}^{n} G(i,j) \tag{8}$$

$$\left(2^N - 1\right)^2 e_\mathrm{r} + \left(2^N - 1\right) f_\mathrm{r} = 2^N - 1, \quad \text{or} \quad \left(2^N - 1\right) e_\mathrm{r} + f_\mathrm{r} = 1 \tag{9}$$

$$\begin{bmatrix} \sum_{i=1}^{m} \sum_{j=1}^{n} R^2(i,j) & \sum_{i=1}^{m} \sum_{j=1}^{n} R(i,j) \\ 2^N - 1 & 1 \end{bmatrix} \begin{bmatrix} e_\mathrm{r} \\ f_\mathrm{r} \end{bmatrix} = \begin{bmatrix} \sum_{i=1}^{m} \sum_{j=1}^{n} G(i,j) \\ 1 \end{bmatrix} \tag{10}$$

$$\begin{bmatrix} \sum_{i=1}^{m} \sum_{j=1}^{n} B^2(i,j) & \sum_{i=1}^{m} \sum_{j=1}^{n} B(i,j) \\ 2^N - 1 & 1 \end{bmatrix} \begin{bmatrix} e_\mathrm{b} \\ f_\mathrm{b} \end{bmatrix} = \begin{bmatrix} \sum_{i=1}^{m} \sum_{j=1}^{n} G(i,j) \\ 1 \end{bmatrix} \tag{11}$$

8.2   Combined Max White and Retinex

This color restoration approach combines the traditional retinex color restoration technique and one of its intrinsic properties, the max white. This approach also assumes that the intensities of the green channel never change. All the changes are introduced to the red and blue channels. Enhancing the performance of this technique is achieved via a set of assumptions for the two founding algorithms.

The hypothesis of the retinex white balancing algorithm requires two equations to adjust the red and blue channels. Also, the max white color restoration technique requires two equations to correct the damaged red and blue channels. The four color mapping constants (e_r, f_r, e_b, f_b) are calculated using Eqs. (12) and (13), and the calculated values of these constants are substituted into Eqs. (6) and (7) for the retrieval of the missing color values.

$$\begin{bmatrix} \max\left\{I_\mathrm{R}^2(x,y)\right\} & \max\left\{I_\mathrm{R}(x,y)\right\} \\ \left(2^N - 1\right) & 1 \end{bmatrix} \begin{bmatrix} e_\mathrm{r} \\ f_\mathrm{r} \end{bmatrix} = \begin{bmatrix} \max\left\{I_\mathrm{G}(x,y)\right\} \\ 1 \end{bmatrix} \tag{12}$$

$$\begin{bmatrix} \max\left\{I_\mathrm{B}^2(x,y)\right\} & \max\left\{I_\mathrm{B}(x,y)\right\} \\ \left(2^N - 1\right) & 1 \end{bmatrix} \begin{bmatrix} e_\mathrm{b} \\ f_\mathrm{b} \end{bmatrix} = \begin{bmatrix} \max\left\{I_\mathrm{G}(x,y)\right\} \\ 1 \end{bmatrix} \tag{13}$$

8.3 Combined Gray World and Retinex

This color restoration approach is more similar to the combined gray world and max white algorithm. In fact, this approach was the first technique to combine two traditional algorithms with a nonlinear mapping for color corrections and adjustments. Similarly, this technique also assumes that the intensities of the green channel never change. The four constants are calculated using Eqs. (14) and (15), and the calculated values of these constants are substituted back into Eqs. (6) and (7) to adjust and correct the faded colors.

$$\begin{bmatrix} \sum\limits_{i=1}^{m} \sum\limits_{j=1}^{n} R^2(i,j) & \sum\limits_{i=1}^{m} \sum\limits_{j=1}^{n} R(i,j) \\ \max\left\{R^2(x,y)\right\} & \max\left\{R(x,y)\right\} \end{bmatrix} \begin{bmatrix} e_r \\ f_r \end{bmatrix} = \begin{bmatrix} \sum\limits_{i=1}^{m} \sum\limits_{j=1}^{n} G(i,j) \\ \max\left\{G(x,y)\right\} \end{bmatrix} \tag{14}$$

$$\begin{bmatrix} \sum\limits_{i=1}^{m} \sum\limits_{j=1}^{n} B^2(i,j) & \sum\limits_{i=1}^{m} \sum\limits_{j=1}^{n} B(i,j) \\ \max\left\{B^2(x,y)\right\} & \max\left\{B(x,y)\right\} \end{bmatrix} \begin{bmatrix} e_b \\ f_b \end{bmatrix} = \begin{bmatrix} \sum\limits_{i=1}^{m} \sum\limits_{j=1}^{n} G(i,j) \\ \max\left\{G(x,y)\right\} \end{bmatrix} \tag{15}$$

8.4 The Overdetermined System

These three combined techniques, which are presented in Sects. 8.1–8.3, can be collected together into a single overdetermined system. The system provides a complete set of the combined color restoration algorithms. Therefore, the three equations, Eqs. (10), (12), and (14), can be combined together into a single equation, Eq. (16). Similarly, a combination of the three equations, Eqs. (11), (13), and (15), provides Eq. (17). The number of the unknown constants is less than the provided number of equations, which makes the system overdetermined. The choice of the best set of equations to solve for the unknown constants is an image-dependent decision, where the color adjustment results may perceptibly vary based on the contents of the image. Implementation is achieved via three main stages, as follows.

Stage 1 includes a decomposition of the input faded image into the three basic colorimetric channels (R, G, and B). The red and blue layers are fed into stage 2, while the green layer is kept without changes until later use in stage 3.

Stage 2 includes all the processing of the red and blue channels and fulfilling the assumptions of the two utilized founding algorithms.

Stage 3 includes three substeps, as follows: In the first substep, a set of the two equations, Eqs. (16) and (17), is used for calculating the mapping constants of the red channel. The blue channel mapping constants are calculated similarly. In the second substep, two different quadratic mappings for each combined technique are used to correct the values of the red and blue channels. The third substep is a simple

Fig. 2 Results from the third solution of the overdetermined color restoration system (combined GWR). (a) Input faded image; (b) output color corrected image; (c), (d), and (e) are zoom in on specific portions of the faded and adjusted images

combination of the original unchanged green channel G with the two corrected red R'' and blue B'' channels.

$$\begin{bmatrix} \sum_{i=1}^{m} \sum_{j=1}^{n} R^2(i,j) & \sum_{i=1}^{m} \sum_{j=1}^{n} R(i,j) \\ 2^N - 1 & 1 \\ \max\left\{R^2(x,y)\right\} & \max\left\{R(x,y)\right\} \end{bmatrix} \begin{bmatrix} e_r \\ f_r \end{bmatrix} = \begin{bmatrix} \sum_{i=1}^{m} \sum_{j=1}^{n} G(i,j) \\ 1 \\ \max\left\{G(x,y)\right\} \end{bmatrix} \tag{16}$$

$$\begin{bmatrix} \sum_{i=1}^{m} \sum_{j=1}^{n} B^2(i,j) & \sum_{i=1}^{m} \sum_{j=1}^{n} B(i,j) \\ 2^N - 1 & 1 \\ \max\left\{B^2(x,y)\right\} & \max\left\{B(x,y)\right\} \end{bmatrix} \begin{bmatrix} e_b \\ f_b \end{bmatrix} = \begin{bmatrix} \sum_{i=1}^{m} \sum_{j=1}^{n} G(i,j) \\ 1 \\ \max\left\{G(x,y)\right\} \end{bmatrix} \tag{17}$$

For a detailed discussion and analysis of the color correction results from the three solutions, the reader might refer to [6–8]. In this study, a few experiments were performed to test the proposed overdetermined system. Figure 2 is a sample of the color correction results using the combined gray world and retinex technique. A test image (old with faded colors) is selected to emphasize the capabilities of this technique. The experiment was repeated on different faded images with different features, and the cast value in each output image was calculated for a numerical

Table 1 Comparison between the three solutions of the overdetermined color restoration system

Corrupted image	Cast measurements			
	Input	MWR	GWR	MWGW
1	18	29	63	52
2	57	76	134	113
3	82	89	161	157
4	32	39	48	43

comparison between the three color restoration schemes. Table 1 displays the comparison results. From the table, we can conclude that the smallest cast values are generally associated with the combined MWR color restoration approach. Therefore, the other two solutions are more recommended for automatic color retrieval of the faded colors. Such conclusion is expected because the MW is an intrinsic property of the retinex approach; however, the combined MWR restores more colors in the dark portions of the damaged image. Also, in some cases, the three algorithms might provide similar casting values, which slightly contradict with the human observation of the corrected image. Therefore, a good and precise evaluation of the color adjustment results should be based on both the human observation and numerical measurements.

9 Machine Learning Techniques

9.1 Colorization

In recent years, machine learning techniques have been improved to recognize complex patterns that were previously difficult. One of the earliest fully automatic methods to colorize grayscale images was introduced by Cheng et al. [50]. This method obtained state-of-the-art performance by employing deep learning to analyze a large amount of training data. In order to further increase the amount of data used in for training, the authors used patch matching noise in relation to the size of the database. Additionally, an automatic random forest method was developed by Deshpande et al. [51]. This technique produced good results when given information about the setting of the image. Another technique utilizing multiple support vector machines (SVM) was developed concurrently by Andress and Zanoci [52]. Each SVM is trained on a different color, classifying whether a pixel contained that SVM's hue. Although this approach presents reasonable performance, it is not fully automatic. For each image colorization, the user must provide a similar input image to the algorithm. Larsson et al. in [53] presented a deep convolutional neural network that could automatically predict per-pixel color histograms. Many different valid color combinations of the image could be generated when applying the predicted histogram. The authors compared different methods with root mean square error. This technique persistently provided significantly better results than all previous methods. Concurrently, Zhang et al. developed a deep convolutional neural network for colorization that adopted classification loss and rebalancing of rare classes [54]. It is shown that the results can be a significant pretext task for feature learning, performed as a cross-channel encoder. Applying ImageNet as training and testing data, this new architecture outperformed all previous work.

9.2   Inpainting

Machine learning techniques have also been employed to perform tasks such as inpainting. Inpainting consists of colorizing portions of an image that have been destroyed. A deep convolutional neural network was demonstrated by Elango and Murugesan in order to tackle this problem [55]. In their solution, an image was divided into cells to estimate the amount of noise in a certain region. Based on the amount of noises seen in a cell, this could indicate the amount of surrounding information to use. This approach is able to successfully correct color images with low or high noise ratios. Xie et al. proposed a denoising auto-encoder architecture to perform inpainting without knowledge of regions where inpainting is needed [56]. This technique obtained results comparable to non-blind inpainting approaches. Kohler et al. introduced a deep neural network utilizing the shape of the damaged region to improve the performance of inpainting [57]. In addition, the authors of this approach found that using extra information about the shape is especially helpful in blind inpainting. A direct comparison of Xie and Kohler's models was never performed. Until 2016, most inpainting procedures were focused on removing patterns such as lines, text, and dots from images. Pathak et al. were the first to introduce a convolutional neural network to fill in large patches from damaged images [58]. Although the patches were not reproduced perfectly, the results were competitive with all other content-aware patch filling techniques. More recently, Yang et al. introduced a deep learning approach that can outperform all other methods of synthesizing content-aware patches [59]. By employing simultaneous optimization of image and texture constraints, a multiscale patch synthesis approach can preserve context and details of the image. The insights of this approach show that using separate networks to extract texture and content information can enhance results.

9.3   Color Constancy

Color constancy is another image processing problem that has been explored by machine learning. The goal of computational color constancy is to accurately estimate the color of light sources. Using this information, images taken in poor light can be corrected. Agarwal et al. [60] compared previous support vector regressions, neural networks, and simple regression methods for color constancy. The results of the comparison show that machine learning techniques were inferior to simple regression tools. Gijsenij and Gevers in [61] recognized that a large amount of color constancy algorithms work well for different kinds of images. Therefore, the authors attempted to create a MoG classifier that can analyze an image and select the best algorithm. Since this process utilizes every state-of-the-art color constancy product, the results outperform the prior methods used by the same authors. Reusing the idea of combining the best of multiple algorithms, Banic and Loncaric introduced a method that increases the accuracy of previous methods by reducing scattering in the chromaticity space [62]. This is implemented by learning the normal distribution of illuminant scattering and applying this knowledge to corrected images. Again, this method reached state-of-the-art accuracy. Bianco et al. in [63] described a convolutional neural network (CNN) that outperforms all state-of-the-art learning and statistical approaches to color constancy. This CNN does not use handcrafted features like many other methods, but instead works in the spatial domain automatically. Lou et al. [64] introduced a deeper convolutional neural network which increases the accuracy of color constancy estimations. The deep CNN is used to estimate the color of light sources by reformulating the problem to CNN-based regression. In addition, this model is fast enough to operate at more than 100 frames per second. Shi et al. proposed a concept where two neural networks are used in synergy [65]. First, a network is used to create many viable hypotheses from a combination of local regions. Following this process, a second network is used to generate confidence values to select the most accurate illuminant value. At its release in 2016, this model reached state-of-the-art performance. Bianco et al. [66] introduced a new 3-stage method in 2017 that first uses a convolutional neural network to estimate local illuminants. The technique then calculates the number of illuminants in the image and aggregates the local regions nonlinearly. An extensive comparison proves the effectiveness of this new method. Many times, patch-based methods for color constancy perform poorly due to the patch's poor representation of the image. To overcome this problem, Hu et al. [67] described a convolutional neural network that weights each patch according to its quality. With this new concept, the model was able to outperform all previous methods while also improving computational efficiency. Most recently, Nikola and Sven has presented an unsupervised learning technique to classify the illumination [68]. Due to the ill-posed nature of the problem, the authors assume that each image will have a warm or cool illumination. This work outperforms all statistical approaches and many learning approaches.

10 Underwater Color Correction Techniques

As the development of underwater cameras has progressed, demand has increased for methods to correct the color of underwater images. Due to different wavelengths of light meeting fluctuating attenuation, images taken underwater have distorted color. One of the first methods to correct the color of these images was introduced in 2005 by Torres-Méndez and Dudek [69]. By forming each image as a Markov random field (MRF), models could learn from training data to probabilistically estimate the correction of color needed. Ancuti et al. presented a fusion-based scheme that relies on user input of four weight maps and two input values representing different versions of the image [70]. One advantage of this technique is that only a single image is required and no knowledge about hardware or underwater conditions are needed. Berman et al. proposed another method that requires the user to estimate the attenuation values of blue-red and blue-green color channels [71]. Using these estimated values, the resulting problem is reduced to simple dehazing on respective color channels. This method exploits that different types of water produce different spectral profiles. Therefore, multiple possible solutions are presented. Chiang and Chen also proposed a dehazing method for underwater color correction [72]. After the algorithm attempts to remove any artificial light in the image, the image is dehazed. Based on the amount of attenuation for each wavelength, each color is corrected. Shamsuddin et al. introduced an automatic and manual method for color correction in underwater images [73]. The automatic method uses histogram clipping while the manual method uses input from the user that sets the darkest and lightest values in the image. Galdran et al. proposed an idea where places in the image corresponding to short wavelengths are recovered [74]. This red channel method enhances the contrast and achieves color correction that is equivalent to all state-of-the-art approaches. An additional method focusing on the red channel correction is presented by Li et al. [75]. First, the blue-green channels are dehazed while the red channel is corrected with the grey world assumption. Afterward, an exposure map is created to correct light and dark regions in the image. Attempting to provide color correction for classification of different species of marine life, Rzhanov et al. combined hyperspectral data, hardware properties, and water conditions to properly correct the color of an image [76]. Due to the ambiguous nature of underwater color correction, this method presents a probabilistic approach that simply corrects and extracts the needed regions of the image. Lu et al. developed a normalized convolution filtering algorithm for enhancing images taken in shallow water [77]. The results show that this method reduces noise, enhances exposure in dark regions, and improves contrast throughout the entire image. In order to correct images from the seafloor derived from an underwater autonomous vehicle (UAV), Bryson et al. exploited structure-from-motion and photogrammetry techniques in order to approximate the true color of an image [78]. This method is used routinely by UAVs in underwater benthic monitoring programs. Li et al. introduced a simple method where a color cast removal step is first performed on the image [79]. Following this step, based on the relationship of the three color channels, a visibility restoration algorithm is applied. This technique provides results comparative and even better than many state-of-the-art underwater image restoration methods. He et al. devised a filter that could smooth an underwater image while still preserving the edges of the image [80, 81]. A big advantage to this method is that it can enhance an underwater image with low computational complexity. Many methods may provide good results, but the computations can take a large amount of time. One of these methods introduced by Schechner and Karpal corrects the colors of an image by using a polarizer with several images of separate orientation in order to improve visibility [82]. Another method with high computational complexity was proposed by Rizzi et al. [83]. This method recovers distortion of color by utilizing partial and global information with unsupervised equalization of color. This process exploits "gray world" and "white patch" for the step of global equalization. On the other hand, when applying local filtering, this chapter takes into account the color spatial distribution. The results of this method are promising for a wide variety of underwater images.

11 Uncategorized Techniques

A virtual retrieval technique for the historical faded art pieces is presented by Pei et al. [25]. This technique introduced a hybrid method including some color contrast enhancement based on saturation and desaturation in the $u'v'Y$ color space. Also, a new patching method using the Markov random field (MRF) model of texture synthesis was introduced. The synthesization process involved three key approaches, weighted mask, annular scan, and auxiliary, with neighborhood searching. This technique maintains the fidelity of complete shapes and prevents the disconnection of edges. The algorithm performs the fine retrieval results at the cost of the computation complexity, which is inherited from the MRF-based neighborhood searching.

Cheng et al. introduced a double-matched method to restore the colors of faded photos [26]. This technique achieves a good color matching via the Euclid distance calculations of all neighboring pixels within small blocks. The luminance and hue components of the faded image and reference image were extensively used in the calculations. Li et al. used the semantic network holding features of objects from faded art pieces [27, 28]. The relationships between these objects were represented by the edges between the nodes of the utilized network. The similarity of the two images represents the weighted average of similarity of their nodes and edges. Two methods for color-based image retrieval were presented, where in the first approach color layers were utilized to represent color information based on color clustering. The similarity between the two input images is calculated based on a new matching method called color layer matching. In the second method, another matching method called color composition matching is utilized to measure the similarity of color composition of the two input images. This method is similar to retrieving the color information of an image from the artist's point of view.

Tarel et al. developed a new algorithm for automatic color retrieval from a single input image based on the median filter [29]. This algorithm is very simple where linear mapping for color correction is utilized. The major advantage is the speed of the algorithm which is a major concern for many image applications including surveillance, intelligent vehicles, and remote sensing. Also, a new filter was introduced to preserve the edges and corners with obtuse angle. Another technique based on the particle swarm optimization with wavelet mutation (WPSO) as the restoration algorithm for half-toned color-quantized images is presented by Yeung et al. [41]. In this algorithm, the number of particles used for searching is increased. Moreover, each particle shares the information in each iteration step, which improves the searching ability and convergence rate. Also, the wavelet mutation reduces the probability of the particles trapped in local optima. Another technique is presented by Wen et al. [42], where the color total variation minimization scheme is utilized to denoise the deblurred color image. A minimization algorithm is employed to solve the total variation minimization problem. The results of this algorithms show the quality improvements of the restored colors. A more recent technique for color correction is presented by Kyung et al. in [44], where the colors of the faded image are separated according to their chromaticity values in the LCybCrg color space. Color corrections are performed based on the gray world assumptions, and weights from the CybCrg are utilized to reduce the contour artifacts.

Lee et al. presented a specialized technique correcting the purple fringing [46]. The corrupted regions are first detected using a chromaticity diagram. Next, these regions are corrected by a colorization process where the colors of the pixels near the purple fringed region are set to some seed pixels. Results from this technique are good, and the corrected colors seem natural.

Jung et al. presented a new class for color restoration techniques using the Mumford-Shah (MS) model and nonlocal image information [47–49]. In this approach, the local Ambrosio-Tortorelli and Shah approximations in MS functional (MS) were extended to new nonlocal formulations. This technique provides good restoration results especially for the fine structures and texture portions of the faded images.

Image mosaicking can be described as a process of stitching separate images together. This process has a large range of applications such as medical, satellite, and text synthesis. To complicate things, many times, images taken in different lighting conditions need to be stitched together. Oloveira et al. introduce a probabilistic approach for color correction in images that are part of a mosaic [93]. This process begins with the original image being segmented into different regions using mean shift. Next, using a fusion algorithm, these regions are extracted and are modeled as groups of Gaussians. With these Gaussians, local mapping of colors is computed. Using these computed mappings, color correction is performed on the entire image. This method is shown to outperform all state-of-the-art image mosaicking color correction methods.

When using a multispectral filter array (MSFA) image sensor, it is essential to remove the infrared cutoff filter (IRCF) to obtain unseen band material. Conversely, without the IRCF, the colors in an image will be desaturated. Park and Kang proposed a method to restore the original color of an image with a signal processing method [94]. In this technique, spectral features are observed from an MSFA sensor. With this information, spectral estimation and decomposition are executed in order to estimate the intensity of the near-infrared band. Once this band is estimated, it can be decomposed along with the visible band to recover the hue and color saturation on each color channel.

Colorizing black and white images or video footage can be invaluable to law enforcement and various other companies. Presently, many machine learning techniques perform very well on this ill-constrained problem. However, Zeng and Jia proposed a matrix equation method that is also very sound [96]. First, by taking photos on a calibration plate, a matrix representing the environment is established. Then, using the calculated matrix, a restored image is estimated. The results indicate that the method is adaptable and has good accuracy.

Another method that uses matrix operations to perform color restoration can be seen in a paper by Sinthanayothin et al. [97]. In this approach, the goal is to correct an object's faded or incorrect colors resulting from unideal lighting conditions. The authors utilize the environment surrounding the object to create six color charts. From these charts, six transformation matrices are calculated and then used to automatically detect the correct color chart. Following this step, the image is divided into several grids, and elements at the grid's corners are evaluated based on their distance from the ideal color chart. Exploiting this information from the corners, each pixel in the segmented grid is adjusted by interpolation from the four corners. The results of this method are accurate and less costly than many state-of-the-art methods.

Augmented reality is a technology that places a computer-generated image in real-world space as seen through a screen. When implementing a product with augmented reality, an image must be captured by multiple cameras and constructed into images that are placed in front of a human's eye. This process usually results in images that have color distortion. Itoh and Amano have created a way to preprocess input colors so that users see output colors as similar [98]. The authors treat this problem as a parametric model where nonlinear color distortion and linear color shift are separated. Through experimentation, calibrated images are adjusted to look like their original counterparts. However, the technology of optical see-through head-mounted displays is limited and shown to pose problems to this technique.

Presently, most digital cameras apply an array of many color filters to capture information about a scene. This color filter array is much different from a full-color camera that applies beam splitters to obtain information. One drawback of a color filter array is that sensors only get information about the intensity of each color channel separately. The result is that two color channels are missing and must be restored. To solve this problem, S. Kim presented a directional interpolation method [99]. When the image is reconstructed, missing information is obtained by interpolation from neighboring pixels. The results of this method are shown to visually outperform other conventional approaches.

When color images are evaluated in quaternion space, image colors are seen as a single unit. Grigoryan et al. described an approach to color enhancement where each color is considered as a vector in quaternion space [100]. The authors proposed using alpha rooting in the frequency spectrum to correct color problems in 3D medical images. Additionally, a second method known as zonal alpha rooting can be employed by dividing the image into separate regions so that different alpha values could be used. By applying these alpha rooting techniques in combination with 2D quaternion discrete Fourier transformations, 3D medical images can be color corrected.

Radiographic testing images taken for manufacturing steam turbines have poor color characteristics. In order to correct the color of these images, Jiang et al.

introduced a pseudo-color enhancement method for images in the HSI color space [101]. This method first transforms each pixel in the HSI space to the value of the original pixel. Afterward, the average intensity of the image is adjusted to half of the original image. After these alterations, the image is transformed back to RGB space. The results of this method are shown to be better than many similar ideas used in industry.

The Terracotta Warriors are an artistic piece that is historically significant to China. These statues have painted surfaces that have slowly shown fading and shedding. Li et al. explored different ideas to restore these treasures and put forward an improved Welsh algorithm [103]. In this technique, the eigenvalues and neighborhood variance are used to form a measurement of mixed distances. This method utilizes the hierarchical color structure of faded paintings and is shown to be better at recoloring gray texture than classical Welsh techniques.

In dental procedures, ceramic restoration of damaged portions of a model can be very difficult. There are many possible shortcomings of human color matching methods, including variable viewing conditions and inadequate observer eyesight. In order to improve the process of matching shades, a procedure that simplifies shade selection to elements of opaque, body, and incisal porcelain is presented by Sorensen and Torres [104]. This system improves communication between dentists and ceramists and solves many of the intrinsic problems of shade selection.

In order to improve the visual appearance of color images, Toet proposed a method that can recover the local contrast of luminance at all levels of resolution [105]. Recognizing that saturation and luminance are often complementary, color contrast can be improved by changing the luminance according to the variations in saturation. Therefore, luminance and saturation are first decomposed into contrast primitives of different scales. Then, by modulating the original luminance at every location and every corresponding saturation contrast primitive, a set of multiscale luminance contrast primitives are constructed. From these multiscale primitives, the image can be reconstructed with enhanced luminance and contrast.

Many times, a group of images about a specific scene or object needs a color correction scheme that balances the color appearances. Zhang et al. attempted to solve this problem with a fast color correction method using principal regions mapped in different color spaces [106]. By selecting principal regions and using them to construct low-degree polynomial mapping functions, the running speed will be low. These functions are run color spaces decorrelated color spaces. After testing with real and synthetic regions, the proposed process is shown to be better than any other solution in 2004.

As old motion picture collections age, it can be observed that much of the color can fade and become damaged. Gangal et al. attempted to solve this problem when films are deteriorated by flashing blotches [107]. This process utilizes five consecutive frames and log-D search with improved multiresolution block matching to restore the damaged color. In addition, a rank-ordered blotch detection scheme with 3D vector median filtering for interpolation was used. It is shown that this product works very efficiently for even severely blotched motion pictures.

Color correction is often evaluated by the human visual system. Using this concept, Huang et al. proposes a novel enhancement method named human visual system controlled color image enhancement and evaluation algorithm (HCCIEE) [108]. This algorithm considers image quality metrics and is based on color processing and luminance in the human visual system. The results of this novel method show that details in the image can be distinguished without any artifacts such as halos and ringing. Due to the lack of artifacts in the results of this method, this application enhances color images better than most state-of-the-art methods.

12   Conclusions

The authors of this chapter surveyed many color restoration techniques, including the single-scale and multiscale retinex, gray world, max white, machine learning, and underwater color correction. In addition to these techniques, the chapter provided a review of numerous unclassified color restoration approaches under the uncategorized section. The chapter focused on the reasons of color fading and the color bleaching models. The survey explained how most of the color restoration/retrieval techniques are built on intuitive assumptions to imitate the human visual system in recognizing different colors. An overdetermined system, joining the three combined gray world, retinex, and max white color retrieval approaches, is presented. A nonlinear mapping process is utilized to correct the faded colors. For future work, the authors will consider a third-order equation for another nonlinear color correction and adjustment. The numerical values of the green layer should also degrade in the proposed future research.

References

1. Land, E. H., & McCann, J. J. (1971). Lightness and retinex theory. *Journal of the Optical Society of America, 61*, 1–11.
2. Hunt, R. (2004). *The reproduction of colour*. Hoboken: John Wiley and Sons.
3. Lam, E. Y. (May 2003). Image restoration in digital photography. *IEEE Transactions on Consumer Electronics, 49*(2), 269–274.
4. Lam, H.-K., Au, O.C., & Wong, C.-W. (2004). Automatic white balancing using standard deviation of RGB components. In *Proceedings of International Symposium on Circuits and Systems, ISCAS* (Vol. 3, pp. 921–924).
5. Starck, J.-L., Murtagh, F., Candès, E. J., & Donoho, D. L. (June 2003). Gray and color image contrast enhancement by the curvelet transform. *IEEE Transactions on Image Processing, 12* (6), 706–717.
6. Lam, E. (2005). Combining gray world and retinex theory for automatic white balance in digital photography. In *Proceedings of 9th International Symposium on Consumer Electronics* (pp. 134–139).
7. Ahmed, A. M. (2009). The max white effect on the gray world white-balancing algorithm. In *Visualization, Imaging and Image Processing Conference, VIIP*, Cambridge, UK.

8. Ahmed, A. M. (2010). Automatic color retrieval using the Retinex and its max white intrinsic property. In *Proceedings of 7th International Conference on Informatics and System, INFOS, Cairo, Egypt* (pp. 1–5).
9. Rosenthaler, L., & Gschwind, R. (2001). Restoration of movie films by digital image processing. In *IEEE Seminar: Digital Restoration of Film and Video Archives* (pp. 6.1–6.5). IEE.
10. Cortelazzo, G. M., Geremia, G. L., & Mian, G. A. (1995). Some results about Wiener-Volterra restoration of the original color quality in old painting imagery. In *Proceedings of IEEE Workshop Nonlinear Signal Image Processing (NSIP'95), Neos Marmaras, Greece* (Vol. I, pp. 86–89).
11. Pappas, M., & Pitas, I. (Feb. 2000). Digital color restoration of old paintings. *IEEE Trans. on Image Processing, 9*(2), 291–294.
12. Gschwind, R. (1989). Restoration of faded color photographs by digital image processing. In *Proceedings of image processing III* (pp. 27–37).
13. Frey, F., & Gschwind, R. (1994). Mathematical bleaching models for photographic three-color materials. *Journal of Imaging Science and Technology, 38,* 513–519.
14. Gschwind, R., & Frey, F. (1994). Electronic imaging, a tool for the reconstruction of faded color photographs. *Journal of Imaging Science and Technology, 38,* 520–525.
15. Gschwind, R., Frey, F. S., & Rosenthaler, L. (1995). Electronic imaging: A tool for the reconstruction of faded color photographs and color movies. In *Proceedings of SPIE image and video processing III* (pp. 57–63).
16. Chambah, M., & Besserer, B. (2000). Digital color restoration of faded motion pictures. In *Proceedings of CGIP Conference* (pp. 338–342).
17. Chambah, M., Besserer, B., & Courtellemont, P. (2002). Approach to automatic digital restoration of faded color film. In *Proceedings of IS&T CGIV 2002* (pp. 613–618).
18. Jobson, D. J., Rahman, Z., & Woodell, G. A. (Mar. 1997). Properties and performance of a center/surround retinex. *IEEE Transactions on Image Processing, 6,* 451–462.
19. Jobson, D. J., Rahman, Z., & Woodell, G. A. (July. 1997). A multiscale retinex for bridging the gap between color images and the human observation of scenes. *IEEE Transactions on Image Processing, 6*(7), 965–976.
20. Chambah, M., Rizzi, A., Gatta, C., Besserer, B., & Marin, D. (2003). Perceptual approach for unsupervised digital color restoration of cinematographic archives. In *Proceedings of SPIE/ IS&T Electronic Imaging* (Vol. 5008, pp. 138–149).
21. Rizzi, A., Chambah, M., Lenza, D., Besserer, B., & Marini, D. (2004). Tuning of perceptual technique for digital movie color restoration. In *Proceedings of SPIE Security, Steganography, and Watermarking of Multimedia Contents VI* (pp. 1286–1294).
22. Rizzi, A., Gatta, C., Slanzi, C., Ciocca, G., & Schettini, R. (2005). Unsupervised color film restoration using adaptive color equalization. In *Proceedings of 8th International Conference on Visual Information and Information Systems* (pp. 1–12).
23. Ahmed, A. M. (2009). Color restoration techniques for faded colors of old photos, printings and paintings. In *IEEE International Conference on Electro/Information Technology, EIT, Windsor, ON, Canada* (pp. 151–156).
24. Strang, G. (2006). *Linear algebra and its applications,* Cengage Learning; 4 edition.
25. Pei, S., Zeng, Y., & Chang, C. (2004). Virtual restoration of ancient Chinese paintings using color contrast enhancement and lacuna texture synthesis. *IEEE Transactions on Image Processing,* Cengage, Boston, MA, USA, *13*(3), 416–429.
26. Cheng, J., Cao, M., & Teng, S. (2008). A double-matched method for color recovery. In *Congress on Image and Signal Processing, CISP* (Vol. 2).
27. Li, X. (1999). *The study of content based image retrieval and image database modeling.* PhD thesis, Dept. of Computer Science and Engineering, Zhejiang University, Hangzhou, China.
28. Li, X., Lu, D., & Pan, Y. (2000). Color restoration and image retrieval for Dunhuang Fresco preservation. *IEEE Multimedia, 7*(2), 38–42.

29. Tarel, J., & Hauti'ere, N. (2009). Fast visibility restoration from a single color or gray level image. In *IEEE 12th International Conference on Computer Vision, ICCV* (pp. 2201–2208).

30. Moradi, M., Nathan, A., Haverinen, H. M., & Jabbour, G. E. (2009). Short channel vertical transistors with excellent saturation characteristics. In *Device Research Conference, DRC* (pp. 171–173).

31. Tye, M. (2002). *Consciousness, color, and content*. Cambridge: MIT Press.

32. Halo Lighting Division, McGraw-Edison Co. (1983). *A complete guide to the language of lighting*. McGraw-Edison Co..

33. Nikitenko, D., Wirth, M., & Trudel, K. (2007). White-balancing algorithms in colour photograph restoration. In *IEEE International Conference on Systems, Man and Cybernetics* Cengage, Boston, MA, USA (pp. 1037–1042).

34. Nikitenko, D., Wirth, M., & Trudel, K. (2008). Applicability of white-balancing algorithms to restoring faded colour slides: An empirical evaluation. *Journal of Multimedia, 3*(5), 9–18.

35. Giakoumis, I., & Pitas, I. (1998). Digital restoration of painting cracks. In *IEEE International Symposium on Circuits and Systems (ZSCAS' 98)*, CA, USA.

36. Han, H., & Sohn, K. (2009). Automatic illumination and color compensation using mean shift and sigma filter. *IEEE Transactions on Consumer Electronics, 55*(3), 978–986.

37. Tao, L., & Asari, V. K. (2005). Adaptive and integrated neighborhood-dependent approach for nonlinear enhancement of color images. *Journal of Electronic Imaging, 14*(4), 043006.

38. Li, Y., He, R. Xu, G., Hou, C., Sun, Y., Guo, L., Rao, L., & Yan, W. I. (2008). Retinex enhancement of infrared images. In *Proceedings of 30th IEEE International Conference on Engineering in Medicine and Biology Society* (pp. 2189–2192).

39. Jie, X., Li-Na, H., Guo-Hua, G., & Ming-Quan, Z. (2009). Real color image enhancement based on the spectral sensitivity of most people vision and stationary wavelet transform. In *Proceedings of 2nd IEEE International Conference on Computer Science and Information Technology, ICCSIT* (pp. 323–328).

40. Wu, J., Wang, Z., & Fang, Z. (2009). Application of retinex in color restoration of image enhancement to night image. In *Proceedings of 2nd International Congress on Image and Signal Processing, CISP*.

41. Yeung, C. W., Ling, S. H., Chan, Y. H., & Leung, F. H. F. (2008). Restoration of half-toned color-quantized images using particle swarm optimization with wavelet mutation. In *IEEE Region 10 Annual International Conference, TENCON* (pp. 1–6).

42. Wen, Y.-W., Ng, M. K., & Huang, Y.-M. (Nov. 2008). Efficient Total variation minimization methods. *IEEE Transactions on Image Processing, 17*(11), 2081–2088.

43. Chambah, M., Besserer, B., & Courtellemont, P. (2001). Recent progress in automatic digital restoration of color motion pictures. In *Proceedings of SPIE Color Imaging: Device-Independent Color, Color Hardcopy, and Applications VII* (pp. 98–109).

44. Kyung, W., Kim, D., Kim, K., & Ha, Y. (2011). Color correction for faded image using classification in LCybCrg color space. In *Proceedings of IEEE Internatoinal Conference on Consumer Electronics, ICCE*, Berlin (pp. 189–193).

45. Ahmed, A. M., & Day, D. D. (2004). Applications of the naturalness preserving transform to image watermarking and data hiding. *Elsevier Digital Signal Processing, 14*, 531–549.

46. Lee, D., Kim, B., & Park, R. (2011). Purple fringing correction using colorization in Yxy color space. In *Proceedings of IEEE International Conference on Consumer Electronics*, ICCE, Las Vegas (pp. 477–478).

47. Jung, M., Bresson, X., Chan, T. F., & Vese, L. A. (2009). Color image restoration using nonlocal Mumford-Shah regularizers. *LNCS, 5681*, 373–387.

48. Jung, M., & Vese, L. A. (2009). Nonlocal variational image deblurring models in the presence of Gaussian or impulse noise. *LNCS, 5567*, 402–413.

49. Jung, M., Bresson, X., Chan, T. F., & Vese, L. A. (2011). Nonlocal Mumford-Shah Regularizers for color image restoration. *IEEE Transactions on Image Processing, 20*(6), 1583–1598.

50. Cheng, Z., Sheng, B., & Yang, Q. (2015). Deep colorization. In *2015 IEEE International Conference on Computer Vision* (pp. 415–423).
51. Deshpande, A., Rock, J., & Forsyth, D. (2015). Learning large-scale automatic image colorization. In *Proceedings of the IEEE International Conference on Computer Vision* (pp. 567–575).
52. Andress, J., & Zanoci, C. (2015). From grayscale to color: Digital image colorization using machine learning.
53. Larsson, G., Maire, M., & Shakhnarovich, G. (2016). Learning representations for automatic colorization. In *European Conference on Computer Vision*. http://cs229.stanford.edu/proj2015/163_report.pdf
54. Zhang, R., Isola, P., & Efros, A. (2016). Colorful image colorization. In B. Leibe, J. Matas, N. Sebe, & M. Welling (Eds.), *Computer Vision—ECCV 2016. ECCV 2016* (Lecture Notes in Computer Science) (Vol. 9907). Cham: Springer.
55. Elango, P., & Murugesan, K. (2010). Image restoration using cellular neural network with contour tracking ideas. *International Journal of Computer Theory and Engineering, 2*(5), 1793–8201.
56. Xie, J., Xu, L., & Chen, E. (2012). Image denoising and inpainting with deep neural networks. In *Proceedings of Neural Information Processing Systems* (pp. 350–358).
57. Köhler, R., Schuler, C., Schölkopf, B., & Harmeling, S. (2014). Mask-specific Inpainting with deep neural networks. In X. Jiang, J. Hornegger, & R. Koch (Eds.), *Pattern recognition. GCPR 2014* (Lecture notes in computer science) (Vol. 8753). Cham: Springer.
58. Pathak, D., Krahenbuhl, P., Donahue, J., Darrell, T., & Efros, A. (2016). Context encoders: Feature learning by inpainting. In *Conference on Computer Vision and Pattern Recognition*.
59. Yang, C., Lu, X., Lin, Z., Shechtman, E., Wang, O., & Li, H. (2016). High-resolution image inpainting using multi-scale neural patch synthesis.
60. Agarwal, V., Gribok, A., & Abidi, M. (2007). Machine learning approach to color constancy. *Neural Networks, 20*(5), 559–563. http://cs229.stanford.edu/proj2015/163_report.pdf
61. Gijsenij, A., & Gevers, T. (2007). Color constancy using natural image statistics. In *Conference on Computer Vision and Pattern Recognition* (pp. 1–8).
62. Banic, N., & Loncaric, S. (2015). Color dog: Guiding the global illumination estimation to better accuracy. In *International Conference on Computer Vision Theory and Applications* (pp. 129–135).
63. Bianco, S., Cusano, C., & Schettini, R. (2015). Color constancy using CNNs. In *Proceedings of the IEEE conference on computer vision and pattern recognition workshops* (pp. 81–89).
64. Lou, Z., Gevers, T., Hu, N., & Lucassen, M. (2015). Color constancy by deep learning. In *British Machine Vision Conference*.
65. Shi, W., Loy, C., & Tang, X. (2016). Deep specialized network for illuminant estimation. In *European Conference* (pp. 371–387).
66. Bianco, S., Cusano, C., & Schettini, R. (2015). Single and multiple illuminant estimation using convolutional neural networks. *IEEE Transactions on Image Processing, 26*(9), 4347–4362.
67. Hu, Y., Wang, B., & Lin, S. (2017). Fully convolutional color Constancy with confidence-weighted pooling. *Computer Vision and Pattern Recognition (CVPR), 2017*, 330–339.
68. Banic, N., & Loncaric, S. (2018). Unsupervised learning for color constancy. In *International Conference on Computer Vision Theory and Applications* (pp. 181–188).
69. Torres-Méndez, L., & Dudek, G. (2005). Color correction of underwater images for aquatic robot inspection. In *Proceedings of the 5th International Conference on Energy Minimization Methods in Computer Vision and Pattern Recognition* (pp. 60–73).
70. Ancuti, C., Ancuti, C. O., Haber, T., & Bekaert, P. (2012). Enhancing underwater images and videos by fusion. In *2012 IEEE Conference on Computer Vision and Pattern Recognition*, Providence, RI (pp. 81–88).
71. Berman, D., Treibitz, T., & Avidan, S. (2017). Diving into haze-lines: Color restoration of underwater images. In *British Machine Vision Conference*.

72. Chiang, J. Y., & Chen, Y. C. (2012). Underwater image enhancement by wavelength compensation and Dehazing. *IEEE Transactions on Image Processing, 21*(4), 1756–1769.
73. Shamsuddin, N., Wan Ahmad, W., Baharudin, B., Kushairi, M., Rajuddin, M., & bt Mohd, F. (2012). Significance level of image enhancement techniques for underwater images. In *2012 International Conference on Computer & Information Science (ICCIS)*, Kuala Lumpeu (pp. 490–494).
74. Galdran, A., Pardo, D., Pic'on, A., & Alvarez-Gila, A. (2015). Automatic red-channel underwater image restoration. *Journal of Visual Communication and Image Representation, 26*, 132–145.
75. Li, C., Quo, J., Pang, Y., Chen, S., & Wang, J. (2016). Single underwater image restoration by blue-green channels dehazing and red channel correction. In *2016 IEEE International Conference on Acoustics, Speech and Signal Processing (ICASSP)* (pp. 1731–1735).
76. Rzhanov, Y., Pe'eri, S., & Šaškov, A. (2016). Probabilistic reconstruction of color for species' classification underwater. In *OCEANS 2015*, Genova (pp. 1–5).
77. Lu, H., Li, Y., & Serikawa, S. (2015). Single underwater image descattering and color correction. In *2015 IEEE International Conference on Acoustics, Speech and Signal Processing* (pp. 1623–1627).
78. Bryson, M., Johnson, M., Pizarro, O., & Williams, S. B. (2016). True color correction of autonomous underwater vehicle imagery. *Field Robot, 33*(6), 853–874.
79. Li, C., Guo, J., Wang, B., Cong, R., Zhang, Y., & Wang, J. (2016). Single underwater image enhancement based on color cast removal and visibility restoration. *Journal of Electron Imaging, 25*(3), 033012.
80. He, K., Sun, J., & Tang, X. (2013). Guided image filtering. *IEEE Transactions on Pattern Analysis Machine Intelligence., 35*(6), 1397–1409.
81. He, K., Sun, J., & Tang, X. Guided image filtering. In *Proceedings of European Conference on Computer Vision* (pp. 1–14).
82. Schechner, Y., & Karpel, N. (2004). Clear underwater vision. In *Proceedings of Conference on Computer Vision and Pattern Recognition* (pp. 536–543).
83. Rizzi, A., Gatta, C., & Marini, D. (2003). A new algorithm for unsupervised global and local color correction. *Pattern Recognition Letters, 24*(11), 1663–1677.
84. Choi, D. H., Jang, I. H., Kim, M. H., & Kim, N. C. (2007). Color image enhancement based on single-scale retinex with a JND-based nonlinear filter. In *2007 IEEE International Symposium on Circuits and Systems* (pp. 3948–3951).
85. Choi, D. H., Jang, I. H., Kim, M. H., & Kim, N. C. (2008). Color image enhancement using single-scale retinex based on an improved image formation model. In *2008 16th European Signal Processing Conference* (pp. 1–5).
86. Wang, W., Li, B., Zheng, J., Xian, S., & Wang, J. (2008). A fast multi-scale Retinex algorithm for color image enhancement. In *2008 International Conference on Wavelet Analysis and Pattern Recognition* (pp. 80–85).
87. An, C., & Yu, M. (2011). Fast color image enhancement based on fuzzy multiple-scale Retinex. In *Proceedings of 2011 6th International Forum on Strategic Technology* (pp. 1065–1069).
88. Hanumantharaju, M. C., Ravishankar, M., Rameshbabu, D. R., & Ramachandran, S. (2011). Color image enhancement using multiscale Retinex with modified color restoration technique. In *2011 Second International Conference on Emerging Applications of Information Technology*, Kolkata (pp. 93–97).
89. Parthasarathy, S., & Sankaran, P. (2012). An automated multi scale Retinex with color restoration for image enhancement. In *2012 National Conference on Communications* (pp. 1–5).
90. Petro, A., Sbert, C., & Morel, J. (2014). Multiscale retinex. *Image Processing, 4*, 71–88.
91. Parthasarathy, S., & Sankaran, P. (2012). Fusion based multi scale RETINEX with color restoration for image enhancement. In *2012 International Conference on Computer Communication and Informatics* (pp. 1–7).

92. Jiang, B., Woodell, G. A., & Jobson, D. (2015). Novel multi-scale retinex with color restoration on graphics processing unit. *Journal of Real-Time Image Processing, 10*(2), 239–253.

93. Oliveira, M., Sappa, A. D., & Santos, V. (2015). A probabilistic approach for color correction in image mosaicking applications. *IEEE Transactions on Image Processing, 24*(2), 508–523.

94. Park, C., & Kang, M. (2016). Color restoration of RGBN multispectral filter array sensor images based on spectral decomposition. *Sensors, 16*, 719.

95. Kim, U. S., Lee, J. H., Park, K. T., & Moon, Y. S. (2012). A novel color restoration method using color projection. In *2012 IEEE International Conference on Consumer Electronics (ICCE)* (pp. 259–260).

96. Zeng, Z., & Jia, H. (2014). Color restoration in the black-and-white video camera. *Optik—International Journal for Light and Electron Optics, 125*(8), 1918–1921.

97. Sinthanayothin, C., Bholsithi, W., & Wongwaen, N. (2016). Color correction on digital image based on reference color charts surrounding object. In *2016 International Symposium on Intelligent Signal Processing and Communication Systems* (pp. 1–6).

98. Itoh, Y., Dzitsiuk, M., Amano, T., & Klinker, G. (2015). Semi-parametric color reproduction method for optical see-through head-mounted displays. *IEEE Transactions on Visualization and Computer Graphics, 21*, 1269–1278.

99. Kim, S. (2017). Full channel color restoration approach using directional interpolation. *International Information Institute Information, 20*(2), 1229–1236.

100. Grigoryan, A., John, A., & Agaian, S. (2017). Color image enhancement of medical images using alpha-rooting and zonal alpha-rooting methods on 2D QDFT. In *Proceedings of SPIE 10136, Medical Imaging 2017: Image Perception, Observer Performance, and Technology Assessment.*

101. Shao, F., Jiang, G., & Yu, M. (2007). Color correction for multi-view images combined with PCA and ICA. *WSEAS Transactions on Biology and Biomedicine, 4*(5), 73–79.

102. Gillespie, A., Kahle, A., & Walker, R. (1986). Color enhancement of highly correlated images, I. Decorrelation and HSI contrast stretches. *Remote Sensing of Environment, 20*(3), 209–235.

103. Li, N., Geng, G., & Wang, K. (2015). The research of terracotta warriors color restoration based on color transfer. *International Journal of Signal Processing, Image Processing and Pattern Recognition, 8*, 89–98.

104. Sorensen, J., & Torres, T. (1987). Improved color matching of metal-ceramic restorations. Part I: A systematic method for shade determination. *The Journal of Prosthetic Dentistry, 58*(2), 133–139.

105. Toet, A. (1992). Multiscale color image enhancement. *Pattern Recognition Letters, 13*(3), 167–174.

106. Zhang, M., & Georganas, N. (2004). Fast color correction using principal regions mapping in different color spaces. *Real-Time Imaging, 10*(1), 23–30.

107. Gangal, A., Kayikçioglu, T., & Dizdaroglu, B. (2004). An improved motion-compensated restoration method for damaged color motion picture films. *Signal Processing: Image Communication, 19*(4), 353–368.

108. Huang, K., Wang, Q., & Wu, Z. (2006). Natural color image enhancement and evaluation algorithm based on human visual system. *Computer Vision and Image Understanding, 103*(1), 52–63.

Part VI
Engineering Mechanics

El Niño (2014–2016) and La Niña (2010–2012): Their Impacts on Water Cycle Components

Muhammed Eltahan ⓘ, Karim Moharm, Mohammed Magooda, and Ahmed H. El-Hennawi

Abstract El Niño and La Niña are considered two of the most important phenomena over the Pacific Ocean that affect the complex earth system and have a great impact and contribution to the weather and climate change. Investigation using spatiotemporal analysis and trend is critical to understand the warm and cold phases of the El Niño–Southern Oscillation (ENSO) which are El Niño and La Niña. In this work, the spatial distribution of temperature, precipitation and evaporation over the Pacific Ocean is investigated during the two different El Niño (2014–2016) and La Niña (2016–2017). Three different data sources were included in this study: temperature from a remote satellite sensor (moderate-resolution imaging spectroradiometer [MODIS]), precipitation from the *Tropical Rainfall Measuring Mission* (TRMM) satellite space mission and evaporation from NASA's numerical climate model, the Modern-Era Retrospective Analysis for Research and Applications (MERRA). This work reveals how El Niño and La Niña events impact the water cycle components.

Keywords EL Niño · La Niña · Climate change · MODIS · TRMM · MERRA

M. Eltahan (✉)
Aerospace Engineering Department, Cairo University, Cairo, Egypt

Institute of Bio-Geosciences (IBG-3 Agrosphere), Jülich, Germany

Centre for High-Performance Scientific Computing, Jülich, Germany
e-mail: m.eltahan@fz-Juelich.de

K. Moharm
Electrical Engineering Department, Alexandria University, Alexandria, Egypt

M. Magooda
Aerospace Engineering Department, Cairo University, Cairo, Egypt

A. H. El-Hennawi
Department of Mechanical Engineering, Ain Shams University, Cairo, Egypt

Mechanical Engineering Department, Techniche Hochschule, luebeck, Germany

© Springer Nature Switzerland AG 2020
M. H. Farouk, M. A. Hassanein (eds.), *Recent Advances in Engineering Mathematics and Physics*, https://doi.org/10.1007/978-3-030-39847-7_25

1 Introduction

El Niño–Southern Oscillation (ENSO) [1] is considered one of the most important irregular cycles that occur in the Pacific Ocean. ENSO affects mainly the wind speed and sea surface temperature [2] which lead to direct negative and positive effects on other meteorological parameters such precipitation and evaporation. ENSO was defined as recurrent variation on sea surface temperature and wind speed which represent frequent climate pattern [21].

ENSO has three modes: neutral, *El Niño and La Niña. Both El Niño and La Niña are opposite of each other. El Niño* is the warmer phase from the ENSO, while *La Niña* is the cooling phase. *Every single ENSO event is not related. All El Niño and La Niña events were monitored and recorded* [13–19]. The main drawbacks of both El Niño and La Niña are the severe storms and precipitation in some places of the world, in addition to droughts in other places. Still the reasons and origin of both El Niño and La Niña are under active research in order to evaluate the impact of these phenomena on agriculture, weather, climate and economy [3–6] and to try to predict these two critical events. It was also shown that every single El Niño and La Niña event has a direct impact on the trend of the global average temperature. Any El Niño and La Niña event generates a short spike (upward and downward, respectively) for a minimum of 1 year on the global temperature trend (heating and cooling effect) [20]. ENSO directly affects both water and energy cycles [10–12]. The importance of the water cycle was studied before in [22, 23].

2 Materials and Methods

2.1 *MODIS [7]*

A level 3 high-resolution sea surface temperature dataset was gathered from moderate-resolution imaging spectroradiometer (MODIS) with 4-km spatial distribution within the defined domain D1 (90.3516, −42.9492, −46.7578, 44.9414) over the Pacific Ocean and surrounding shores as shown in Fig. 1. D1 is defined in Fig. 1a.

2.2 *TRMM [8]*

Precipitation rate (in mm/h) over the defined domain D1 in Fig. 1a from the *Tropical Rainfall Measuring Mission* (TRMM) space mission is used in this study. Precipitation rate with spatial resolution of 0.25 degrees was used in this work within the El Niño event (2014–2016).

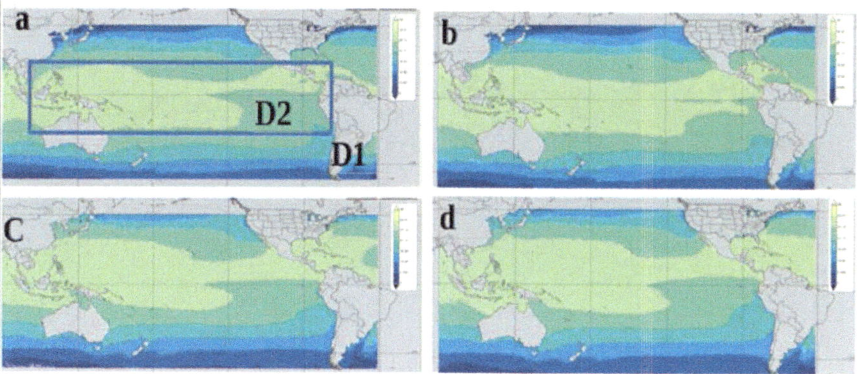

Fig. 1 Seasonal spatial SST maps over the Pacific Ocean for the El Niño event (2014–2016). (**a**) DJF, (**b**) MAM, (**c**) JAJ, (**d**) SON

2.3 MERRA [9]

Evaporation in (kg s/m^2) over the defined D1 from the global climate numerical model Modern-Era Retrospective Analysis for Research and Applications (MERRA) was selected for this work. The spatial resolution was 0.5 × 0.667 degrees.

3 Results and Discussion

In this section, SST seasonal spatial maps over the Pacific Ocean for an El Niño event are shown in Fig. 1. It is shown that the maximum temperature in the core of the Pacific Ocean reached 38 °C, showing most of the coasts in the Asian continent especially Indonesia, Philippines, Taiwan, Singapore, Malaysia, Thailand and Vietnam, coastal parts of north and east of Australia, coastal areas of southern part of North America, Mexico, Guatemala, Salvador, Costa Rica and Panama. Figure 2 shows the average SST interannual monthly changes for El Niño (2014–2016) within the defined domain D2 (95.4609, −21.4336, −75.3984, 28.4883) as shown in Fig. 1a. Most months show high averages in both 2015 and 2016. Only 4 months show a negative average trend (September, October, November, December) in year 2016 which was the end of the El Niño event (2014–2016). On the other hand, Fig. 3 shows the seasonal spatial SST for the La Niña event (2010–2012). The interannual monthly trends are shown in Fig. 4 over domain D7 which is defined in Fig. 4a. It is clearly shown that there is a decrease on the average SST on most of the months in year 2011 and then an increase again in year 2012 on most of the months.

El Niño and La Niña have measurable and noticeable pattern and impact on both water and energy cycles. With respect to precipitation as shown in Figs. 5 and 6, the

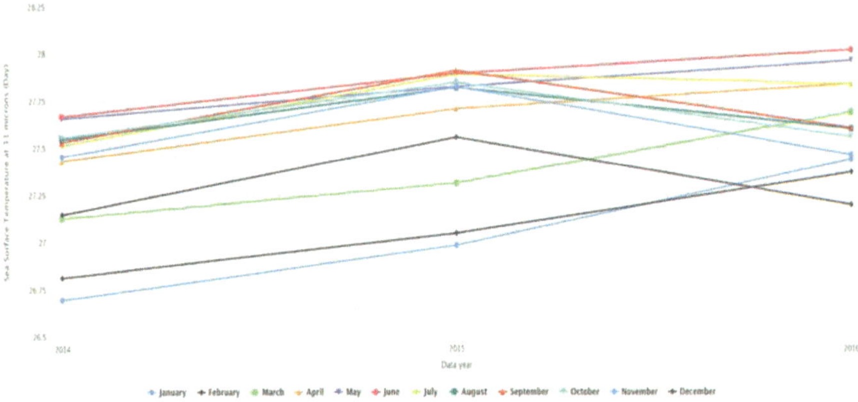

Fig. 2 Average SST from MODIS over D2 defined in Fig. 1a during the El Niño event

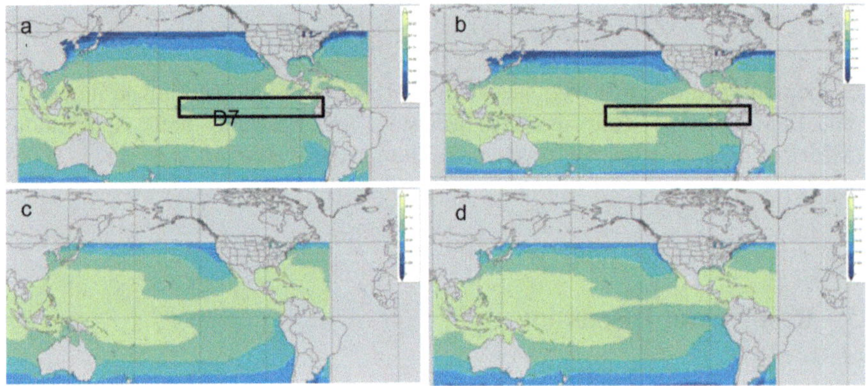

Fig. 3 Seasonal spatial SST maps over the Pacific Ocean for La Niña event (2010–2012). (a) DJF, (b) MAM, (c) JAJ, (d) SON

spatial seasonal maps for the precipitation rate from TRMM correlated with the spatial temperature pattern from MODIS. For MAM season with high temperature, the precipitation rate increases as shown in Figs. 1b, 5b, and 6b.

For the second main component in the water cycle which is evaporation, it is clear that the JAJ in the El Niño (2014–2016) season has the highest spatial distribution of evaporation which mainly focused around coasts of Australia. As shown in the three defined domains, D3, D4 and D5, in Fig. 7b, it reached maximum values in the JAJ season. From both Figs. 7 and 8, it is shown that evaporation during El Niño (2014–2016) is higher than during La Niña (2010–2012).

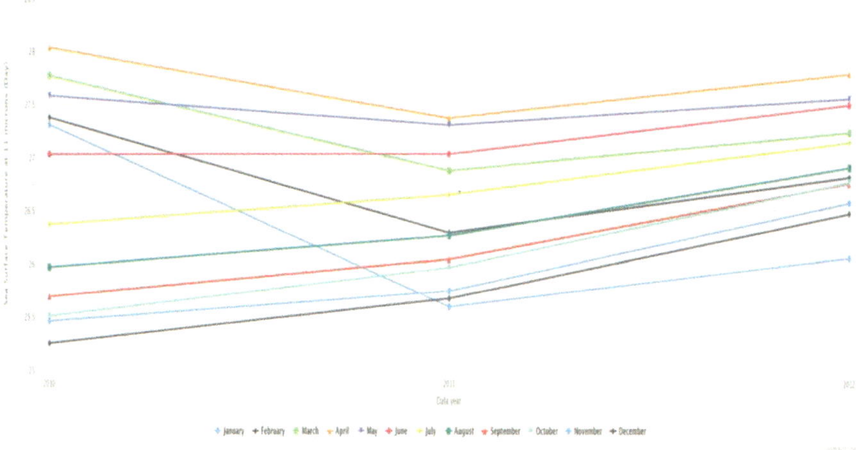

Fig. 4 Average SST from MODIS over D7 defined in Fig. 3a during the La Niña event

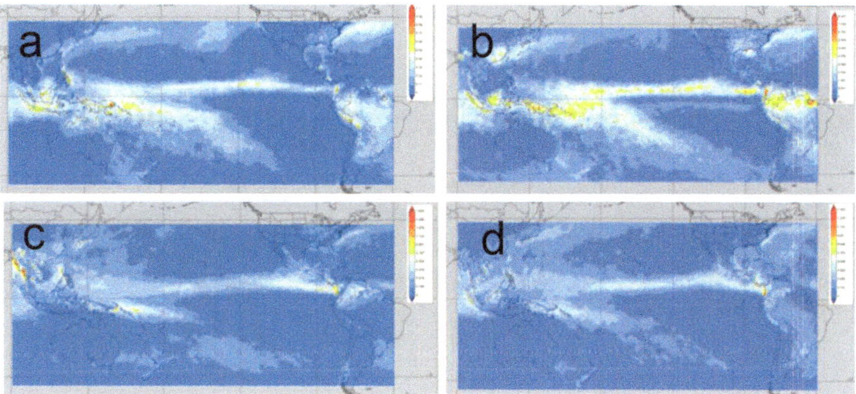

Fig. 5 Seasonal spatial precipitation rate (in mm/h) maps over the Pacific Ocean during El Niño. (**a**) DJF, (**b**) MAM, (**c**) JAJ, (**d**) SON

4   Conclusion

In this work, the spatial distribution of the water budget components, precipitation and evaporation, during the two different phases of ENSO, the warm El Niño phase (2014–2016) and the cold La Niña phase, was presented. There are three main meteorological variables from different data sources. Temperature, precipitation and evaporation from both satellite remote sensing and numerical models, MODIS, TRMM and MERRA, respectively, are the core of this study. It is shown that there is correlation between the spatial pattern of increasing temperature due to El Niño and both precipitation and evaporation.

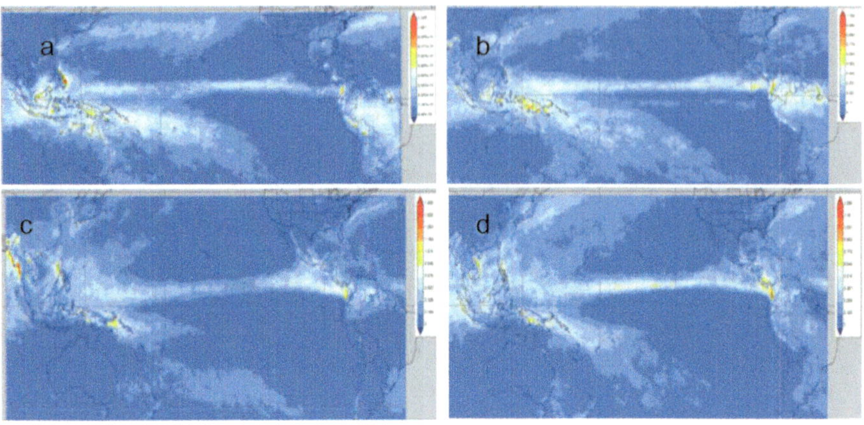

Fig. 6 Seasonal spatial precipitation rate (in mm/h) maps over the Pacific Ocean during La Niña. (**a**) DJF, (**b**) MAM, (**c**) JAJ, (**d**) SON

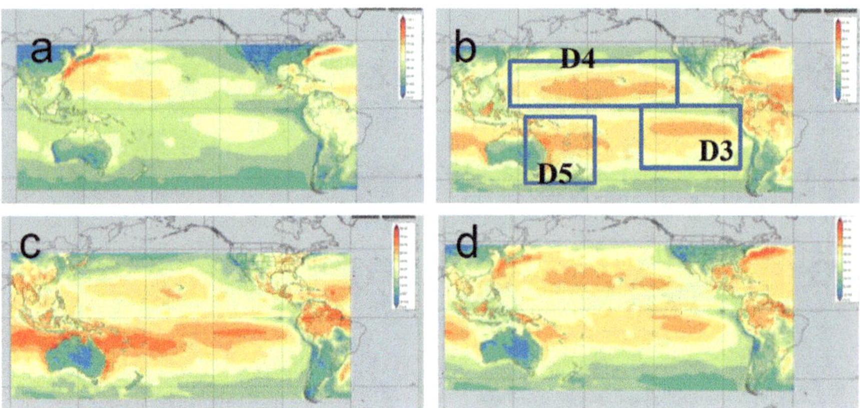

Fig. 7 Seasonal spatial evaporation (in kg s/m^2) maps over the Pacific Ocean during El Niño, (**a**) DJF, (**b**) MAM, (**c**) JAJ, (**d**) SON

During El Niño (2014–2016), temperature was highest in MAM and corresponding higher precipitation was observed at the same period, while JAJ has the most evaporation. On the other hand, during La Niña (2010–2012), the spatial pattern for the three meteorological variables is shown.

Evaluation of water and energy cycle components over different El Niño and La Niña events using both satellite and numerical climate models could lead to more understanding for such critical and vital event. Building correlations between different modes of ENSO using recent techniques of deep learning may also lead to a descriptive and quantitative model that improves our understanding of the ENSO phenomena.

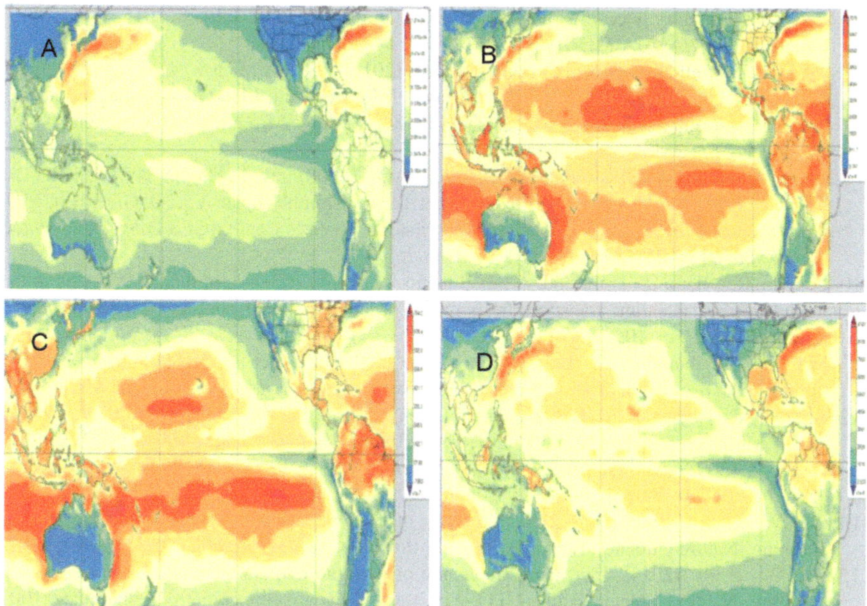

Fig. 8 Seasonal spatial evaporation (in kg s/m^2) maps over the Pacific Ocean during La Niña, (**a**) DJF, (**b**) MAM, (**c**) JAJ, (**d**) SON

References

1. Cai, W., Wang, G., Dewitte, B., Wu, L., Santoso, A., Takahashi, K., Yang, Y., Carréric, A., & McPhaden, M. J. (2018). Increased variability of eastern pacific el niño under greenhouse warming. *Nature, 564*, 201–206.
2. Timmermann, A., An, S.-I., Kug, J.-S., Jin, F.-F., Cai, W., Capotondi, A., Cobb, K., Lengaigne, M., McPhaden, M. J., Stuecker, M. F., Stein, K., Wittenberg, A. T., Yun, K.-S., Bayr, T., Chen, H.-C., Chikamoto, Y., Dewitte, B., Dommenget, D., Grothe, P., Guilyardi, E., Ham, Y.-G., Hayashi, M., Ineson, S., Kang, D., Kim, S., Kim, W., Lee, J.-Y., Li, T., Luo, J.-J., McGregor, S., Planton, Y., Power, S., Rashid, H., Ren, H.-L., Santoso, A., Takahashi, K., Todd, A., Wang, G., Wang, G., Xie, R., Yang, W.-H., Yeh, S.-W., Yoon, J., Zeller, E., & Zhang, X. (2018). El Niño– Southern Oscillation complexity. *Nature, 559*, 535–545. https://doi.org/10.1038/s41586-018- 0252-6pmid:30046070
3. Ropelewski, C. F., & Halpert, M. S. (1987). Global and regional scale precipitation patterns associated with the El Niño/Southern Oscillation. *Monthly Weather Review, 115*, 1606–1626.
4. Glynn, P. W., & DE Weerdt, W. H. (1991). Elimination of two reef-building hydrocorals following the 1982–83 El Niño warming event. *Science, 253*, 69–71.
5. Bove, M. C., O'Brien, J. J., Eisner, J. B., Landsea, C. W., & Niu, X. (1998). Effect of El Niño on US landfalling hurricanes, revisited. *Bulletin of the American Meteorological Society, 79*, 2477–2482.
6. Vincent, E. M., et al. (2011). Interannual variability of the South Pacific Convergence Zone and implications for tropical cyclone genesis. *Climate Dynamics, 36*, 1881–1896.
7. NASA Goddard Space Flight Center, Ocean Ecology Laboratory, Ocean Biology Processing Group. (2014). *MODIS-Aqua Ocean Color Data*; NASA Goddard Space Flight Center, Ocean

Ecology Laboratory, Ocean Biology Processing Group. Retrieved July 28, 2015, from https://doi.org/10.5067/AQUA/MODIS_OC.2014.0

8. Tropical Rainfall Measuring Mission [TRMM]. (2011). TRMM Microwave Imager Precipitation Profile L3 1 month 0.5 degree x 0.5 degree V7, Greenbelt, MD, Goddard Earth Sciences Data and Information Services Center (GES DISC). Retrieved from https://disc.gsfc.nasa.gov/datacollection/TRMM_3A12_7.html

9. Global Modeling and Assimilation Office [GMAO]. (2008). tavgM_2d_int_Nx: MERRA 2D IAU Diagnostic, Vertical Integrals and Budget Terms, Monthly Mean V5.2.0, Greenbelt, MD, USA, Goddard Earth Sciences Data and Information Services Center (GES DISC). https://doi.org/10.5067/JBJDIWKAEU03

10. Castillo, R., Nieto, R., Drumond, A., & Gimeno, L. (2014). The role of the ENSO cycle in the modulation of moisture transport from major oceanic moisture sources. *Water Resources Research, 50,* 1046–1058.

11. Mo, K. C., Schemm, J.-K. E., & Yoo, S.-H. (2009). Influence of ENSO and the Atlantic multidecadal oscillation on drought over the United States. *Journal of Climate, 22,* 5962–5982.

12. Barlow, M., Nigam, S., & Berbery, E. H. (2001). ENSO, Pacific decadal variability, and US summertime precipitation, drought, and stream flow. *Journal of Climate, 14,* 2105–2128.

13. Davis, M. (2001). *Late Victorian holocausts: El Niño famines and the making of the third world* (p. 271). London: Verso.

14. Cai, W., et al . (2015). Increased frequency of extreme La Niña events under greenhouse warming. Nature Climate Change 5:132–137. (https://www.nature.com/articles/nclimate2492)

15. Cai, W., et al . (2014). Increasing frequency of extreme El Niño events due to greenhouse warming. Nature Climate Change 4:111–116. (https://www.nature.com/articles/nclimate2100)

16. Climate Prediction Center. (2015, 4 November). Monitoring & Data: ENSO Impacts on the U. S.—Previous Events. Cpc.noaa.gov. Retrieved January 3, 2017.

17. ENSO Diagnostic Discussion. (n.d.). NOAA's Climate Prediction Center. Retrieved January 10, 2011.

18. NOAA's Climate Prediction Center. (2011, 8 September). *La Niña is back*. Retrieved December 10, 2011.

19. Nicholls, N. (2008). Recent trends in the seasonal and temporal behaviour of the El Niño Southern Oscillation. *Geophysical Research Letters, 35*(19), L19703. Bibcode:2008GeoRL..3519703N. https://doi.org/10.1029/2008GL034499

20. Brown, P. T., Li, W., & Xie, S.-P. (2015). Regions of significant influence on unforced global mean surface air temperature variability in climate models. *Journal of Geophysical Research: Atmospheres, 120*(2), 2014JD022576. https://doi.org/10.1002/2014JD022576

21. Santoso, A., McPhaden, M. J., & Cai, W. (2017). The defining characteristics of ENSO extremes and the strong 2015/16 El Niño. *Reviews of Geophysics, 55,* 1079–1129.

22. Oki, T. (1999). The global water cycle. In K. A. Browning & R. J. Gurney (Eds.), *Global energy and water cycles* (pp. 10–29). Cambridge: Cambridge University Press.

23. Dai, A., & Trenberth, K. E. (2002). Estimates of freshwater discharge from continents: latitudinal and seasonal variations. *Journal of Hydrometeorology, 3,* 660–687. https://doi.org/10.1175/1525-7541(2002)003<0660:EOFDFC>2.0.CO;2

Frequency Scaling in a Sweeping-Jet Fluidic Oscillator Working at Low Reynolds Numbers: A Multiple-Relaxation Time LBM Model

Mohammed A. Boraey

Abstract The operation of a sweeping-jet fluidic oscillator (FO) is numerically investigated at low Reynolds number range using the multiple-relaxation time lattice Boltzmann method (MRT-LBM). The scaling between the normalized oscillation frequency (i.e., Strouhal number) and the operating Reynolds number is established. The lower limit of the investigated range of the Reynolds number corresponds to the onset of oscillation (i.e., below which steady-state operation exists). The oscillation frequency and amplitude were found to scale linearly with the Reynolds number. The results are useful in applications where the maximum flow rate is limited due to other design constraints and hence the cyclic behavior of FO is required at a lower Reynolds number range.

Keywords Fluidic oscillator · Low Reynolds number · Multiple-relaxation time lattice Boltzmann method

1   Introduction

Fluidic oscillators (FO) are devices which are capable of generating oscillating fluid motion without the need for moving parts. Due to the absence of any moving parts, FO are maintenance-free and very cheap to manufacture and operate. The main idea in FO is to guarantee the existence of a feedback mechanism between two geometrically identical parts of the FO that will generate a dynamic behavior if a minimum value of the flow rate is maintained [1, 2].

According to the feedback mechanism, FO can be classified into two categories: the wall attachment (commonly called sweeping-jet) FO and jet interaction FO [1]. The most popular type is the sweeping-jet FO where two feedback loops ensure

M. A. Boraey (✉)
Mechanical Power Engineering Department, Zagazig University, Zagazig, Egypt

Smart Engineering Systems Research Center, Nile University, Giza, Egypt
e-mail: boraey@ualberta.ca

© Springer Nature Switzerland AG 2020
M. H. Farouk, M. A. Hassanein (eds.), *Recent Advances in Engineering Mathematics and Physics*, https://doi.org/10.1007/978-3-030-39847-7_26

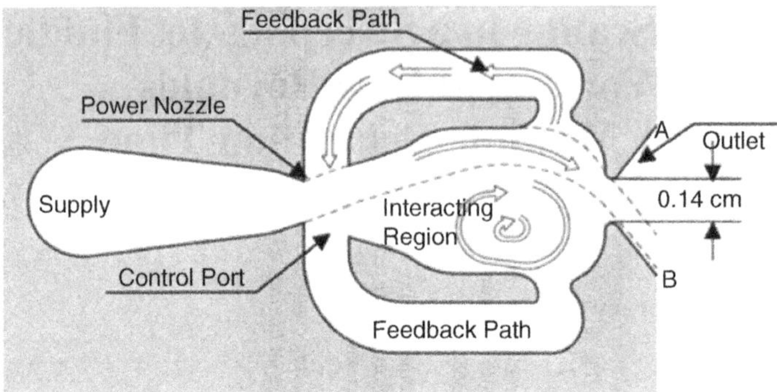

Fig. 1 A sample sweeping-jet fluidic oscillator [4]

the cyclic operation through the Coanda effect [3]. Figure 1 [4] shows a schematic diagram of a typical sweeping-jet FO.

Due to its unique characteristics, FO found many applications like medical applications [5], combustion control [6], bubble generation [7, 8], flow control [2, 9, 10], micromixers [11], and film cooling applications [12]. For each of these applications, the minimum flow rate to keep the dynamic instability has to be maintained. Most of the time, the operating flow rate is much higher than the required minimum and cyclic operation is always guaranteed.

However, some applications like micromixers put a constraint on the size and flow rate and hence approach the lower limit of the cyclic operation mode. For each FO design, there is a lower limit on the flow rate to ensure operation in the oscillating mode (similar to the critical Reynolds number for the onset of vortex shedding for the flow around cylinders), and the characteristics of operation close to this limit are lacking in literature.

The goal of the present study is to numerically investigate the operation of FO at low Reynolds number range and find the relation between the oscillation frequency and Reynolds number with the lowest Reynolds number corresponding to the onset of oscillations.

2   Problem Description

The considered FO is a 2D sweeping-jet FO with the geometry and relative dimensions shown in Fig. 2 [3]. The flow is from left to right. The Reynolds number is defined as follows:

$$Re = \frac{uL}{\nu} \tag{1}$$

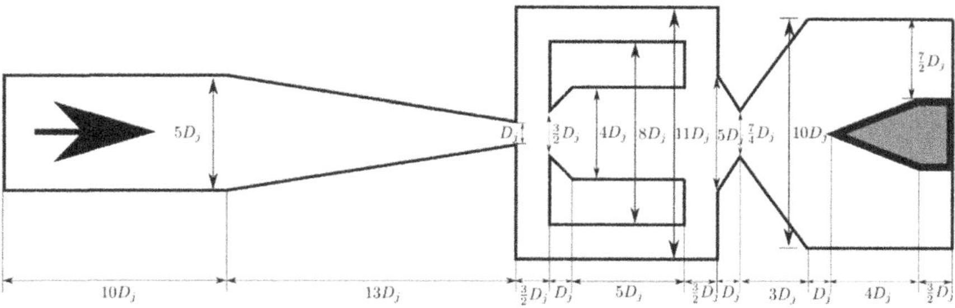

Fig. 2 The geometry and relative dimensions of the used fluidic oscillator [3]

where u and $L = 5D_j$ are the inlet velocity and FO inlet width, ν is the used fluid kinematic viscosity, and D_j is the power jet diameter.

The fluid is assumed to be viscous, Newtonian, and a constant-property fluid while the flow is unsteady, incompressible, and laminar. The tested range of *Re* number ranges from 180 to 300 and was changed by altering the inlet flow velocity u.

Since the tested range of *Re* number is very low, a turbulence model is not needed. The tested range corresponds to the lowest *Re* number which results in a hydrodynamic instability and hence the cyclic behavior of the FO. Below this range, the operation of the FO is steady with symmetric and time-independent flow field. As will be shown in the results section, the linear relation between the *Re* number and the oscillation frequency f confirms the validity of the reported results as compared to other computational [13] and experimental [3] investigations.

For each case, the oscillation frequency is calculated in addition to the velocity field. The flow field is predicted using the multiple-relaxation time lattice Boltzmann method.

Velocity inlet boundary condition is used for the left inlet boundary and outflow boundary condition is used for the right boundary [14, 15]. No-slip boundary condition was used for all other boundaries.

3 Numerical Model

In order to predict the hydrodynamic field inside the studied sweeping-jet FO, the multiple-relaxation time lattice Boltzmann method (MRT-LBM) is used [15–17].

The LBM is a robust numerical simulation method that is capable of dealing with computational fluid dynamics problems with different degrees of complexity in terms of the geometry and variable material properties [18, 19]. LBM offers many afdvantages like straightforward implementation, easy incorporation of additional physics, lower computational cost, and high parallelizability potential [16, 20].

The method solves a discretized version of the Boltzmann equation [14, 16]:

$$f_i(x + c_i \Delta t, t + \Delta t) = f_i(x, t) + \Omega(x, t) \tag{2}$$

where f is the particle probability density function, c_i is the velocity in the ith direction, and $\Omega(f)$ is a collision operator. The present study uses the *D2Q9* lattice configuration which has the following weights w_i and speeds:

$$w_i = \begin{cases} \dfrac{4}{9} & i = 0 \\[2mm] \dfrac{1}{9} & i = 1, 2, 3, 4 \\[2mm] \dfrac{1}{36} & i = 5, 6, 7, 8 \end{cases} \tag{3}$$

$$c_i = \begin{cases} (0, 0) & i = 0 \\ (\pm c, 0), (0, \pm c) & i = 1, 2, 3, 4 \\ (\pm c, \pm c) & i = 5, 6, 7, 8 \end{cases} \tag{4}$$

$c = \Delta x / \Delta t$ is the lattice speed in LBM units.

For the MRT-LBM, the collision operator Ω is expressed as [15, 21]:

$$\Omega(f) = -M^{-1} S[m - m^{\mathrm{eq}}] \tag{5}$$

M is a transformation matrix from the velocity space to the moment space and m^{eq} is the transformed equilibrium distribution function:

$$m = M \cdot f \tag{6}$$

$$m^{\mathrm{eq}} = M \cdot f^{\mathrm{eq}} \tag{7}$$

For the *D2Q9* lattice configuration, M is given by:

$$M = \begin{bmatrix} 1 & 1 & 1 & 1 & 1 & 1 & 1 & 1 & 1 \\ -4 & -1 & -1 & -1 & -1 & 2 & 2 & 2 & 2 \\ 4 & -2- & -2 & -2 & -2 & 1 & 1 & 1 & 1 \\ 0 & 1 & 0 & -1 & 0 & 1 & -1 & -1 & 1 \\ 0 & -2 & 0 & 2 & 0 & 1 & -1 & -1 & 1 \\ 0 & 0 & 1 & 0 & -1 & 1 & 1 & -1 & 1 \\ 0 & 0 & -2 & 0 & 2 & 1 & 1 & -1 & -1 \\ 0 & 1 & -1 & 1 & -1 & 0 & 0 & 0 & 0 \\ 0 & 0 & 0 & 0 & 0 & 1 & -1 & 1 & -1 \end{bmatrix} \tag{8}$$

The relaxation matrix S is given by:

$$S = \text{diag}(0, s_1, s_2, 0, s_4, 0, s_6, s_\nu, s_\nu) \tag{9}$$

The kinematic viscosity ν is related to the relaxation parameter s_ν by the following relation [17, 22]:

$$\nu = c_s^2 \left[\frac{1}{s_\nu} - \frac{1}{2} \right] \tag{10}$$

The equilibrium particle probability density function f^{eq} is given by:

$$f_i^{\text{eq}}(x, t) = w_i \rho \left[1 + \frac{u \cdot c_i}{c_s^2} + \frac{(u \cdot c_i)^2}{2c_s^4} - \frac{u \cdot u}{2c_s^2} \right] \tag{11}$$

$c_s = \sqrt{\frac{1}{3}} \frac{\Delta x}{\Delta t}$ is the lattice sound speed. The density and velocity are calculated from f as follows:

$$\rho = \sum_{i=0}^{8} f_i \tag{12}$$

$$u_\alpha = \frac{1}{\rho} \sum_{i=0}^{8} c_{i\alpha} f_i \tag{13}$$

4   Results and Discussion

In this section, the results of the numerical simulation for the operation of a sweeping-jet FO at low Reynolds number range are presented. The investigated range for Re number is 180–300.

Figure 3 shows the normalized velocity contours for the case of $Re = 300$ at different phase angles θ of the oscillation cycle. The graph shows how the central jet sweeps the two sides of the FO in a complete cycle. It also shows that the sweeping jet has the highest velocity magnitude in the whole FO and that the velocity at the two side feedback loops is very small.

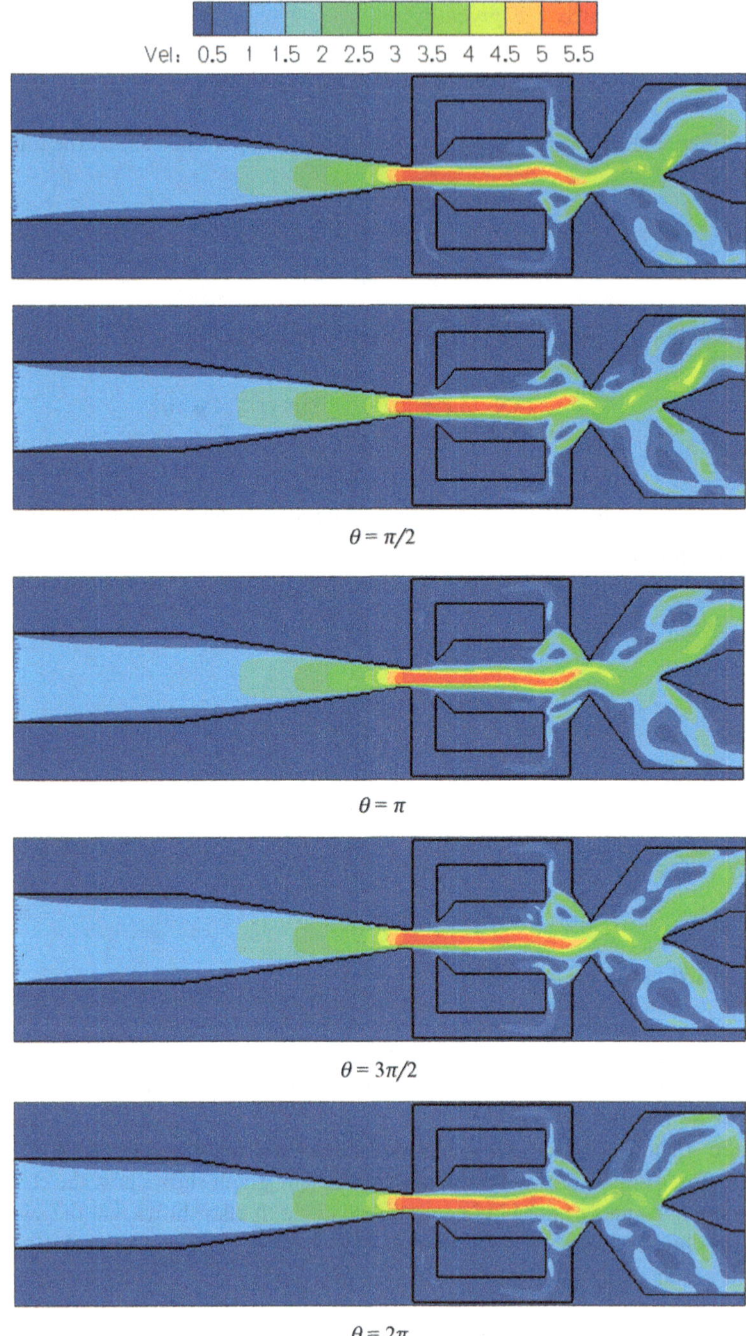

Fig. 3 Normalized velocity contours for $Re = 300$ at different phase angels θ

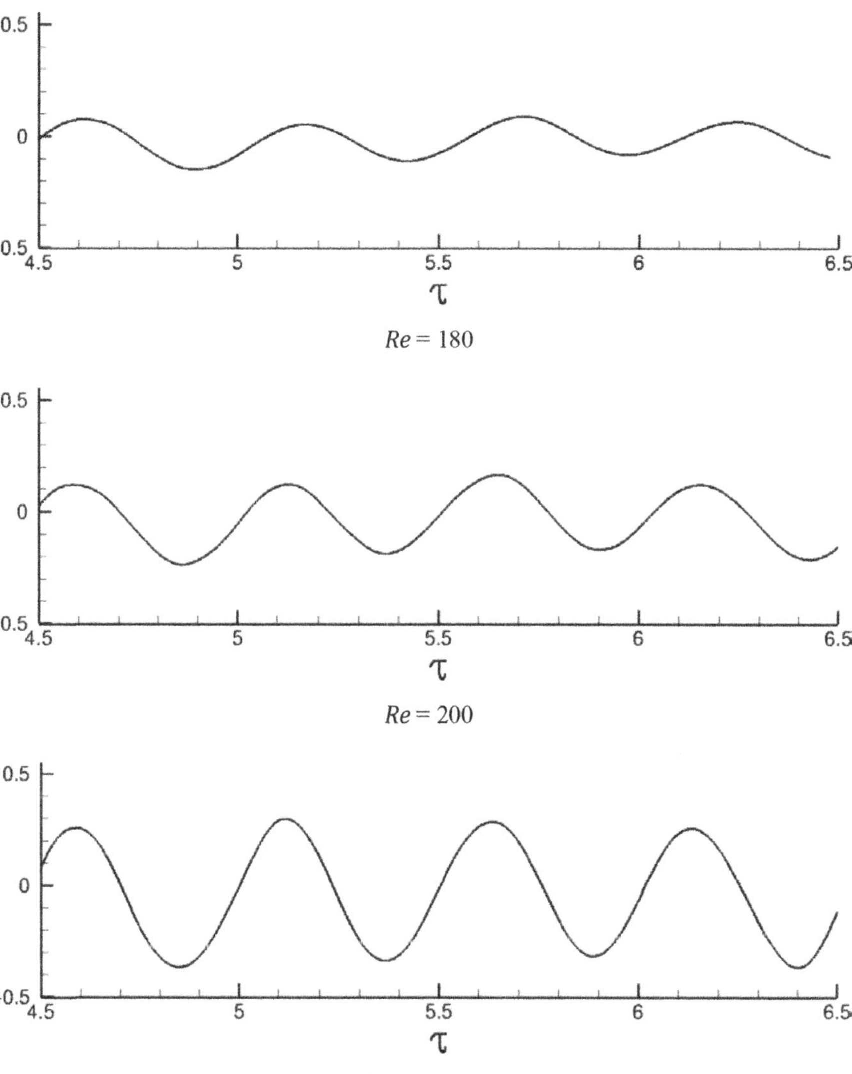

Fig. 4 Variation of the normalized vertical velocity for different *Re* numbers

Figure 4 shows the change of the normalized vertical velocity component at the centerline of the FO at $6D_j$ downstream of the power jet with the normalized time τ defined as:

$$\tau = \frac{tu}{L} \tag{14}$$

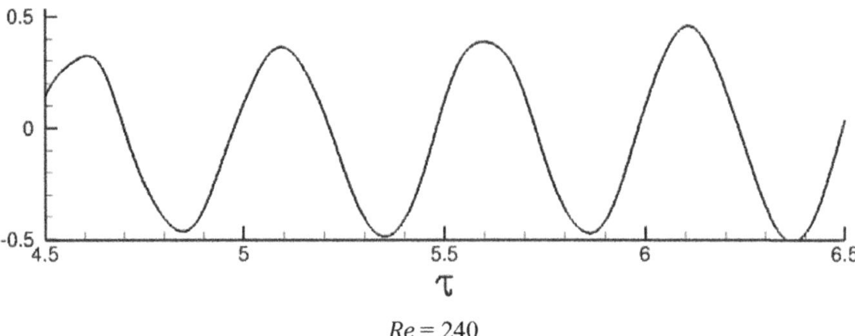

$Re = 240$

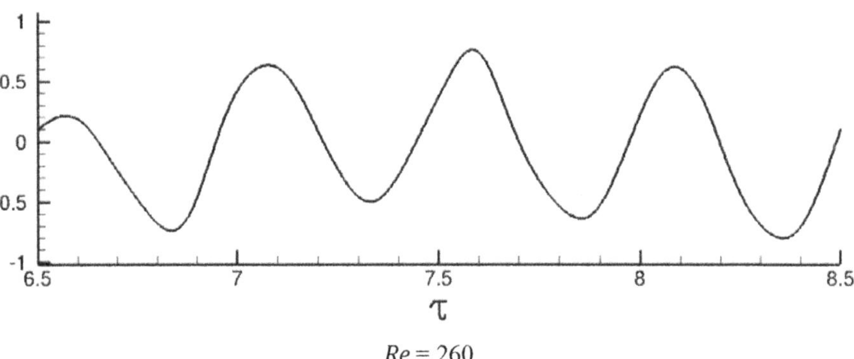

$Re = 260$

Fig. 4 (continued)

Fig. 5 Variation of the oscillation amplitude with the *Re* number

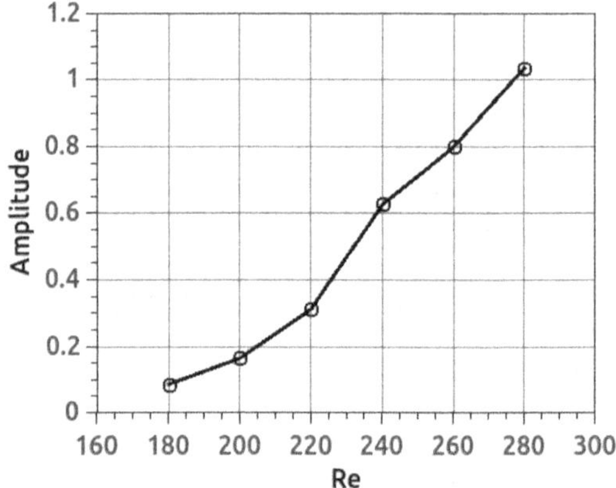

The figure shows an increase in the frequency and amplitude of the oscillations with the increase in the *Re* number.

These results can be summarized in Fig. 5 which shows the variation of the oscillation amplitude (for the normalized vertical velocity component) for different

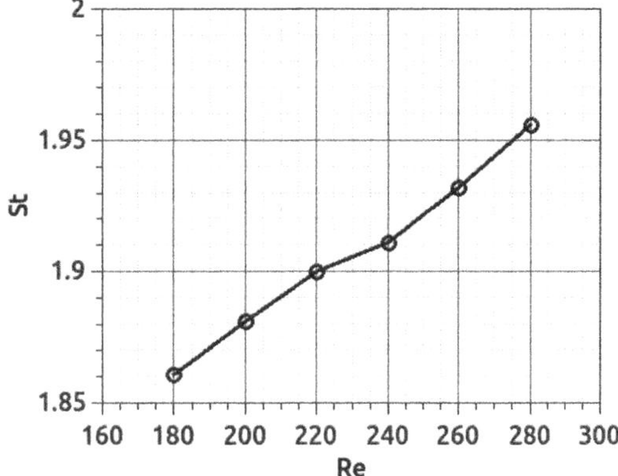

Fig. 6 Variation of the *St* number with the *Re* number

Re numbers. There is an almost-linear dependence between the *Re* number and the oscillation amplitude in this range.

The variation of the Strouhal, *St*, number with the *Re* number is shown in Fig. 6. For the current study, the *St* number is defined as follows:

$$St = \frac{fL}{u} \tag{15}$$

where f is the oscillation frequency. It is also clear that the relation between *Re* and *St* is very close to being linear. This trend agrees well with the published and numerical results [3, 13].

The increase in the *Re* number results in an increase in both the amplitude and frequency and hence a better mixing potential for applications like micromixers.

5   Conclusions

The operation and performance of a sweeping-jet FO are numerically investigated at the low *Re* range. The MRT-LBM is used due to its superior ability in dealing with problems with geometric complexity. The hydrodynamic flow field is predicted for a range of *Re* number from 180 to 300.

It is shown that both the oscillation amplitude and frequency increase linearly with the *Re* number in the low *Re* range with the lower limit of *Re* corresponding to the onset of oscillation.

The importance of the presented results comes from the need for some FO applications to work with very small flow rates while keeping the advantages of the dynamic characteristics of FO.

References

1. Gregory, J., & Tomac, M. N. (2013). A review of fluidic oscillator development. In: *43rd AIAA fluid dynamics conference* (p. 2474).
2. Raghu, S. (2013). Fluidic oscillators for flow control. *Experiments in Fluids, 54*(2), 1455.
3. Bobusch, B. C., Woszidlo, R., Bergada, J., Nayeri, C. N., & Paschereit, C. O. (2013). Experimental study of the internal flow structures inside a fluidic oscillator. *Experiments in Fluids, 54*(6), 1559.
4. Raman, G., & Raghu, S. (2004). Cavity resonance suppression using miniature fluidic oscillators. *AIAA Journal, 42*(12), 2608–2612.
5. Gebhard, U., Hein, H., Just, E., & Ruther, P. (1997). Combination of a fluidic micro-oscillator and micro-actuator in liga-technique for medical application. In *Proceedings of international solid state sensors and actuators conference (Transducers' 97)* (Vol. 2, pp. 761–764). IEEE.
6. Guyot, D., Bobusch, B., Paschereit, C. O., & Raghu, S. (2008). Active combustion control using a fluidic oscillator for asymmetric fuel flow modulation. In *44th AIAA/ASME/SAE/ASEE Joint Propulsion Conference & Exhibit* (p. 4956).
7. Zimmerman, W. B., Tesăr, V., & Bandulasena, H. H. (2011). Towards energy efficient nanobubble generation with fluidic oscillation. *Current Opinion in Colloid & Interface Science, 16*(4), 350–356.
8. Zimmerman, W. B., Hewakandamby, B. N., Tesǎr, V., Bandulasena, H. H., & Omotowa, O. A. (2009). On the design and simulation of an airlift loop bioreactor with microbubble generation by fluidic oscillation. *Food and Bioproducts Processing, 87*(3), 215–227.
9. Cerretelli, C., & Kirtley, K. (2009). Boundary layer separation control with fluidic oscillators. *Journal of Turbomachinery, 131*(4), 041001.
10. Tesǎr, V., Zhong, S., & Rasheed, F. (2012). New fluidic-oscillator concept for flow-separation control. *AIAA Journal, 51*(2), 397–405.
11. Tesǎr, V. (2009). Oscillator micromixer. *Chemical Engineering Journal, 155*(3), 789–799.
12. Hossain, M. A., Prenter, R., Lundgreen, R. K., Ameri, A., Gregory, J. W., & Bons, J. P. (2017). Experimental and numerical investigation of sweeping jet film cooling. *Journal of Turbomachinery, 140*(3), 031009–031009-13.
13. Gebhard, U., Hein, H., & Schmidt, U. (1996). Numerical investigation of fluidic microoscillators. *Journal of Micromechanics and Microengineering, 6*(1), 115.
14. Mohamad, A. A. (2011). *Lattice Boltzmann method. Fundamentals and engineering applications with computer codes*. New York: Springer.
15. Krüger, T., Kusumaatmaja, H., Kuzmin, A., Shardt, O., Silva, G., & Viggen, E. (2017). *The lattice Boltzmann method: Principles and practice*. Cham: Springer.
16. Boraey, M. A. (2017). Simulation of the lid-driven cavity flow at Reynolds numbers between 100 and 1000 using the multi-relaxation-time lattice boltzmann method. *Mugla Journal of Science and Technology, 3*(2), 110–115.
17. Succi, S. (2001). *The lattice Boltzmann equation: for fluid dynamics and beyond*. Oxford: Clarendon Press.
18. Boraey, M. A., & Epstein, M. (2010). Lattice Boltzmann modeling of viscous elementary flows. *Advances in Applied Mathematics and Mechanics, 2*(4), 467–482.
19. Guo, Z., & Shu, C. (2013). *Lattice Boltzmann method and its applications in engineering* (Vol. 3). Singapore: World Scientific.
20. Boraey, M. A. (2019). Thermal optimization of square pin-fins in crossflow using the lattice boltzmann method with quadratic thermal equilibrium. *Physica A: Statistical Mechanics and its Applications, 532*, 121880.
21. Boraey, M. A. (2018). A hydro-kinematic approach for the design of compact corrugated plate interceptors for the de-oiling of produced water. *Chemical Engineering and Processing—Process Intensification, 130*, 127–133.
22. Boraey, M. A. (2019). An asymptotically adaptive successive equilibrium relaxation approach for the accelerated convergence of the lattice Boltzmann method. *Applied Mathematics and Computation, 353*, 29–41.

Water-Aluminum Oxide Nano-Fluid Nusselt Number Enhancement and Neural Network Accelerated Prediction

Omar Sallam, Adel M. El-Refaey, and Amr Guaily

Abstract The fluid flow and heat transfer of nano-fluids, water based with suspended aluminum oxide nanoparticles, are numerically simulated using the Galerkin/least-squares finite element method. The predicted Nusselt number values show an enhancement with increase of both the Reynolds number and the volume fraction concentration of the nanoparticles that improve the thermal properties of the base fluid. The increase in the Nusselt number is as high as 27.75% at 200 Re and 0.15 nanoparticle concentrations compared to pure water. Our results show a linear relation between the Nusselt number and nanoparticle concentration for a given Reynolds number, and the thermal boundary layer gradient shows more inclination with nanoparticle concentration loading, which plays a vital role in convective heat transfer improvement. A two-layer neural network with Bayesian regularization technique is trained by 41 training samples with 9 inputs based on the numerical results to predict the Nusselt number of the nano-fluid. The output shows satisfactory predictions with 0.1088 RMSE compared to 1.3212 without regularization.

Keywords Nusselt number FEM · Nano-fluid · Neural network · Convective heat transfer

O. Sallam (✉)
School of Engineering and Applied Sciences, Nile University, Giza, Egypt
e-mail: okhaled@nu.edu.eg

A. M. El-Refaey
Basic and Applied Science Department, College of Engineering and Technology, Arab Academy for Science, Technology and Maritime Transport, Cairo, Egypt
e-mail: adel_elrefaey@aast.edu

A. Guaily
Faculty of Engineering, Cairo University, Giza, Egypt

Smart Engineering Systems Research Center (SESC), Nile University, Giza, Egypt
e-mail: aguaily@nu.edu.eg

M. H. Farouk, M. A. Hassanein (eds.), *Recent Advances in Engineering Mathematics and Physics*, https://doi.org/10.1007/978-3-030-39847-7_27

1   Introduction

Convective heat transfer enhancement study is one of the most common and attractive research tracks in past decades and nowadays, as it plays a vital role in excelling the operation of thermo-fluid equipment, electric/electronic devices' cooling, or buildings' air-conditioning systems. Nano-fluids' effects, characteristics, and limitations are considered as a spirited research area in the thermo-fluid field, which can enhance the convection heat transfer by improving thermal properties of the base working fluids by adding suspended solid particles in nano size with specific thermophysical properties.

Ting [1] numerically studied the laminar forced convection of nano-fluid water based with suspended copper oxide and aluminum oxide nanoparticles in a triangle cross-section duct subjected to constant heat flux. Ting realized an increase in the Nusselt number with both Peclet number and particle volume concentration. The numerical results showed an increase in the Nusselt number to 35% when the aluminum oxide particle volume concentration is 2% compared to the pure water case. Li [2] experimentally studied the convective heat transfer in a tube for both regimes, laminar and turbulent, in a wide range of Reynolds numbers for copper water nano-fluid. Li found an increase in the Nusselt number to 60% at nanoparticle concentration of 2%; he also realized that at small values of concentration, the increase in pressure drop is very small to be considered. Vijjha et al. [3] numerically studied laminar flow in a flat tube automobile radiator for different base fluids and nanoparticles. Their results show enhancement of the Nusselt number by 94% when loading the base fluid with 0.1 volume fraction concentration of aluminum oxide, while 89% increase in the Nusselt number when loading the base fluid with 0.06 volume fraction concentration of copper oxide. At the same Nusselt number of pure base fluid, aluminum oxide nano-fluid reduced the volume flow rate of working fluid; hence, the pumping power savings reached 82%, while copper oxide case reduction is found to be 77%.

Leong et al. [4] studied theoretically the effect of ethylene glycol base fluid with copper nanoparticles on automotive cooling system; the results of Nusselt number predictions show an enhancement by 3.8% at nanoparticle volume concentration of 0.02 at the same radiator air frontal area. At the original cooling rate, the air frontal area can be reduced by 18.7% with nanoparticle loading, while the pumping power increased by 12.12%.

Despite CFD being sometimes easy and a cheap tool to predict the thermal and hydrodynamic fields of thermo-fluid applications as it works as a virtual lab to calculate and visualize the fluid mechanics behavior and characteristics, generally CFD is still computationally expensive and time consuming to predict the results of solving the complex nonlinear partial differential equations that describe the fluid and heat flow phenomena. This urges many researchers to reduce the computational cost of thermo-fluid result predictions by using artificial intelligence techniques to predict similar results that CFD predicts.

Beigzadeh et al. [5] trained a neural network and genetic algorithm for predicting the Nusselt number and friction factor of forced convection flow over plate fins; the training dataset is extracted from CFD results validated with experimental data. The inputs to algorithms are the Reynolds number and spanwise spacing ratio of the fins. Despite genetic algorithm being much simpler than neural network in developing a direct algebraic correlation for both outputs, neural network shows higher accuracy in predicting the results with smaller error percentage.

Zendehboudi and Saidur [6] designed a multilayer perceptron-artificial neural network to predict the thermal conductivity of 26 different nano-fluids, with the input data to the network based on 993 experiments. Neural network with relative root mean squared error 1.534% assures that other literatures' empirical correlations of thermal conductivity predictions overestimate or underestimate the right values.

2 Mathematical Modeling

In this section, the governing equations are presented in the nondimensional form.

Finite element approximation of governing equations is derived based on the Galerkin/least-squares method (GLS) and then a strong to weak form is presented. Validation of a finite element code is conducted with previous studies for the continuum model. Nano-fluid thermophysical effective property formulas are presented with respect to previous literatures. Also neural network design and implementation are discussed.

2.1 Governing Equations

The two-dimensional planar governing equations in its nondimensional form are presented below to model the incompressible fluid flow with heat transfer between the walls and fluid domain.

Continuity equation

$$\frac{\partial u}{\partial x} + \frac{\partial v}{\partial y} = 0 \tag{1}$$

Momentum equation in x direction

$$\left(\frac{\partial u}{\partial t} + u\frac{\partial u}{\partial x} + v\frac{\partial u}{\partial y}\right) + \frac{\partial P}{\partial x} - \frac{1}{Re}\left(\frac{\partial^2 u}{\partial x^2} + \frac{\partial^2 u}{\partial y^2}\right) = 0 \tag{2}$$

Momentum equation y direction

$$\left(\frac{\partial v}{\partial t} + u\frac{\partial v}{\partial x} + v\frac{\partial v}{\partial y}\right) + \frac{\partial P}{\partial y} - \frac{1}{Re}\left(\frac{\partial^2 v}{\partial x^2} + \frac{\partial^2 v}{\partial y^2}\right) = 0 \qquad (3)$$

Energy equation

$$\left(\frac{\partial T}{\partial t} + u\frac{\partial T}{\partial x} + v\frac{\partial T}{\partial y}\right) - \frac{1}{Re\ Pr}\left(\frac{\partial^2 T}{\partial x^2} + \frac{\partial^2 T}{\partial y^2}\right) = 0 \qquad (4)$$

The modified continuity equation

The pressure stabilization technique introduced by El Hanafy et al. [7], which is a simplified version of the original technique devised by Peeters et al. [8], is used to restore the link between the continuity equation and the momentum equation by adding a second-order derivative term of pressure field, so pressure can be computed from the modified continuity Eq. (5).

$$\frac{\partial u}{\partial x} + \frac{\partial v}{\partial y} = \varepsilon\left(\frac{\partial^2 p}{\partial x^2} + \frac{\partial^2 p}{\partial y^2}\right) \qquad (5)$$

where u, v, T, and P are the nondimensional velocity in the x direction, velocity in the y direction, temperature, and pressure.

ε is the pressure dissipation parameter with a small value, order of the time step, used to control the amount of added dissipation.

Dimensionless numbers used are shown as below.

Reynolds number	$Re = \frac{\rho_{nf} U_\infty D}{\mu_{nf}}$
Prandtl number	$Pr = \frac{cp_{nf}\,\mu_{nf}}{k_{nf}}$

where ρ_{nf}, U_∞, D, μ_{nf}, cp_{nf}, and k_{nf} are the density, inlet velocity, cylinder diameter, dynamic viscosity, specific heat, and thermal conductivity, respectively.

2.2 Finite Element Approximation

Finite element numerical technique weighted residual method [10] is used to solve the nonlinear partial differential equations in its nondimensional form. The method aims to minimize the error or the residuals of the equation over the whole domain as shown in Eq. (6).

$$\iint w_i(x, y)\, R(x, y) \cdot \mathrm{d}A = 0 \tag{6}$$

where w_i and R are the weights of weighted residual method and residual equation to be solved respectively.

Galerkin Least-Squares Method

As Galerkin method adopts symmetric weight functions through the element as its weights are the trial functions $W_i = N_i$, the method is not stable for partial differential equations that contain advection terms representing wave propagation in the domain, e.g., the momentum and energy equations. Galerkin/least-squares method (GLS) introduced by Hughes et al. [9] is used to modify the weights of advection terms in momentum and energy equations to give biasing to the weight in the direction of wave flow as shown in Eq. (7).

$$W_i = N_i + \tau\, L(N_i) \tag{7}$$

where τ is the relaxation factor defined by:

$$\tau = \frac{1}{\sqrt{\left(\frac{2}{\Delta t}\right)^2 + \left(\frac{2C}{\Delta x}\right)^2 + \left(\frac{4D}{\Delta x^2}\right)^{2.}}} \tag{8}$$

C and D are the advection and diffusion term coefficients as shown in Table 1.

Weak Form of Governing Equations

Continuity equation

$$\iint N_i \left(\frac{\partial u}{\partial x} + \frac{\partial v}{\partial y}\right) \mathrm{d}A + \iint \varepsilon N_i \left(\left(\frac{\partial N_i}{\partial x} + \frac{\partial N_i}{\partial y}\right)\left(\frac{\partial p}{\partial x} + \frac{\partial p}{\partial y}\right)\right) \mathrm{d}A$$

$$- \oint \varepsilon N_i \frac{\partial p}{\partial n}\, \mathrm{d}S = 0 \tag{9}$$

Table 1 Advection/diffusion coefficient

Equation	Convection coefficient, C	Diffusion coefficient, D
Momentum	$\sqrt{u^2 + v^2}$	$\frac{1}{Re}$
Energy	$\sqrt{u^2 + v^2}$	$\frac{1}{Re\, Pr} = \frac{1}{Pe}$

x momentum equation

$$\iint N_i \frac{\partial u}{\partial t}\, \mathrm{d}A + \iint \left(u_g \frac{\partial u}{\partial x} + v_g \frac{\partial u}{\partial y} \right) W_i \mathrm{d}A + \iint N_i \frac{\partial p}{\partial x}\, \mathrm{d}A$$

$$+ \left(\frac{1}{Re} \iint \cdot \left(\frac{\partial N_i}{\partial x} \frac{\partial u}{\partial x} + \frac{\partial N_i}{\partial y} \frac{\partial u}{\partial y} \right) \right) \mathrm{d}A - \oint N_i \frac{\partial u}{\partial n}\, \mathrm{d}S = 0 \qquad (10)$$

y momentum equation

$$\iint N_i \frac{\partial u}{\partial t}\, \mathrm{d}A + \iint \left(u_g \frac{\partial v}{\partial x} + v_g \frac{\partial v}{\partial y} \right) W_i \mathrm{d}A + \iint N_i \frac{\partial p}{\partial y}\, \mathrm{d}A$$

$$+ \left(\frac{1}{Re} \iint \cdot \left(\frac{\partial N_i}{\partial x} \frac{\partial v}{\partial x} + \frac{\partial N_i}{\partial y} \frac{v}{\partial y} \right) \right) \mathrm{d}A - \oint N_i \frac{\partial u}{\partial n}\, \mathrm{d}S = 0 \qquad (11)$$

Energy equation

$$\iint N_i \frac{\partial T}{\partial t}\, \mathrm{d}A + \iint \left(u_g \frac{\partial T}{\partial x} + v_g \frac{\partial T}{\partial y} \right) W_i \mathrm{d}A$$

$$+ \left(\frac{1}{Re\ Pr} \iint \cdot \left(\frac{\partial N_i}{\partial x} \frac{\partial T}{\partial x} + \frac{\partial N_i}{\partial y} \frac{T}{\partial y} \right) \right) \mathrm{d}A - \oint N_i \frac{\partial T}{\partial n} = 0 \qquad (12)$$

2.3 Nano-Fluid Effective Thermophysical Properties

Effective nano-fluid thermophysical property formulas used in this work are adopted from previous literatures' correlations that vary between theoretical derivations and experimental results.

Mixture theory [11] is used to determine the effective density of the nano-fluid (Eq. 13). The theory is used by most researchers who deal with discrete flow applications.

$$\rho_{\mathrm{nf}} = (1 - \varphi)\rho_{\mathrm{bf}} + \varphi\rho_{\mathrm{p}} \qquad (13)$$

Effective thermal conductivity is determined by Eq. (14) which is derived theoretically by Maxwell [11] for spherical solid particles suspended in base fluid.

Table 2 Thermophysical properties of base fluid and nanoparticles

	Base fluid	Nanoparticle
Property	Water	Al_2O_3
Specific heat (J/kg K)	4182	765
Density (kg/m^3)	1000	3950
Thermal conductivity (W/mK)	0.6	35
Dynamic viscosity (Pa s)	0.001003	

$$k_{nf} = k_f \left[\frac{\left(k_p + 2k_f\right) - 2\varphi\left(k_f + k_p\right)}{\left(k_p + 2k_f\right) + \varphi\left(k_f - k_p\right)} \right] \tag{14}$$

The specific heat of the nano-fluid is calculated from correlation (Eq. 15) taken from Esfe's experimentation [12] on the water base fluid with suspended nanoparticles.

$$cp_{nf} = \frac{(1 - \varphi)\left(\rho c_p\right)_{bf} + \varphi\left(\rho c_p\right)_p}{\rho_{nf}} \tag{15}$$

$$\mu_{nf} = \mu_{bf}\left(123^2\varphi^2 + 7.3\,\varphi + 1\right) \tag{16}$$

Dynamic viscosity correlation (Eq. 16) derived by Maiga [13] is considered; the experimentation was set on water as base fluid with aluminum oxide nanoparticles (Table 2).

3 Physical Domain and Boundary Conditions

Figure 1 shows the domain dimensions with respect to cylinder diameter. The boundary conditions set for the case are the cylinder with nonslip condition and fixed temperature and walls with zero shear and zero temperature gradient that represent the slip and adiabatic conditions, respectively. Pressure is set to zero at the domain outlet and inlet velocity is +1 in the x direction with inlet fixed temperature.

The H grid is generated with quad elements as shown in Fig. 2; the elements are clustered to be finer near the cylinder walls and diverge away with 1.1 transition ratio; the total number of elements is 7600 with 7875 nodes.

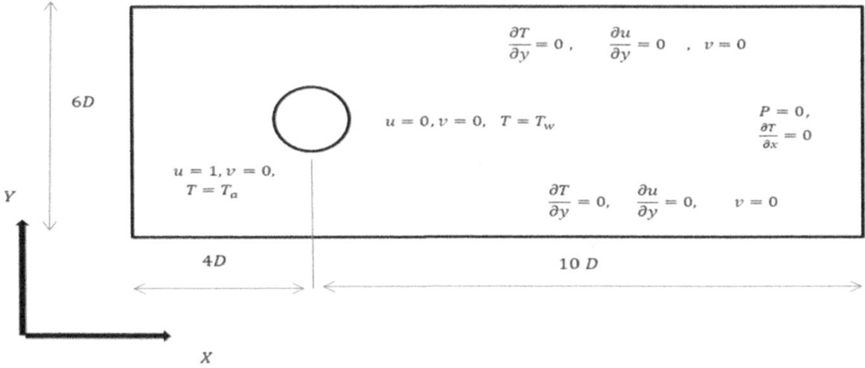

Fig. 1 Physical domain and boundary conditions

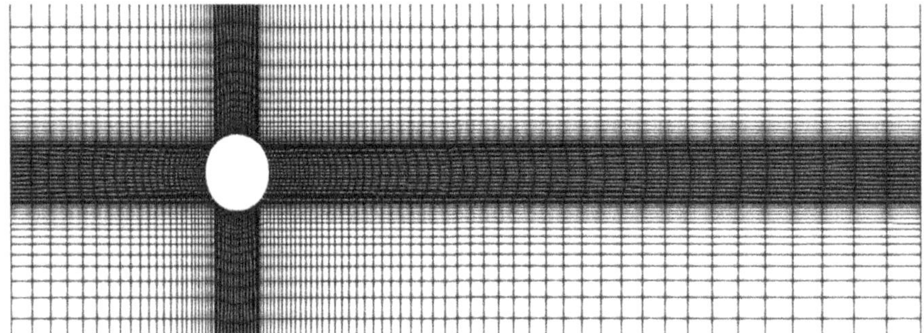

Fig. 2 Clustered H grid

4 Code Validation

In this section, the home-built finite element solver code is validated with previous experimental literatures' results for continuity, momentum and energy equations.

4.1 Continuity/Momentum Equation Validation

Continuity and momentum equations are validated for external flow over the cylinder benchmark problem after reaching steady state for two cases, namely, the 20 and 40 Reynolds numbers where two symmetric vortices appear downstream of the cylinder. Figure 3 shows the streamlines and symmetric vortices at 20 Reynolds number. The validation is considering against Huang [14] results of the downstream vortices' length with respect to the cylinder diameter that depends on Reynolds number. Table 3 shows how the present work's nondimensional vortex length agrees with the Huang results.

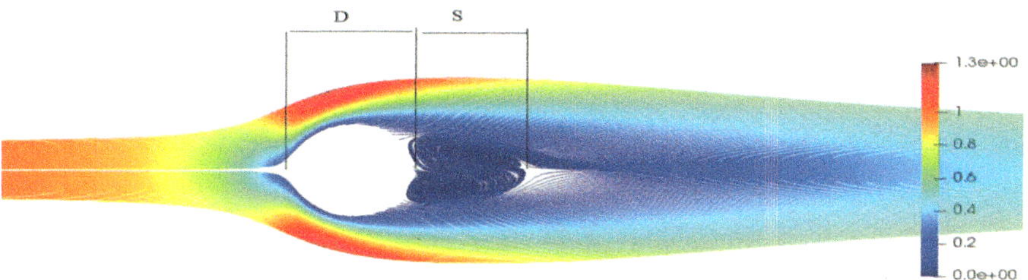

Fig. 3 Streamlines of 20 *Re*

Table 3 Vortex length to cylinder diameter with Reynolds number dependency validation

Re	*S/D*, Present work	*S/D*, Huang et al. [14]	Relative error percentage %
20	0.935	0.991	5.6
40	1.82	1.94	6.18

Table 4 Nusselt number corellation for external flow over cylinder

Reynolds number	Nusselt number and Zakaukas correlation [15]
4–40	$Nu = 0.911\ Re^{0.385} Pr^{\frac{1}{3}}$
40–4000	$Nu = 0.638\ Re^{0.466} Pr^{\frac{1}{3}}$

4.2 Energy Equation Validation

The energy equation is validated by numerical Nusselt number (Eq. 17) and compared with Zakaukas [15] correlation based on his experimentations in 1972 [15]. Equations set by Zakaukas shown in Table 4 are functions of Reynolds number and Prandtl number with numeric constants that depend on the Reynolds number range and flow regime. Figure 4 shows how the present work's results meet the experimental results of Nusselt numbers with RMSE 0.1859. The Peclet number is fixed at 50 for all simulation cases.

$$Nu = \left| \frac{\partial T}{\partial \mathbf{n}} \right| \tag{17}$$

where **n** is the normal unit vector to a cylinder surface.

5 Neural Network Implementation

A two hidden layer neural network with five hidden neurons is shown in Fig. 5, which is designed to predict a single output feature (Nusselt number) with respect to nine input features (represented by Reynolds number, particle concentration, solid

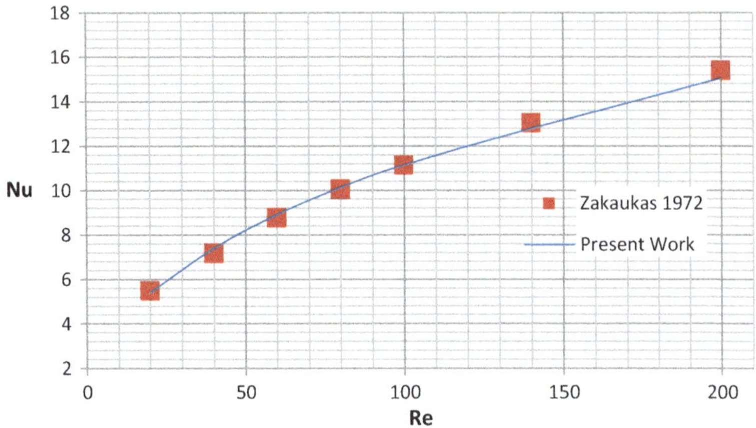

Fig. 4 Nusselt and Reynolds numbers' dependency validation

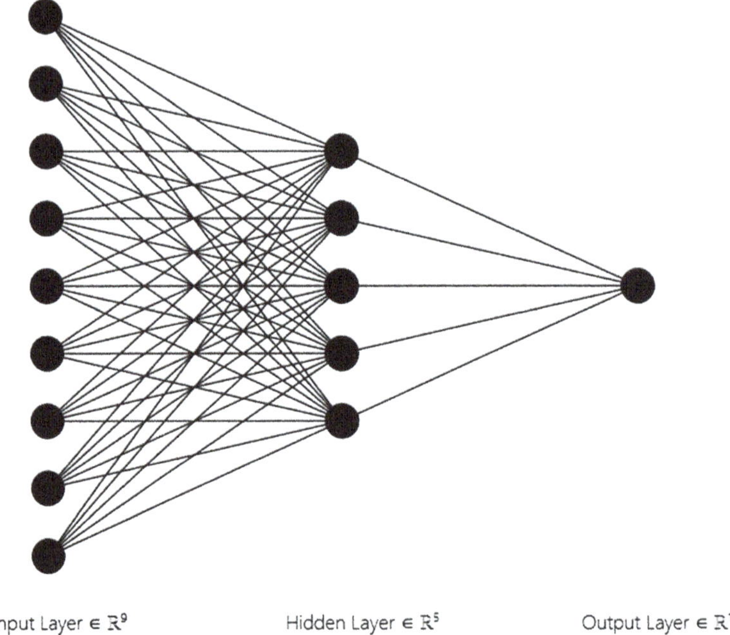

Fig. 5 Two-layer neural network

particle density, base fluid density, solid particle specific heat, base fluid specific heat, solid particle thermal conductivity, base fluid thermal conductivity, and base fluid dynamic viscosity). Input/output data extracted from the CFD numerical solution of Nusselt number are used as a training dataset for the network with 41 training samples m.

5.1   Back Propagation

Mean square error (Eq. 18) is used as a minimization objective function for bias (b_i) and weight (w_i) determination of the network.

$$J = \frac{1}{m} \sum_{1}^{m} (y^- - y)^2 \tag{18}$$

where J, m, y, and y^- are the cost function, dataset samples number, predicted output, and correct output, respectively.

The updated weights and biases can be calculated as follows:

$$w_i^n = w_i^{n-1} - \vartheta \, \frac{\mathrm{d}J}{\mathrm{d}w_i^{n-1}} \tag{19}$$

$$b_i^n = b_i^{n-1} - \vartheta \, \frac{\mathrm{d}J}{\mathrm{d}b_i^{n-1}} \tag{20}$$

where n and ϑ are the iteration number and learning rate, respectively.

5.2   Overfitting Elimination

One of the most common errors of curve fitting in a neural network process is overfitting that causes noisy predictions by the network. The reason of curve overfitting is that some input neurons have typically higher weights compared to other neurons causing the same output effect but with lower weights, which leads the neural network to translate the delicate increase or decrease of higher-weight neurons as responsible for the output variation.

To avoid overfitting of neural network results, the training dataset samples shall be increased and better predictions can be attained. In some applications, increasing the training dataset is difficult or time consuming as CFD analysis requires a high computational power to be performed in addition to the time required accessing its results.

Neural Network Regularization

Bayesian regularization is used for overfitting elimination by adding a new term J_w to the original objective function J_0 (Eq. 21) that is used to penalize large neuron weights of the network.

$$J = \beta J_0 + \alpha J_w \qquad (21)$$

where J_0 is the original objective function and J_w the Frobenius norm for the weight matrix and α and β are the objective function parameters to be optimized with the Bayesian framework described by Mackay [16] and Okut [17] in their literatures. In this framework, the weights and biases are initialized randomly with Gaussian distribution assumption and objective function parameters α and β are calculated via statistical methods.

6 Results and Discussion

The effect of different volume fraction concentrations of aluminum oxide nanoparticles on Nusselt number is studied. The CFD simulation is set for a range of Reynolds number 20–200 and nanoparticle volume fraction concentration 0–0.15. The thermal boundary layer is the main cause of the convection heat transfer phenomena shown in Fig. 6; the contours illustrate the temperature gradient in the flow from the high temperature at the cylinder wall to the free stream fluid temperature. Also the thermal boundary layer is calculated for the given range of nanoparticle volume concentration and plotted with respect to the nondimensional vertical height from the cylinder wall as shown in Fig. 7. The graph shows that the thermal boundary layer gradient increases with increase of nanoparticle volume concentration that means enhancement of convection heat transfer due to increase of Prandtl number as an effect of particle loading. Figure 8 shows an enhancement of Nusselt number with increase of both Reynolds number and nanoparticle volume concentration, as Nusselt number at 200 Re and concentration of 0.15 is increased by 27.75% compared to pure water at the same Reynolds number (Fig. 9). A linear

Fig. 6 Temperature contours for thermal boundary layer at 20 Re

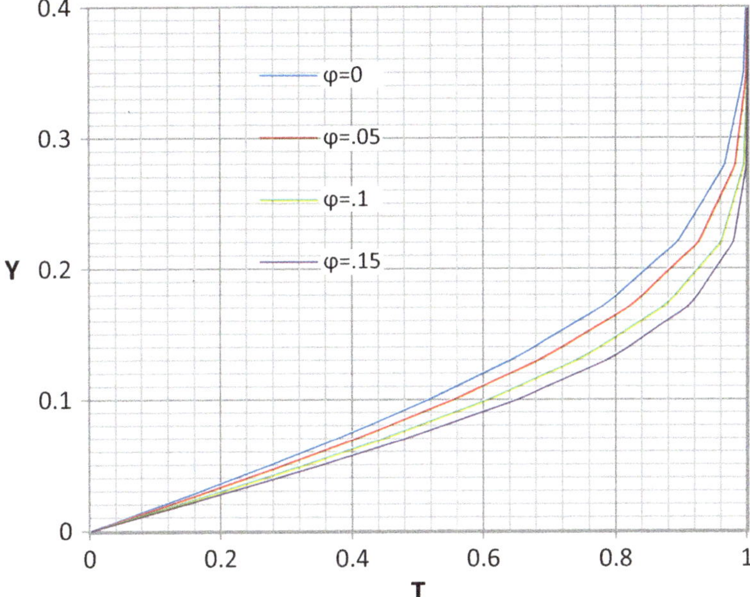

Fig. 7 Thermal boundary layer at 20 *Re* for different particle volume concentrations

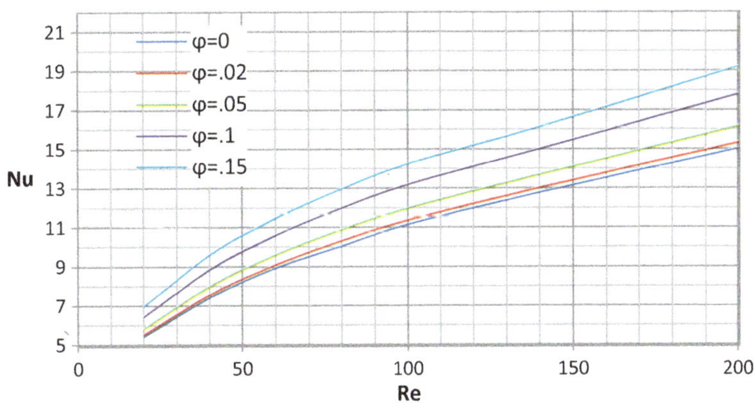

Fig. 8 Nanoparticle concentration effect on water-aluminum oxide nano-fluid

proportionality behavior of Nusselt number is observed with nanoparticle concentration at constant Reynolds number as shown in Fig. 6.

Gradient descent neural network algorithm is tested by a CFD dataset with 41 training samples; the dataset is validated by the CFD Nusselt number predictions and plotted as shown in Fig. 10; the validation plot shows a noisy curve fitting and deviation of the neural network predictions from the CFD results with RMSE 1.3212. Bayesian regularization neural network technique is used to avoid the

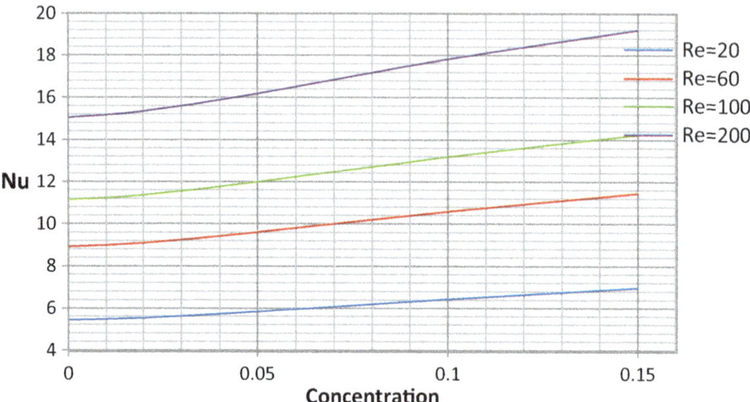

Fig. 9 Nusselt number linear relation with water-aluminum oxide concentration

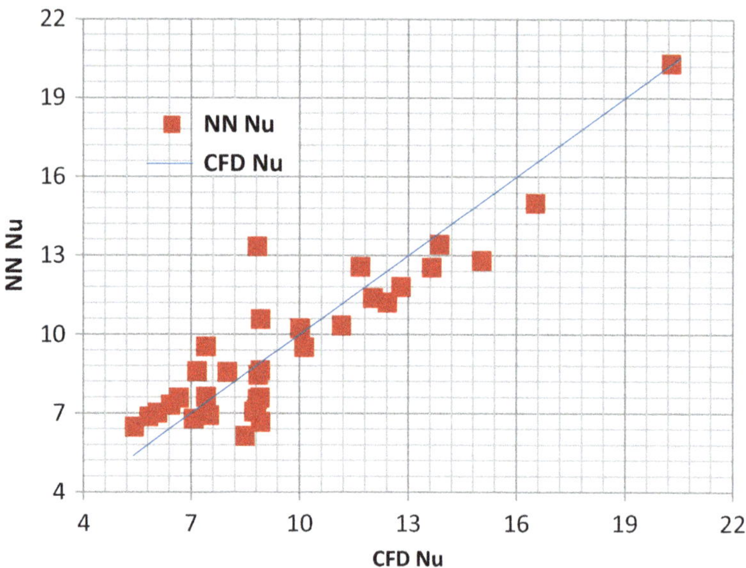

Fig. 10 Gradient descent neural network validation result

overfitting and noisy predictions, and more accurate Nusselt number predictions are attained with RMSE to 0.0188 and limit the variation between the predicted Nusselt number and CFD results as shown in Fig. 11. The neural network with weights and biases obtained by Bayesian regularization neural network technique is fed forward with a dataset of input cases not included in the training dataset to test the reliability and robustness of a neural network in Nusselt number result prediction as shown in Fig. 12 with RMSE 0.4806.

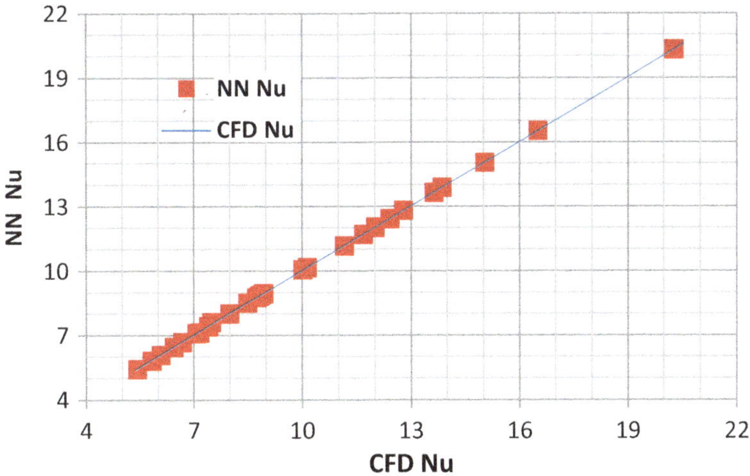

Fig. 11 Regularized neural network validation result

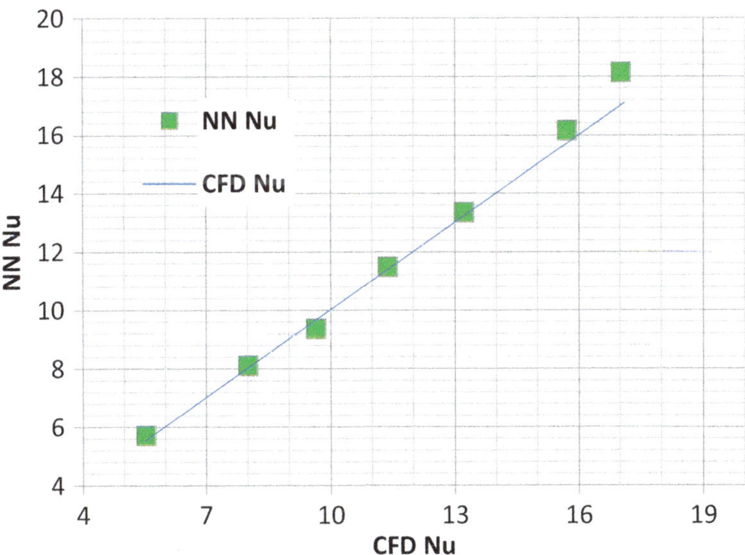

Fig. 12 Regularized neural network tesiting results

7 Conclusion

Convective heat transfer for external flow over a cylinder is studied with water-based aluminum oxide nanoparticles. Nusselt number prediction shows a proportional dependency with both Reynolds number and nanoparticle volume fraction concentration; at 200 *Re* and 0.15 volume concentration, the Nusselt number enhanced with 27.75% compared to the pure water case. A linear proportional trend is observed between Nusselt number and nanoparticle concentration at constant Reynolds number. The thermal boundary layer gradient shows an increase with nanoparticle loading that enhances convective heat transfer and Nusselt number. Bayesian regularization neural network succeeded to predict Nusselt number with nine input features with RMSE 0.0188 and 0.4806 for validated and tested data inputs, respectively.

References

1. Ting, H., & Hou, S. (2016). Numerical study of laminar flow and convective heat transfer utilizing nanofluids in equilateral triangular ducts with constant heat flux. *Materials, 9*(7), 576.
2. Li, Q., & Xuan, Y. (2002). Convective heat transfer and flow characteristics of Cu-water nanofluid. *Science in China Series E: Technological Science, 45*(4), 408–416.
3. Vajjha, R. S., Das, D. K., & Namburu, P. K. (2010). Numerical study of fluid dynamic and heat transfer performance of Al_2O_3 and CuO nanofluids in the flat tubes of a radiator. *International Journal of Heat and Fluid Flow, 31*(4), 613–621.
4. Leong, K., Saidur, R., Kazi, S., & Mamun, H. (2010). Performance investigation of an automotive car radiator operated with nanofluid-based coolants (nanofluid as a coolant in a radiator). *Applied Thermal Engineering, 30*(17–18), 2685–2692.
5. Beigzadeh, R., Rahimi, M., Jafari, O., & Alsairafi, A. A. (2016). Computational fluid dynamics assists the artificial neural network and genetic algorithm approaches for thermal and flow modeling of air-forced convection on interrupted plate fins. *Numerical Heat Transfer Part A: Applications, 70*(5), 546–565.
6. Zendehboudi, A., & Saidur, R. (2019). A reliable model to estimate the effective thermal conductivity of nanofluids. *Heat and Mass Transfer, 55*(2), 397–411.
7. Elhanafy, A., Guaily, A., & Elsaid, A. (2017). Pressure stabilized finite elements simulation for steady and unsteady Newtonian fluids. *Journal of Applied Mathematics and Computational Mechanics, 16*(3), 17.
8. Peeters, M. F., Habashi, W. G., & Nguyen, B. Q. (1991). Finite element solution of the incompressible Navier-Stokes equations by a Helmholtz velocity decomposition. *International Journal for Numerical Methods in Fluids, 13*(2), 135–144.
9. Hughes, T. J., Franca, L. P., & Balestra, M. (1986). A new finite element formulation for computational fluid dynamics: V. Circumventing the Babuška-Brezzi condition: A stable Petrov-Galerkin formulation of the Stokes problem accommodating equal-order interpolations. *Computer Methods in Applied Mechanics and Engineering, 59*(1), 85–99.
10. Hutton, D. V. (2004). *Fundamental of finite element analysis.* McGraw-Hill: Elizabeth A. Jonse Lt.
11. Drew, D. A., & Passman, S. L. (1999). *Theory of multicomponent fluids.* Berlin: Springer.
12. Esfe, M. H., Saedodin, S., & Mahmoodi, M. (2014). Experimental studies on the convective heat transfer performance and thermophysical properties of MgO–water nanofluid under turbulent flow. *Experimental Thermal and Fluid Science, 52*, 68–78.
13. Maïga, S. E. B., Nguyen, C. T., Galanis, N., & Roy, G. (2004). Heat transfer behaviours of nanofluids in a uniformly heated tube. *Superlattices and Microstructures, 35*(3–6), 543–557.
14. Huang, Y. Q., Deng, J., & Ren, A. L. (2003). Research on lift and drag in unsteady viscous flow around circular cylinders. *Journal-Zhejiang University Engineering Science, 37*(5), 596–601.
15. Žukauskas, A. (1972). Heat transfer from tubes in crossflow. *Advances in Heat Transfer, 8*, 93–160.

16. MacKay, D. J. (1992). A practical Bayesian framework for backpropagation networks. *Neural Computation, 4*(3), 448–472.
17. Okut, H. (2016). Bayesian regularized neural networks for small n big p data. In *Artificial neural networks-models and applications* (pp. 21–23)

Variants of the Finite Element Method for the Parabolic Heat Equation: Comparative Numerical Study

Ahmed A. Hamada, Mahmoud Ayyad, and Amr Guaily

Abstract Different variants of the weighted residual finite element method are used to obtain a solution for the parabolic heat equation, which is considered to be the model equation for the steady-state Navier-Stokes equations. Results show that the collocation and the least-squares variants are more suitable for first-order systems. Results also show that the Galerkin/least-squares method is more diffusive than other methods and hence gives stable solutions for a wide range of Péclet numbers.

Keywords Galerkin · Least-squares · Collocation · Finite element method · Parabolic heat equation

A. A. Hamada (✉)
Civil, Environmental and Ocean Engineering, Schaefer School of Engineering and Science, Stevens Institute of Technology, Hoboken, NJ, USA

Aerospace and Aeronautical Engineering Department, Faculty of Engineering, Cairo University, Giza, Egypt
e-mail: AHamada@stevens.edu

M. Ayyad
Civil, Environmental and Ocean Engineering, Schaefer School of Engineering and Science, Stevens Institute of Technology, Hoboken, NJ, USA

Engineering Mathematics and Physics Department, Faculty of Engineering, Cairo University, Giza, Egypt

A. Guaily
Engineering Mathematics and Physics Department, Faculty of Engineering, Cairo University, Giza, Egypt

Smart Engineering Systems Research Center (SESC), School of Engineering and Applied Sciences, Nile University, Shaikh Zayed City, Egypt

© Springer Nature Switzerland AG 2020

M. H. Farouk, M. A. Hassanein (eds.), *Recent Advances in Engineering Mathematics and Physics*, https://doi.org/10.1007/978-3-030-39847-7_28

1   Introduction

The behavior of different variants of the finite element method is examined using the second-order differential equation (DE) represented by the parabolic heat equation [1]. The most important features of the finite element method (FEM) [2] can be summarized that FEM can approximate complicated geometrical boundaries easily; also it accounts for boundary conditions in an easy, straightforward manner; and FEM is modular where the problem may be changed to a more sophisticated problem without changing the program structure. Thus, the method used in this study is the finite element method (FEM).

The weighted residual FEM [3–6] succeeded to solve boundary value problems for parabolic differential equations (DE). The objective of this paper is to study the behavior of different finite element techniques with respect to the second-order ordinary DE from two points of view: first, the inherent numerical viscosity and, second, the stability of the method with convection dominated flows.

The used FEM and their acronyms used throughout the paper are:

- SG: standard Galerkin method [7].
- C: collocation method [8–10].
- LS: least-squares method [11–16].
- GLS: Galerkin/least-squares method [17–19].
- CG: collocation/Galerkin method [20–22].
- CLS: collocation/least-squares method [23–25].
- CGLS: collocation/Galerkin/least-squares method.

The SG and the C methods both depend on the weighted residual method [6]. In the SG method, the weight function is chosen to be a basis function. While the weight function of the C method is the Dirac delta. The C method is used at least since the 30s [10] by Lanczos [8]. The C method is well known in chemical engineering [3, 6]. The numerical results from the C method are found to be sensitive to the chosen collocation points. The roots of the Legendre polynomials are found to be the optimum collocation points [26–28].

The LS method is considered as an alternative to the SG formulation. The weight function in the LS is the differential operator of the DE applied on the basis function. The LS FEM is receiving high attention in different communities [29–33]. In 1988, Hughes and Shakib [17–19] have presented the GLS method to solve the advective diffusive systems. The GLS is based on adding the differential operator, introduced in the LS, to the SG method to improve the stability of the advective diffusive equation [34].

The paper is organized as follows. The differential equation is described in Sect. 2. Then, the exact and the numerical solutions, using the finite element method, of the differential equation are discussed in Sect. 3. Finally, Sect. 4 presents the discussion of the results and the conclusion.

2 Problem Statement

Common examples of one-dimensional (1D) second-order differential equations is the parabolic heat equation. The DE is solved numerically using the weighted residual finite element method (FEM). The nondimensional DE is expressed in the Cartesian coordinate as

$$T_{xx}(x) = P_e T_x(x) \tag{1}$$

where $T(x)$ is the temperature at distance x, the subscript x denotes the derivative with respect to x, and P_e is the Péclet number ($P_e = \frac{Lu}{\alpha}$; L is the characteristic length of the rod, u is the local flow velocity, and α is the thermal diffusivity). Dirichlet boundary conditions are used and are discussed later in Sect. 3.2.2.

3 Solving the Differential Equation

In this section, the exact and numerical solution of the parabolic heat equations (Eq. (1)) is discussed.

3.1 The Exact Solution

The exact solution of the parabolic heat equation (Eq. (1)) is found for the used boundary conditions; $T(1) = 0$ and $T(2) = 1$, using the principle of superposition,

$$T(x) = \frac{e^{P_e(x-1)} - 1}{e^{P_e} - 1} \tag{2}$$

3.2 The Finite Element Solution

The Weak Form

The weak forms of the parabolic heat Eq. (1) are

$$\int_{-1}^{1} [(P_e T_x(x) - T_{xx}(x))w_i(x)]\mathrm{d}x = 0 \tag{3}$$

where $w_i(\cdot)$ are the weight functions. Integration by parts is applied to the higher-order derivative term in Eq. (3). The generated term, product of functions, is neglected because of using Dirichlet boundary condition which yields to

$$\int_{-1}^{1} [P_e T_x(x) w_i(x) + T_x(x) w_{ix}(x)] dx = 0 \tag{4}$$

The linear shape functions, $N_j(\cdot)$, are used to approximate the dependent quantity $T(x) = N_j(x) T_j$ between nodes, where T_j are the nodal values. Then, the weak form of Eq. (4) can be rewritten in the indicial notation as

$$\int_{-1}^{1} \left[P_e N_{jx}(x) w_i(x) + N_{jx}(x) w_{ix}(x)\right] dx T_j = 0 \tag{5}$$

The Weight Function

The used weight function depends on the used FE technique. As mentioned in the introduction, Sect. 1, different techniques are used to compare the performance of the FE techniques. The weight function in the case of the SG is the shape functions themselves, Eqs. (6a); in the C method is the Dirac delta function, Eq. (6b); and in the LS method is a differential operator of the DE, Eq. (6c). The GLS, CG, CLS, and CGLS are combinations of the main three methods, SG, C, and LS, as shown in Eqs. (6d), (6e), (6f), and (6g), respectively.

$$w_i(x) = N_i(x) \tag{6a}$$

$$w_i(x) = \delta_i(x) \tag{6b}$$

$$w_i(x) = L(N_i(x)) \tag{6c}$$

$$w_i(x) = N_i(x) + \tau L(N_i(x)) \tag{6d}$$

$$w_i(x) = \delta_i(x) + N_i(x) \tag{6e}$$

$$w_i(x) = \delta_i(x) + \tau L(N_i(x)) \tag{6f}$$

$$w_i(x) = \delta_i(x) + N_i(x) + \tau L(N_i(x)) \tag{6g}$$

where $\delta_i(x)$ is the Dirac delta, τ is the stabilization parameter, and $L(N_i(x))$ is the differential operator. τ is the stabilization parameter given by [18]

$$\tau = \left(\left(\frac{2P_e}{l_e}\right)^2 + \left(\frac{4}{l_e^2}\right)^2\right)^{-0.5} \tag{7}$$

where l_e is the length of the element in a uniform grid. $L(N_i(x))$ is the differential operator given by

$$L(N_i(x)) = P_e N_{ix}(x) - N_{ixx}(x) \tag{8}$$

The integration is calculated using the Gaussian quadrature points [35]. The first derivative of the Dirac delta function, using its properties, is

$$\delta_x(x) = \frac{-\delta(x)}{x} \tag{9}$$

4 Results and Discussion

4.1 The Exact Solution

The exact solution is a straight line at a zero Péclet number (a pure diffusion). As the Péclet number increases, the exact solution of the parabolic heat equation deviates from the straight line, as a consequence of the equation being convection dominated, as shown in Fig. 1. As a result, a boundary layer starts to form and the higher the Péclet number, the thinner the boundary layer.

4.2 The Finite Element Solution

The numerical solution of the parabolic heat equation (Eq. (5)) is conducted for different numbers of elements: 25, 50, 75, and 100 at different Péclet numbers and 1, 50, 100, and 500 by the techniques mentioned previously. The adopted procedure, integration by parts, that deals with the higher derivative term in the DE produces a singular matrix in the C method with a very small condition number. Thus, the solution from the C method ceased to exist. Figures 2, 3, 4, 5, 6, and 7 show the effect of changing the Péclet numbers for each method. Figures 2 and 3 show that the SG and CG methods become unstable as the Péclet number increases because the effect of the dispersive term increases. Considering linear shape functions for the second-order DE leads to the conclusion that the LS method became a pure diffusive

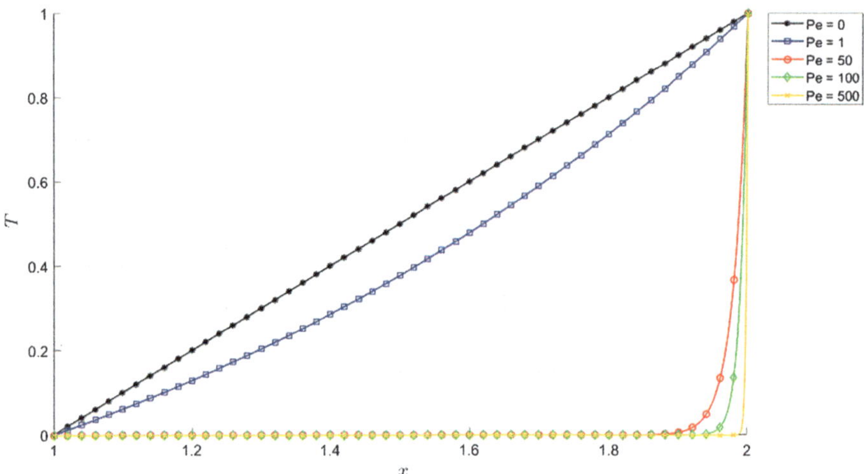

Fig. 1 Exact solutions of the parabolic heat equation at different Péclet numbers

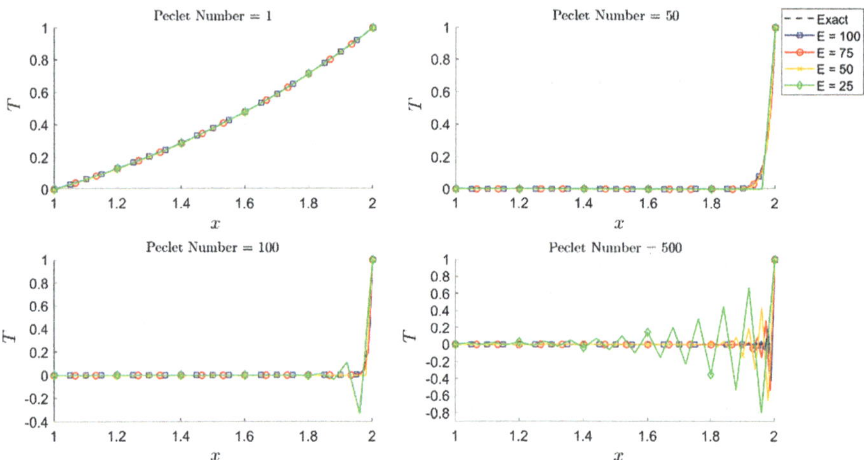

Fig. 2 Exact and numerical solutions using the SG FEM for the parabolic heat equation at different numbers of elements and different Péclet numbers

method whatever the value of the Péclet number as shown in Fig. 4. The reasoning behind this behavior for the LS method is that the weight function is the derivative of the residual of the equation which contains the second derivative of the linear shape functions and consequently vanishes. The remedy is to use the shape functions of the same order as the highest derivative appearing in the DE.

The CLS becomes more stable as the Péclet number increases, as shown in Fig. 5, because the effect of the diffusive term from the LS increases. Figures 6 and 7 show that the GLS method is always stable whatever the value of the Péclet number, while the CGLS method becomes unstable at higher Péclet number because the effect of

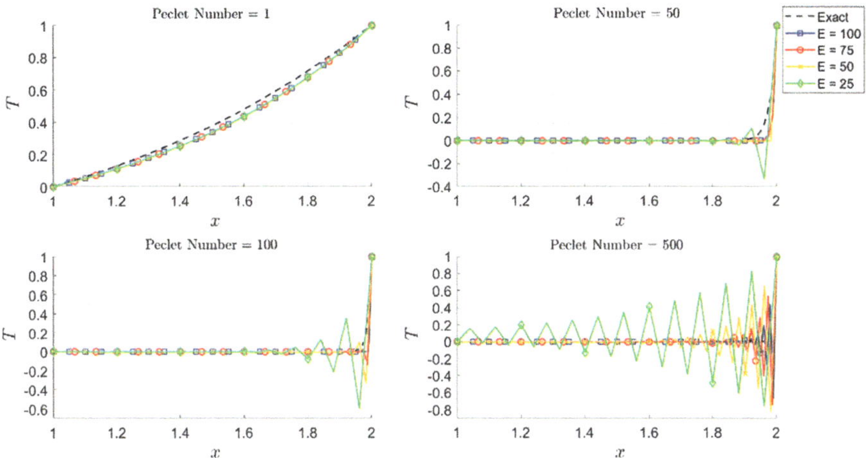

Fig. 3 Exact and numerical solutions using the CG FEM for the parabolic heat equation at different numbers of elements and different Péclet numbers

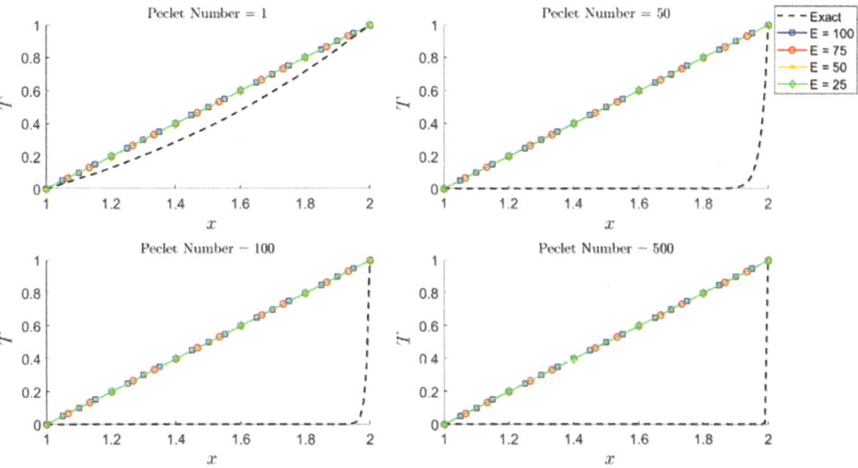

Fig. 4 Exact and numerical solutions using the LS FEM for the parabolic heat equation at different numbers of elements and different Péclet numbers

the convection term in the SG part is higher than the effect of the diffusivity in the LS part. Moreover, Fig. 8 shows that the GLS is the most accurate diffusive method. Hence, the results proved that GLS is the most recommended method to solve second-order DE, such as Navier-Stokes equations, with high accuracy [36, 37]. However, solving the first-order DE, such as Euler equation or Navier-Stokes equations formulated as first-order systems [38], with LS and C gives a solution with higher accuracy than other methods [10].

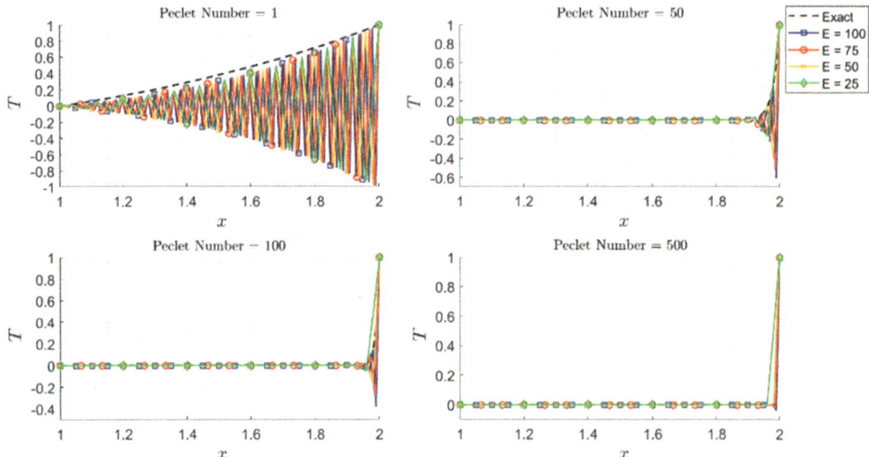

Fig. 5 Exact and numerical solutions using the CLS FEM for the parabolic heat equation at different numbers of elements and different Péclet numbers

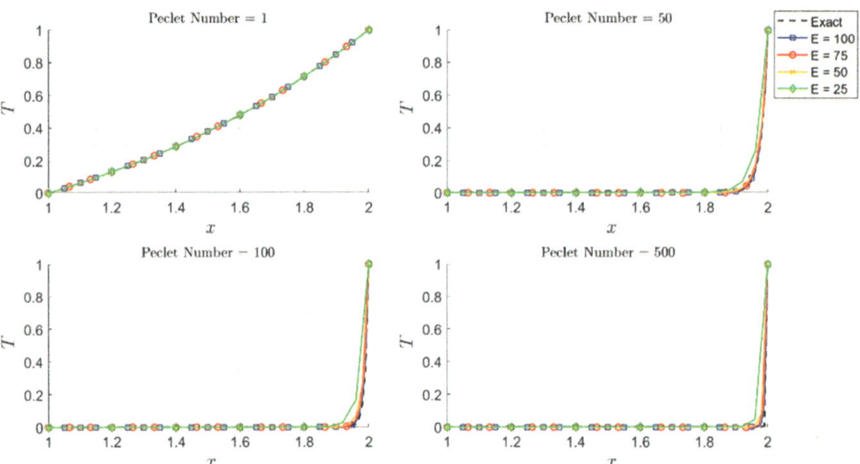

Fig. 6 Exact and numerical solutions using the GLS FEM for the parabolic heat equation at different numbers of elements and different Péclet numbers

Figures 9 and 10 show the accuracy and the convergence of the FE methods using the absolute relative error:

$$\text{Absolute Relative Error} = \left| \frac{\text{Exact Solution} - \text{Numerical Solution}}{\text{Exact Solution}} \right| \tag{10}$$

The GLS method gives a high accuracy solution regardless of the value of the Péclet number and converges by increasing the number of elements. The solution

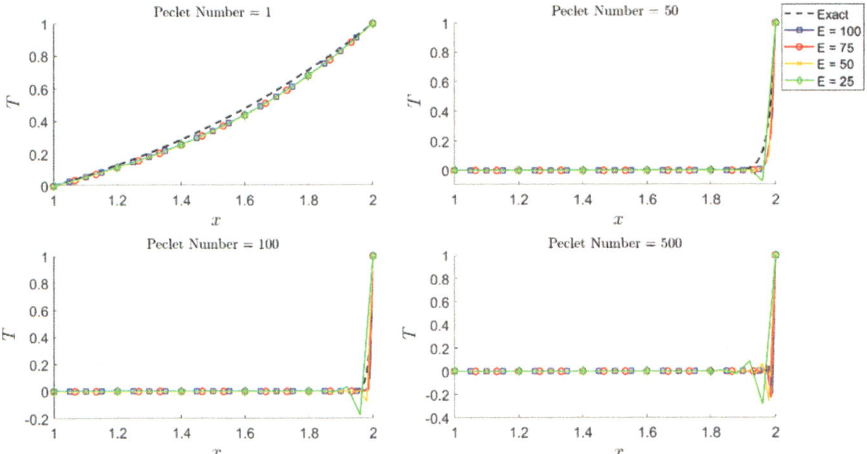

Fig. 7 Exact and numerical solutions using the CGLS FEM for the parabolic heat equation at different numbers of elements and different Péclet numbers

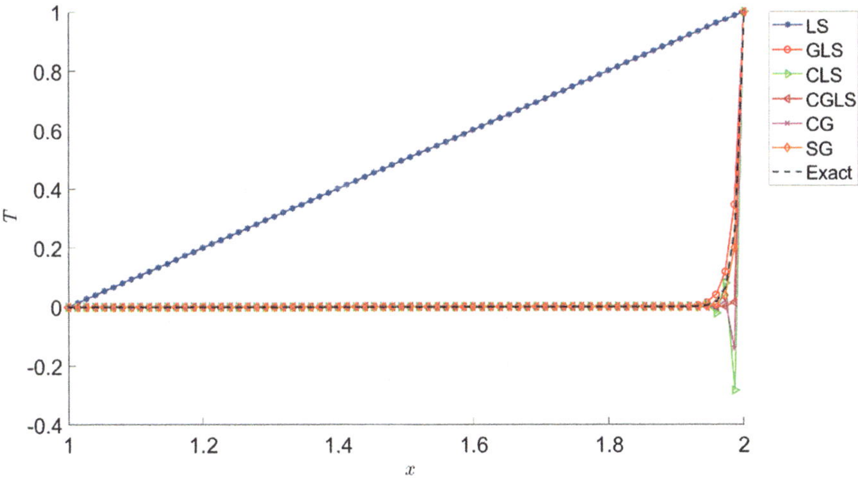

Fig. 8 The exact solution at Péclet number 100 and the numerical solutions using 75 elements

from the SG method is promising at small values of Péclet number, and the errors decrease with the increase of number of elements. However, the solution from the SG methods diverges when the value of Péclet number is increased. Moreover, the methods CLS, CG, and CGLS have nearly the same error because, as discussed previously, the dispersive term effects are higher than the diffusion term effects. Finally, these figures emphasize that the LS method fails to solve the second-order DE with linear shape functions.

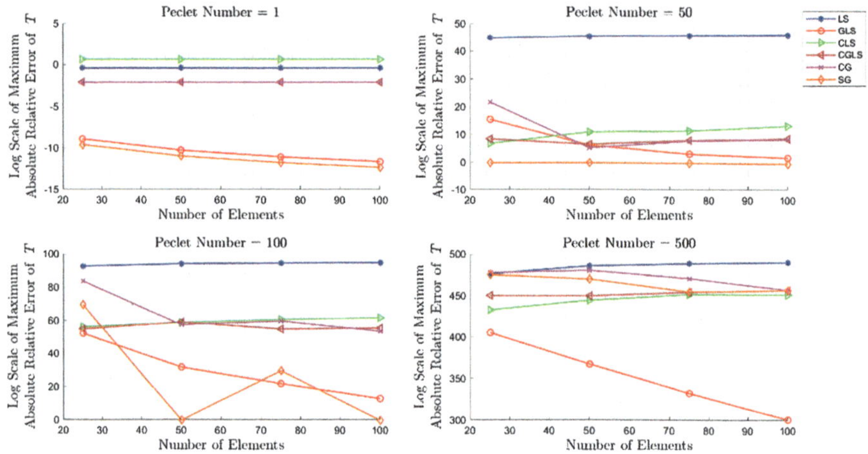

Fig. 9 The logarithmic scale of maximum absolute relative error of the numerical solutions at different numbers of elements and different Péclet numbers

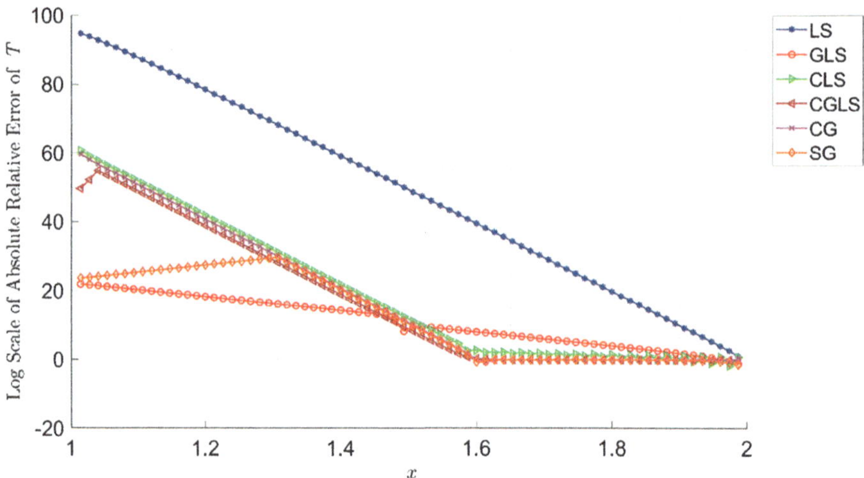

Fig. 10 The logarithmic scale of absolute relative error of the numerical solutions at Péclet number 100 using 75 elements

5 Conclusion

The parabolic heat equation is used to study the behavior of different versions of the weighted residual method. The Péclet number is varied from 1, which is considered to be diffusion dominated case, till a very high value of about 500, reflecting the case of convection dominated case. The GLS method turned out to be the most stable technique for the case at hand, while the CLS method is believed to be suitable for very high Péclet number case or in general for the inviscid flows which correspond to very high Reynolds number flows.

References

1. Burgers, J. M. (1995). Hydrodynamics—Application of a model system to illustrate some points of the statistical theory of free turbulence. In *Selected papers of JM Burgers* (pp. 390–400). Dordrecht: Springer.
2. Zienkiewicz, O. C., Taylor, R. L., Nithiarasu, P., & Zhu, J. Z. (1977). *The finite element method* (Vol. 3). London: McGraw-Hill.
3. Villadsen, J., & Michelsen, M. L. (1978). *Solution of differential equation models by polynomial approximation* (Vol. 7). Englewood Cliffs, NJ: Prentice-Hall.
4. Demkowicz, L., Oden, J. T., Rachowicz, W., & Hardy, O. (1991). An hp Taylor-Galerkin finite element method for compressible Euler equations. *Computer Methods in Applied Mechanics and Engineering, 88*(3), 363–396.
5. Zienkiewicz, O. C., Taylor, R. L., & Zhu, J. Z. (2005). *The finite element method: Its basis and fundamentals*. Oxford: Elsevier.
6. Finlayson, B. A. (2013). *The method of weighted residuals and variational principles* (Vol. 73). Philadelphia: SIAM.
7. Thomée, V. (1984). *Galerkin finite element methods for parabolic problems* (Vol. 1054). Berlin: Springer.
8. Lanczos, C. (1938). Trigonometric interpolation of empirical and analytical functions. *Journal of Mathematics and Physics, 17*(1–4), 123–199.
9. Pinder, G. F., Frind, E. O., & Celia, M. A. (1978, January). Groundwater flow simulation using collocation finite elements. In *Proceedings of the 2nd International Conference on Finite Elements in Water Resources*.
10. Frind, E. O., & Pinder, G. F. (1979). A collocation finite element method for potential problems in irregular domains. *International Journal for Numerical Methods in Engineering, 14*(5), 681–701.
11. Hall, T. (1970). *Carl Friedrich Gauss. A biography* (Vol. 175). Cambridge, MA: The MIT Press.
12. Eason, E. D. (1976). A review of least-squares methods for solving partial differential equations. *International Journal for Numerical Methods in Engineering, 10*(5), 1021–1046.
13. Jiang, B. N. (1998). *The least-squares finite element method: Theory and applications in computational fluid dynamics and electromagnetics*. Berlin: Springer Science & Business Media.
14. Deang, J. M., & Gunzburger, M. D. (1998). Issues related to least-squares finite element methods for the Stokes equations. *SIAM Journal on Scientific Computing, 20*(3), 878–906.
15. Bochev, P. B., & Gunzburger, M. D. (1998). Finite element methods of least-squares type. *SIAM Review, 40*(4), 789–837.
16. Gerritsma, M. I., & Proot, M. M. (2002). Analysis of a discontinuous least squares spectral element method. *Journal of Scientific Computing, 17*(1–4), 297–306.
17. Hughes, T. J. R., & Shakib, F. (1988). *Computational aerodynamics and the finite element method*. Rep. AIAA.
18. Hughes, T. J., Franca, L. P., & Hulbert, G. M. (1989). A new finite element formulation for computational fluid dynamics: VIII. The Galerkin/least-squares method for advective-diffusive equations. *Computer Methods in Applied Mechanics and Engineering, 73*(2), 173–189.
19. Shakib, F. (1989). *Finite element analysis of the compressible Euler and Navier-Stokes equations* [PhD thesis]. Stanford, CA: Stanford University.

20. Dunn, R. J., Jr., & Wheeler, M. F. (1976). Some collocation-Galerkin methods for two-point boundary value problems. *SIAM Journal on Numerical Analysis, 13*(5), 720–733.
21. Rogers, J. M., & McCulloch, A. D. (1994). A collocation-Galerkin finite element model of cardiac action potential propagation. *IEEE Transactions on Biomedical Engineering, 41*(8), 743–757.
22. Klinkel, S., Chen, L., & Dornisch, W. (2015). A NURBS based hybrid collocation–Galerkin method for the analysis of boundary represented solids. *Computer Methods in Applied Mechanics and Engineering, 284*, 689–711.
23. Moritz, H. (1978). Least-squares collocation. *Reviews of Geophysics, 16*(3), 421–430.
24. Hein, G. W. (1986). A model comparison in vertical crustal motion estimation using leveling data. National Oceanic and Atmospheric Administration (NOAA) Technical Report NOS 117 NGS 35, Rockville, Md.
25. Laible, J. P., & Pinder, G. F. (1989). Least squares collocation solution of differential equations on irregularly shaped domains using orthogonal meshes. *Numerical Methods for Partial Differential Equations, 5*(4), 347–361.
26. Prenter, P. M. (2008). *Splines and variational methods*. New York: Courier Corporation.
27. Villadsen, J. V., & Stewart, W. E. (1967). Solution of boundary-value problems by orthogonal collocation. *Chemical Engineering Science, 22*(11), 1483–1501.
28. De Boor, C., & Swartz, B. (1973). Collocation at Gaussian points. *SIAM Journal on Numerical Analysis, 10*(4), 582–606.
29. Aziz, A. K., & Liu, J. L. (1991). A weighted least squares method for the backward-forward heat equation. *SIAM Journal on Numerical Analysis, 28*(1), 156–167.
30. Bell, B. C., & Surana, K. S. (1994). A space–time coupled p-version least-squares finite element formulation for unsteady fluid dynamics problems. *International Journal for Numerical Methods in Engineering, 37*(20), 3545–3569.
31. Bochev, P. B. (1997). Analysis of least-squares finite element methods for the Navier–Stokes equations. *SIAM Journal on Numerical Analysis, 34*(5), 1817–1844.
32. Bochev, P., Cai, Z., Manteuffel, T. A., & McCormick, S. F. (1996, September). First-order system least-squares for the Navier-Stokes equations. In *NASA Conference Publication* (pp. 41–56).
33. Bochev, P., Manteuffel, T. A., & McCormick, S. F. (1999). Analysis of velocity-flux least-squares principles for the Navier–Stokes equations: Part II. *SIAM Journal on Numerical Analysis, 36*(4), 1125–1144.
34. Shakib, F., & Hughes, T. J. (1991). A new finite element formulation for computational fluid dynamics: IX. Fourier analysis of space-time Galerkin/least-squares algorithms. *Computer Methods in Applied Mechanics and Engineering, 87*(1), 35–58.
35. Isaacson, E., & Keller, H. B. (2012). *Analysis of numerical methods*. New York: Courier Corporation.
36. Elhanafy, A., Guaily, A., & Elsaid, A. (2019). Numerical simulation of Oldroyd-B fluid with application to hemodynamics. *Advances in Mechanical Engineering, 11*(5), 168781401985284.
37. Elhanafy, A., Guaily, A., & Elsaid, A. (2019). Numerical simulation of blood flow in abdominal aortic aneurysms: Effects of blood shear-thinning and viscoelastic properties. *Mathematics and Computers in Simulation, 160*, 55–71.
38. Abohadima, S., & Guaily, A. (2016). Hyperbolic model for the classical Navier-Stokes equations. *The Canadian Journal of Chemical Engineering, 94*(7), 1396–1401.

An Intelligent IoT-Based Wearable Health Monitoring System

Ahmed Kassem, Mohamed Tamazin, and Moustafa H. Aly

Abstract Due to the increasing usage of wireless technologies and the miniaturization of electronic sensors, progress in wearable health monitoring technologies has been improved drastically, with strong potential to alter the future of healthcare services by using Internet of Things (IoT) active health monitoring sensors for omnipresent monitoring of patients and athletes through their regular daily routines. Medical applications such as remote monitoring, biofeedback and telemedicine create an entirely new base of medical quality and cost management. The objective of this work is to develop a low-cost, high-quality multipurpose wearable smart system for healthcare monitoring of patients with heart diseases and fitness athletes. In this chapter, we discuss the three phases of our proposed system. In the first phase, we use the Raspberry-Pi as an open-source microcontroller with a HealthyPi hat acting as a medium between the Raspberry-Pi and the biomedical sensors connected to the HealthyPi hat, with various parameters such as temperature, ECG, heartbeat, and oximetry. We began our experiment using 15 test subjects with different genders, age and fitness level. We placed the proposed wearable device and collected the readings' data for each test subject while resting, walking and running. The second phase is connecting our system to an open-source IoT platform to represent the data through a graphical IoT dashboard to be viewed by doctors remotely, as well as implementing action rules that send alarms to patient and doctor in case of problem detection. In the third phase, we designed and tested a fuzzy logic system that inputs the accelerometer, gyroscope, heart rate and blood oxygen level data collected from the experiments and provides the physical state (resting, walking or running) as output, which helps in determining the health status of the patient/ athlete. The obtained results of the proposed method show a successful remote health status monitoring of test subjects through the IoT dashboard in real time and detection of abnormalities in their health status, as well as efficient detecting of the physical motion mode using the proposed fuzzy logic system design.

A. Kassem (✉) · M. Tamazin · M. H. Aly
Electronics and Communications Engineering Department, Arab Academy for Science, Technology and Maritime Transport, Alexandria, Egypt

© Springer Nature Switzerland AG 2020

M. H. Farouk, M. A. Hassanein (eds.), *Recent Advances in Engineering Mathematics and Physics*, https://doi.org/10.1007/978-3-030-39847-7_29

Keywords Raspberry-Pi · Internet of Things · Fuzzy · ECG · Telemedicine · Biofeedback · Accelerometer · Gyroscope · Wireless

1 Introduction

A great aim is to provide a multipurpose system that not only can be used to monitor heart disease patients or fitness athletes' performance but also used in rural areas and countries with low healthcare capabilities and budgets.

The areas of IoT, telemedicine and biomedical sensors were researched to reach the design and method of implementing a multipurpose wireless monitoring system for patients and athletes [1, 2].

The importance of a wireless monitoring system for patients and athletes is outlined in identifying the patient's essential health parameters and activities remotely with the assistance of sensors situated on the human body.

This chapter proposes a real-time multipurpose system based on IoT that can share real-time medical data between the patients and doctors. This proposed system has an extensive application area, which includes but is not limited to the management of disease like heart disease, where the patient needs to be continuously monitored, or the monitoring of an athlete's health status and fitness level.

Now, we are highlighting some of the many benefits of the proposed system:

- Cost-effective: the patient or athlete can be monitored remotely from any location. This minimizes the travelling cost, hospital bill and time wastage for multiple visits.
- Fast services: the system enables immediate assistance to the patient by the healthcare takers and doctors.
- Management on real-time basis: this enables the patient to get necessary treatment immediately, which helps in preventing further complications.
- Improvement of life quality: the proposed method can also help ageing people, as well as chronically ill people, to improve their life quality with the assistance of health experts who will be monitoring the patient's health status and receive notifications of any abnormalities.

Further data analysis is performed by designing a fuzzy logic system using the Mamdani method to detect the type of physical motion taking place, enabling doctors to fully understand all aspects of the patient's health status. This is in addition to provide fitness athletes assurance over their health status and enable them to strive for improvement in their fitness levels [3–6].

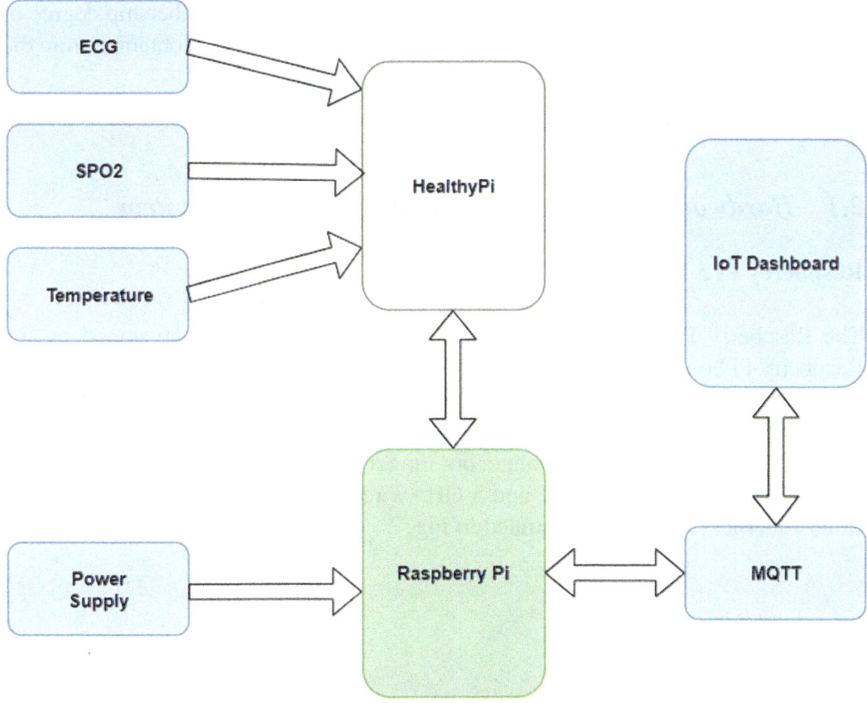

Fig. 1 Multipurpose IoT-ready health monitoring proposed system block diagram

2 Proposed Method

Figure 1 shows the block diagram of our proposed system which is divided into three phases.

In the first phase, the ECG sensor, the temperature sensor and the SPO2 sensor begin data collection and transmit it to the HealthyPi controller. The controller performs the configured digital signal processing in its programming code on the raw input data and provides a clear, understandable output that is transmitted to the Raspberry Pi [7, 8]. In the second phase, the Raspberry Pi immediately begins transmission of the output data to the selected IoT dashboard using the MQTT protocol. In the third phase, we use the provided output data to perform further analysis and enable detection of physical and health states of the patient/athlete. The output data are used as inputs in our fuzzy logic system which we designed using the Mamdani fuzzy inference method. Five inputs are added to the system: accelerometer in x-direction, accelerometer in y-direction, gyroscope in y-direction, heart rate (HR) and oxygen saturation.

Membership functions are determined for each sensor and each motion type output. Then, we populated the inference engine with our if-then rules to relate the inputs to outputs using the data readings from the experiments. The inference engine obtains the degree of fulfilment for each input and infers the membership degree of each output from the fuzzy rules. The most probable output is then obtained from the membership degrees of each output through defuzzification.

2.1   Hardware Components Used in the Proposed System

Raspberry Pi 3 B+

The Raspberry Pi is a series of small single-board computers developed by the Raspberry Pi Foundation to provide accessible teaching of basic computer science in schools and research areas.

As per the cited datasheet, the Raspberry Pi 3 Model B+ is one of the latest products in the Raspberry Pi 3 range, boasting a 64-bit quad-core processor running at 1.4 GHz, dual-band 2.4-GHz and 5-GHz wireless LAN, Bluetooth 4.2/BLE, and faster Ethernet [9]. This is illustrated in Fig. 2.

HealthyPi 3 Hat

The HealthyPi, Fig. 3, is the first fully open-source, full-featured vital sign monitor. Using the Raspberry Pi as its computing and display platform, the HealthyPi add-on HAT turns the Raspberry Pi into a vital sign monitoring system. It includes the following sensors: ECG and respiration, TI ADS1292R; pulse oximetry, TI

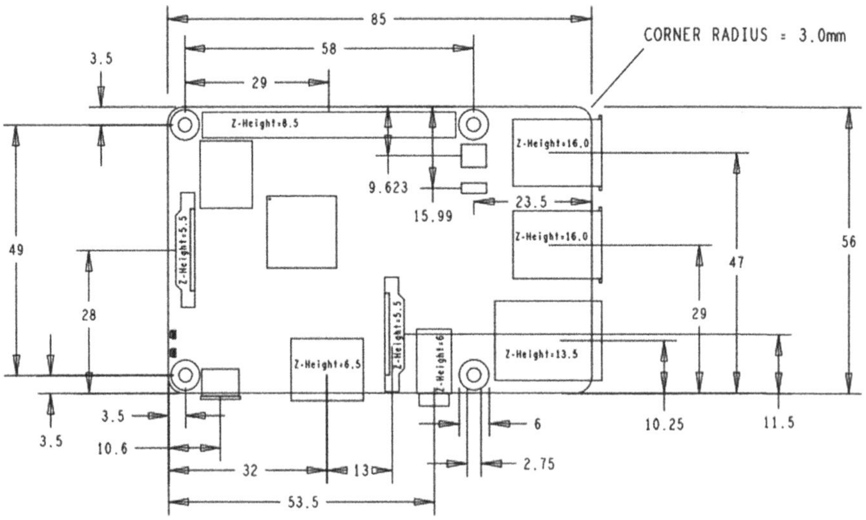

Fig. 2 Raspberry Pi 3 B+ specs [9]

Fig. 3 HealthyPi v3 Hat [10]

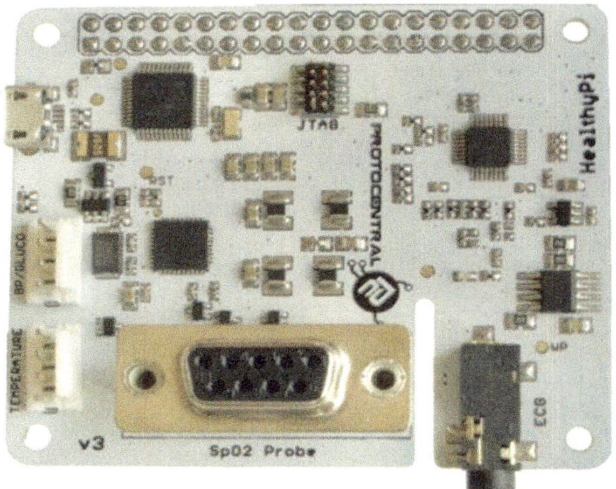

AFE4400; temperature, Maxim MAX30205; microcontroller, Atmel ATSAMD21; and programmability, Arduino Zero Bootloader [10].

Electrocardiogram (ECG) Sensor

As per the cited datasheet, the ADS1292R, Fig. 4, is a multichannel, simultaneous sampling, 24-bit, delta-sigma ($\Delta\Sigma$) analogue-to-digital converter (ADC) with a built-in programmable gain amplifier (PGA), internal reference and an on-board oscillator. The ADS1292R incorporates all features commonly required in portable, low-power medical electrocardiogram (ECG), sports and fitness applications, with high levels of integration and exceptional performance. The ADS1292R enables the creation of scalable medical instrumentation systems at significantly reduced size, power and overall cost [11].

Pulse Oximeter Sensor

As per the cited datasheet, the AFE4400, Fig. 5, is a fully integrated analogue front-end (AFE) ideally suited for pulse oximeter applications. It consists of a low-noise receiver channel with an integrated analogue-to-digital converter (ADC), an LED transmit section and diagnostics for sensor and LED fault detection. The device is a very configurable timing controller. This flexibility enables the user to have complete control of the device timing characteristics. To ease clocking requirements and provide a low-jitter clock to the AFE4400, an oscillator is also integrated that functions from an external crystal. The AFE4400 communicates to an external microcontroller or host processor using an SPI™ interface [12].

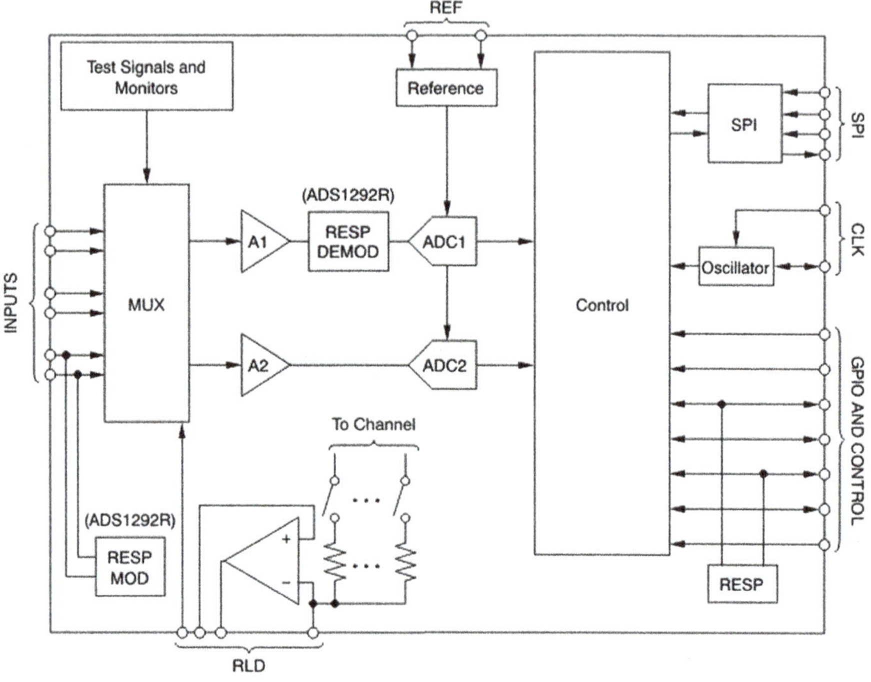

Fig. 4 ADS1292R ECG sensor layout [11]

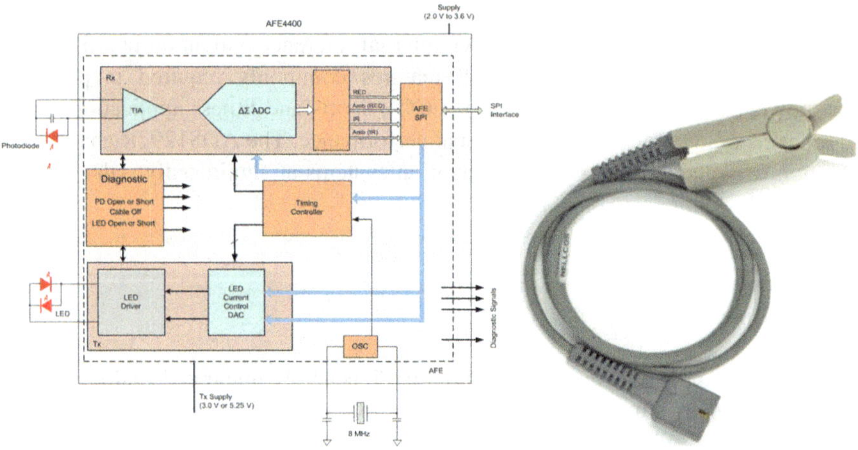

Fig. 5 AFE4400 pulse oximeter layout and cable [12]

Body Temperature Sensor

The MAX30205, Fig. 6, temperature sensor accurately measures temperature and converts the temperature measurements to a digital form using a high-resolution,

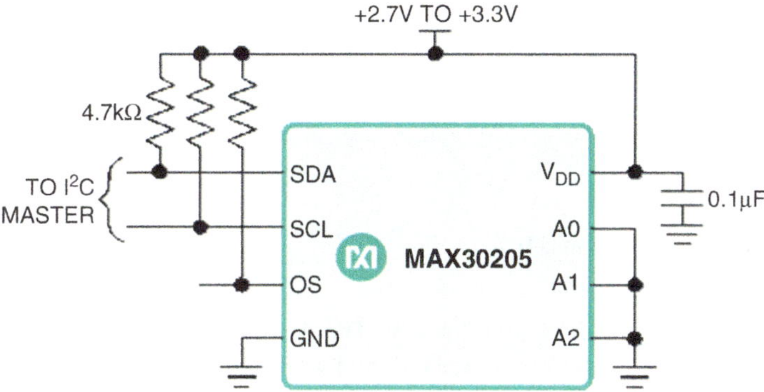

Fig. 6 MAX30205 body temperature layout [13]

sigma-delta, analogue-to-digital converter (ADC). The clinical thermometry accuracy specification is met when the sensor is soldered directly to the HealthyPi. The communication is performed through an I2C-compatible two-wire serial interface, which accepts standard write, read, send and receive byte commands to read the temperature data.

The sensor has a 2.7-V to 3.3-V supply voltage range, low 600-µA supply current and a lock-up-protected I2C-compatible interface that makes them ideal for wearable fitness and medical applications [13].

2.2 Software Components Used in the Proposed System

IoT Dashboard

The IoT dashboards monitor and control boards and sensors. IoT dashboards are filled with graphs, charts and many widgets; they are the digital tools we use to visualize and organize the data that is coming from the physical sensors to our computers. They are cloud-based and global. The IoT usage is in part, thanks to the evolution of cloud computing and its proficient data collection, analysis and processing capabilities. With the grown accessibility of cloud data-storage platforms, businesses or private users no longer need server rooms to store data nor the IT engineer needed to run it. The IoT dashboards can be accessed simply with a URL and any web browser or mobile application, globally anywhere in the world. In our work, we used an open-source IoT dashboard io.adafruit.com.

MQTT Protocol

The MQTT, Fig. 7, stands for MQ Telemetry Transport. It is a publish/subscribe, simple and lightweight messaging protocol, designed for small devices, unreliable

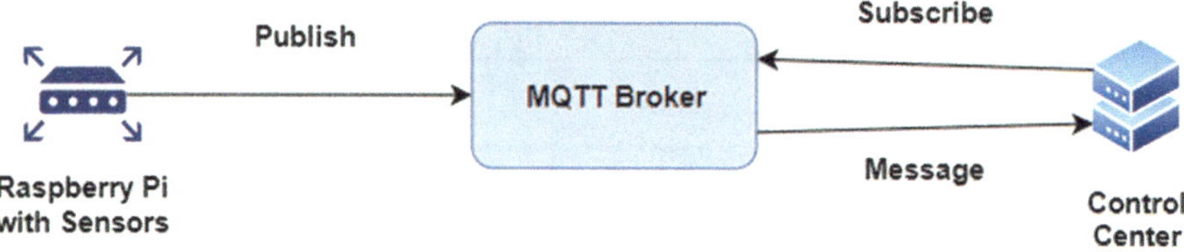

Fig. 7 MQTT block diagram [14]

network, low bandwidth or high latency. The design principles minimize the network bandwidth and device resource required while also attempts to ensure reliability and assurance of delivery. These principles assist in making the protocol ideal to the ever-growing "machine-to-machine" (M2M) or "Internet of Things" (IoT) world of connected devices, where bandwidth and battery power are at a premium [14, 15].

Andy Stanford-Clark of IBM and Arlen Nipper of Arcom invented the MQTT. The standard ports for MQTT are TCP/IP port 1883 which is reserved with IANA for use with MQTT. The TCP/IP port 8883 is also registered for using MQTT over SSL [14, 15].

Android Application

In IoTool, Fig. 8, the API allows IoT researching and fast prototyping with minimal costs in the areas of IoT, eHealth, sports and wellness.

As per the cited datasheet, currently, IoTool supports more than 100 different sensors with more than 258 sensor readings, 50 actuators and different types of triggers connected to an ordinary smartphone through a flexible extension system. The IoTool works on Android devices. The IoTool on a smartphone can process, collect, show values and diagrams, encrypt, store and sync to cloud [16].

3 Fuzzy Logic Algorithm

The fuzzy logic is a many-valued logic, where fuzzy variables' value ranges from 0 to 1. It tries to model human reasoning and relativity of opinion. The membership function is a curve that defines how each input space is mapped to a degree of membership between 0 and 1. The fuzzy input parameters, such as the numerical value of heart rate, is represented by a fuzzy membership function. There are many types of membership functions: triangle, trapezoidal, bell-shaped, etc. [17]. The fuzzy inference system tries to define the fuzzy membership functions to feature vector variables and classes and deduce fuzzy rules to relate feature vector inputs to classes [18, 19].

The steps of fuzzy classification are shown below in Fig. 9:

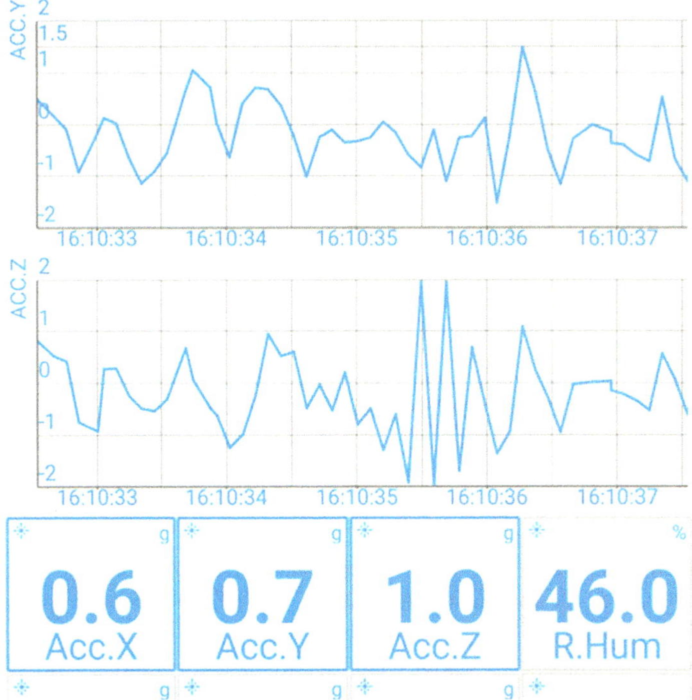

Fig. 8 IoTool interface showing accelerometer readings [16]

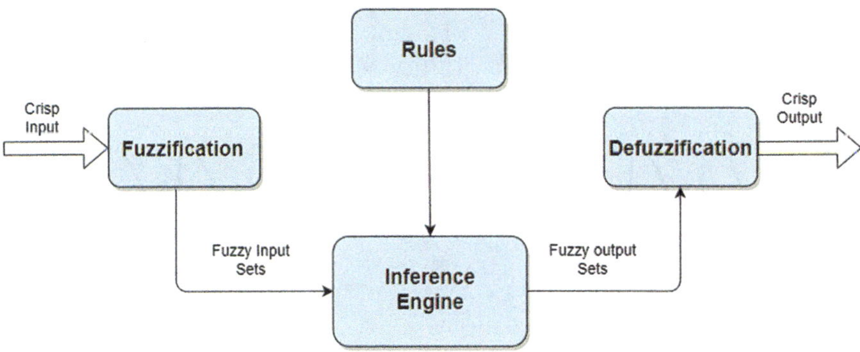

Fig. 9 Fuzzy logic system block diagram [20]

- Input/output variable definition: the set of sensors is defined to be the inputs, and the set of motion modes are defined to be the outputs.
- Membership function determination: for each sensor input and each motion mode output, a set of membership functions are defined to associate an input feature value to sets such as "high", "medium" and "low".

- Fuzzy rule generation: the fuzzy *if-then* rules are defined to relate inputs to outputs using statistical data readings obtained from the test experiments.
- Infer output: the degree of fulfilment (DOF) is obtained for each input and then the membership degree of each output is inferred from the fuzzy rules.
- Defuzzification: the most probable output is obtained from the membership degrees of each output [18–20].

4 Experimental Work

We started our implementation by collecting readings using the ECG, temperature and oximeter sensors connected to the HealthyPi hat which is connected to the Raspberry Pi. These readings were collected from 15 different test subjects with varying age, gender and fitness levels. Subjects were tested in three physical motion types—resting, walking and running—which provided 45 unique readings to use in our system [21–23] (Fig. 10).

The readings (results) were collected and summarized in Tables 1, 2, 3, and 4.

Then we were able to transmit this data to our IoT dashboard using the MQTT protocol and a built-in MQTT client. This enabled us to monitor and view a live stream of the test subject's health status including ECG graph, heart rate, body temperature and blood oxygen level. This is explained in Fig. 11. We configured our IoT dashboard to perform further processing on the received data and send alerts to

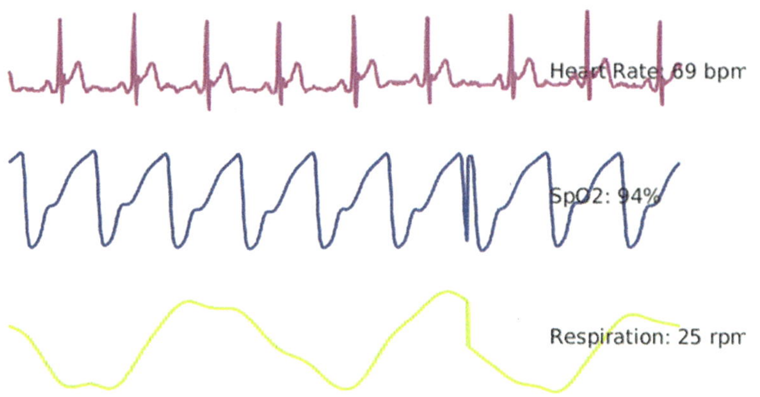

Fig. 10 Real-time data on screen

Table 1 Resting mode readings

Sensor	Ax	Ay	HR (BPM)	O_2	Temperature (°C)
Min	−0.01	−0.14	68	96	36.2
Max	0.35	0.58	80	100	37.0

Table 2 Walking mode readings

Sensor	Ax	Ay	HR (BPM)	O_2	Temperature (°C)
Min	−0.4	−0.45	84	97	36.4
Max	0.66	0.58	100	100	37.1

Table 3 Running mode readings

Sensor	Ax	Ay	HR (BPM)	O_2	Temperature (°C)
Min	−0.6	−0.5	110	96	36.5
Max	0.81	0.95	190	100	37.3

Table 4 All motion modes' readings combined

Sensor	Ax	Ay	HR (BPM)	O_2	Temperature (°C)
Min	−0.01	−0.14	68	96	36.2
Max	0.81	0.95	190	100	37.3

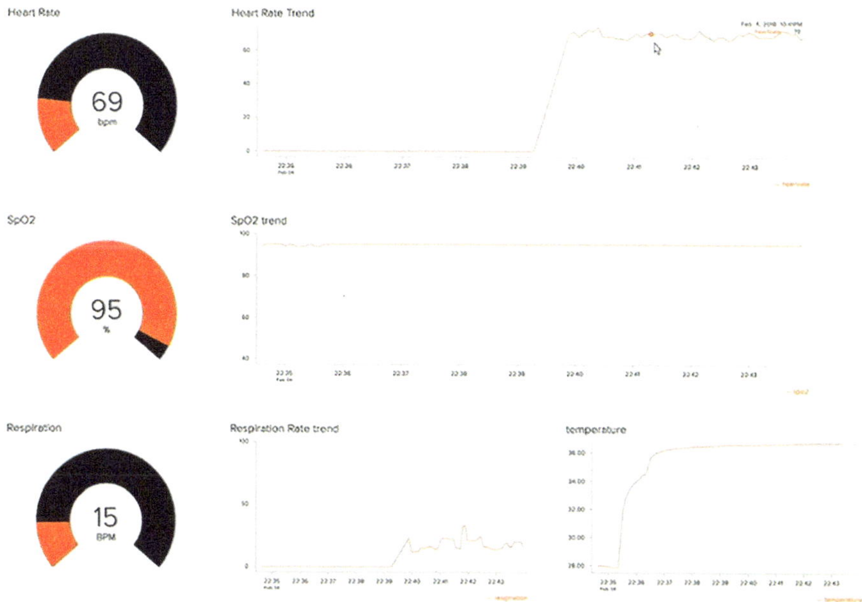

Fig. 11 IoT dashboard real-time data

the patient and doctor if a sensor reading level decreases or increases beyond a single point.

Finally, we created a fuzzy logic system designed using the Mamdani method with triangle membership functions. We used five inputs to our system: accelerometer in x-direction, accelerometer in y-direction, gyroscope in y-direction, heart rate and blood oxygen Level.

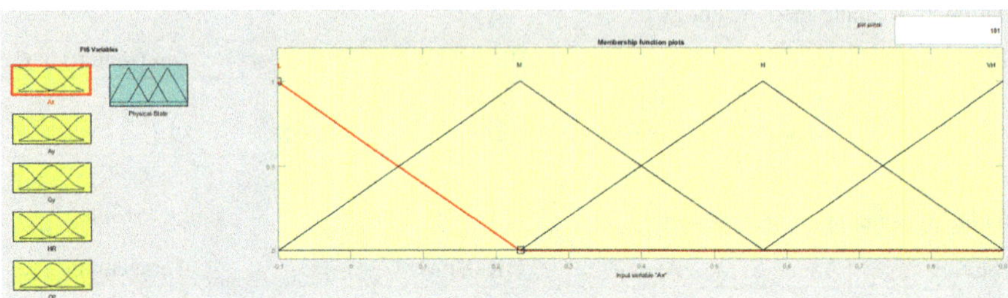

Fig. 12 Membership function for accelerometer in *x*-direction

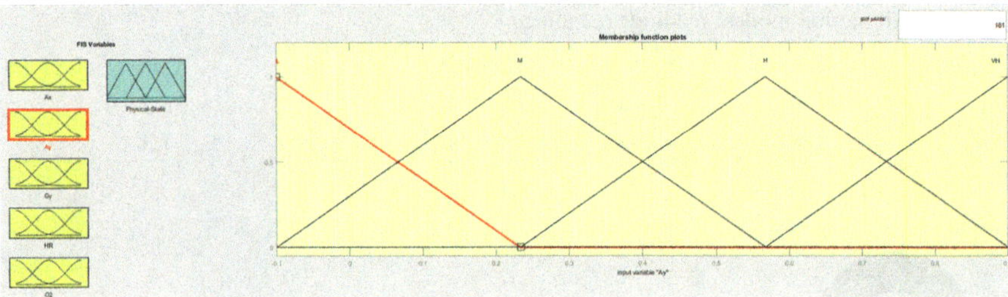

Fig. 13 Membership function for accelerometer in *y*-direction

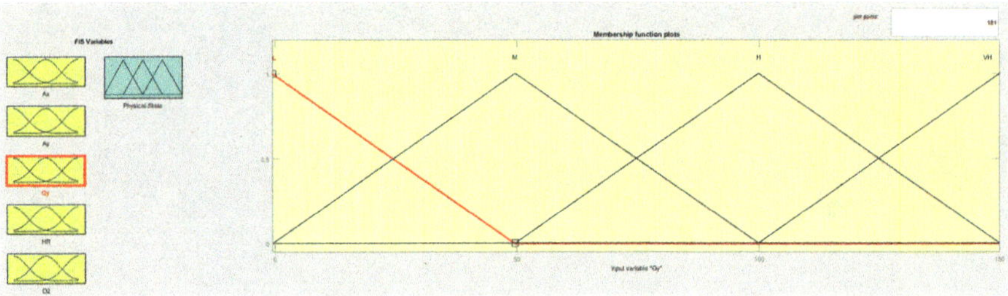

Fig. 14 Membership function for gyroscope in *y*-direction

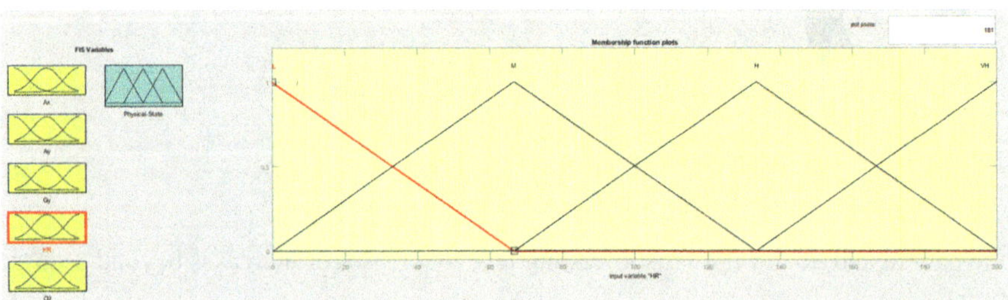

Fig. 15 Membership function for heart rate

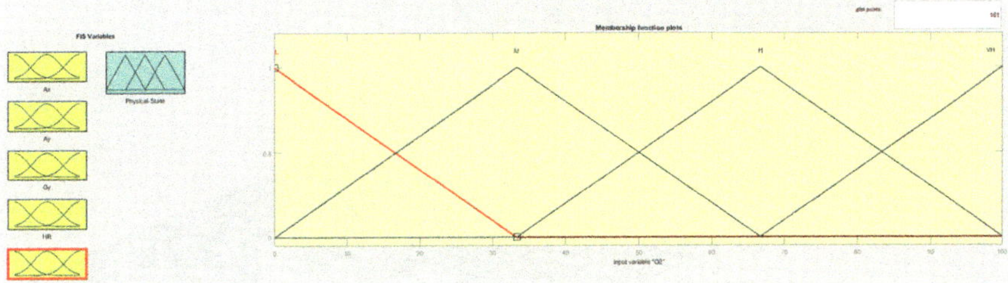

Fig. 16 Membership function for blood oxygen level

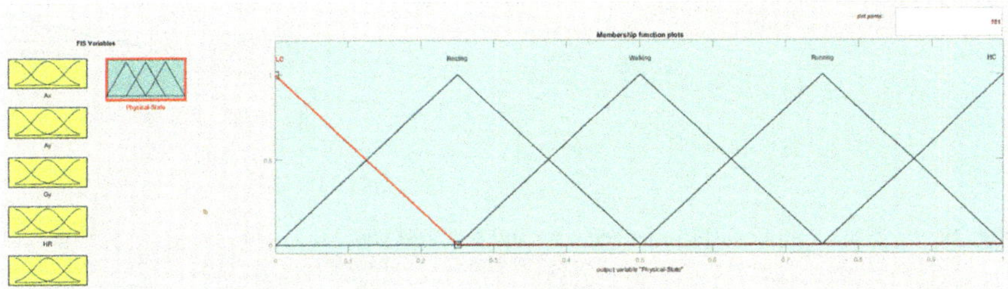

Fig. 17 Membership function for motion mode output

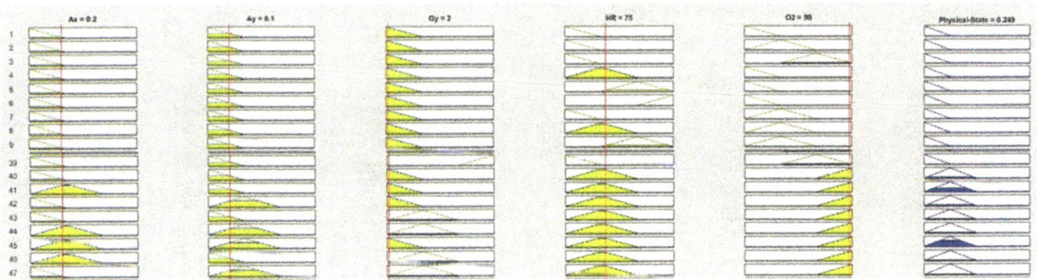

Fig. 18 Output result inferred from rules showing resting mode

Figures 12, 13, 14, 15, and 16 show the input membership functions, while Fig. 17 shows the output membership function.

5 Results and Discussion

A successful implementation of our system enabled us to monitor the test subjects in real time, to detect any abnormalities in their health status and to perform further analysis on their collected readings' data with our designed fuzzy logic system and detect the type of physical motion occurring.

Figures 18, 19, 20, 21, 22, and 23 show the output detection of different motion modes.

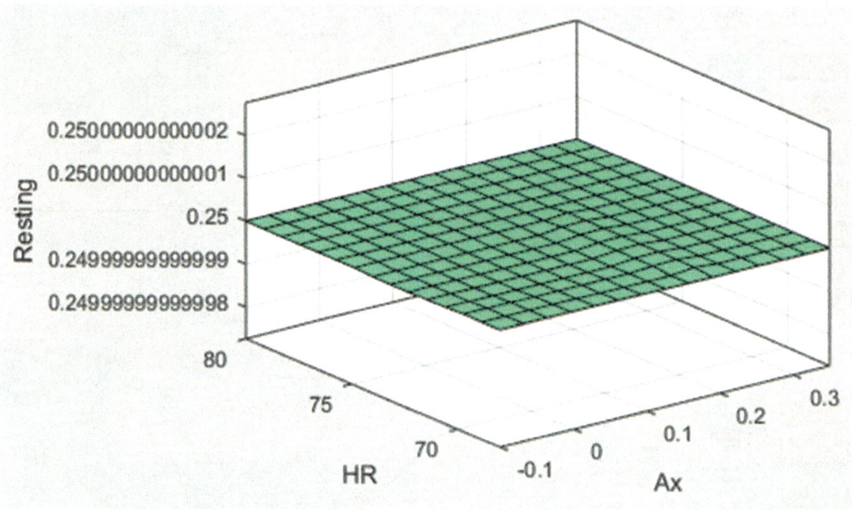

Fig. 19 Predicted output surface between Ax and HR: rest state

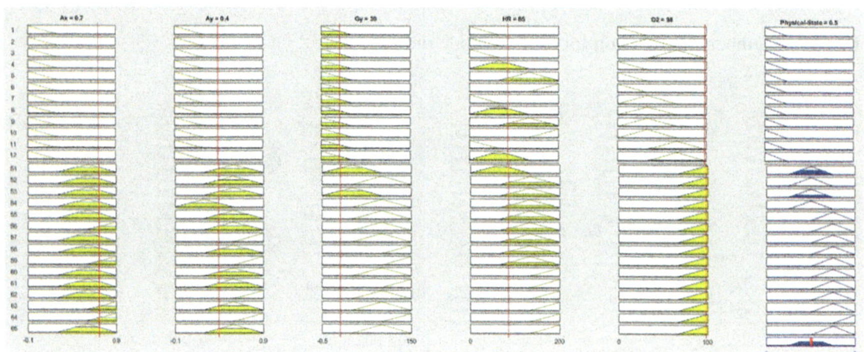

Fig. 20 Output result inferred from rules showing: walking mode

We were also able to provide immediate notification to test subjects and their doctors once any issue or abnormality in their health status readings was detected using the IoT dashboard io.adafruit.com.

An observation was made that body temperature within normal ranges does not differ in the detection of physical motion type, as the body temperature difference is minute and is considered negligible in our experiment.

Table 5 shows the count of successful detection on motion type and unsuccessful detection as well as the error percentage after performing 45 test cases.

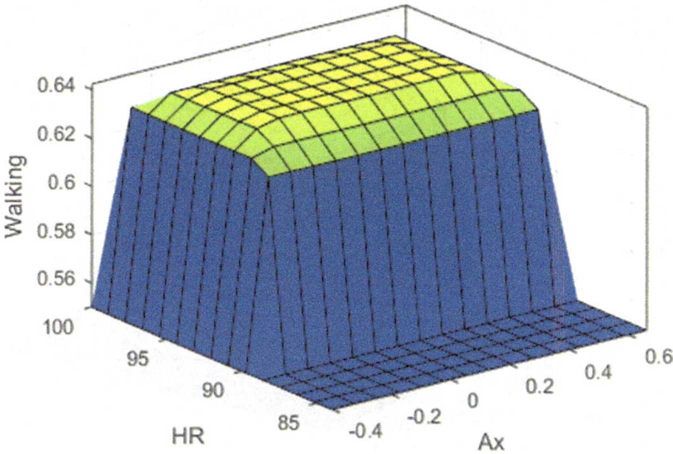

Fig. 21 Predicted output surface between Ax and HR: walking state

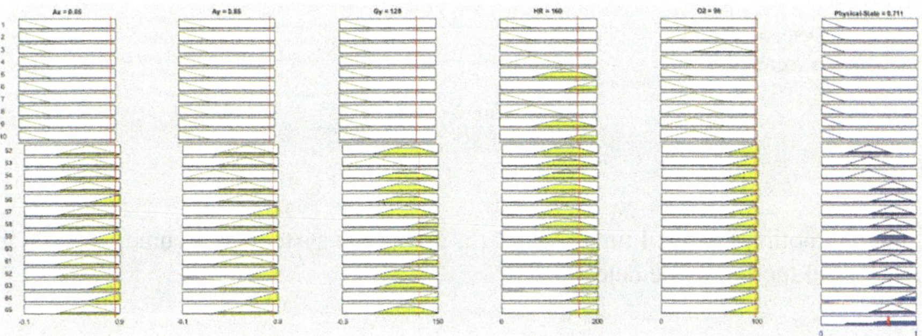

Fig. 22 Output result inferred from rules showing running mode

6   Conclusion

The importance of having a multipurpose IoT-ready health monitoring system is declared in the areas of healthcare and fitness sports, as well as highlighting the many benefits of such system in providing a balance between cost, quality and manageability for patients, athletes, healthcare centres and doctors.

We were able to accurately collect the health status readings using three smart wearable sensors and a smartphone. Remote monitoring of the health vital signs of patients and transmitting the readings in real time to the IoT dashboard to be viewed by doctors was achieved successfully.

The proposed fuzzy logic system is able to detect the correct physical motion mode with high accuracy. Using the 45 test cases from our experiments, the proposed fuzzy logic system was able to successfully detect the resting motion mode 14 times out of 15, the walking motion mode 13 times out of 15 and the

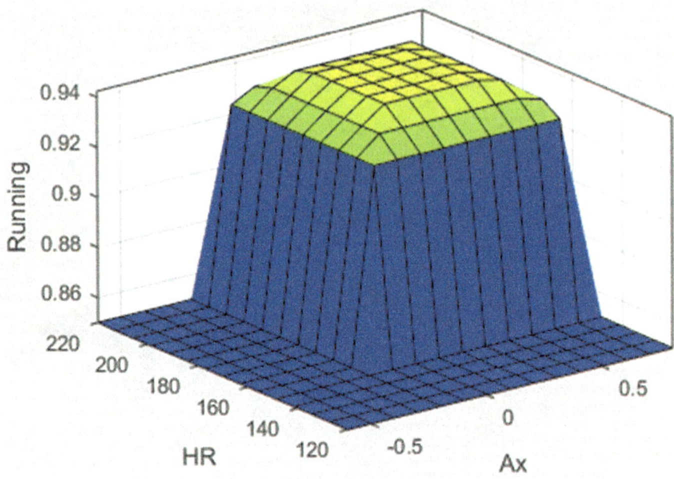

Fig. 23 Predicted output surface between Ax and HR: running state

Table 5 Detection results and error percentage

Detection	Resting	Walking	Running
Successful	14	13	14
Unsuccessful	1	2	1
Accuracy	92%		

running motion mode 14 times out of 15, giving the system an accuracy of 92% in successful motion mode detection.

References

1. Mahfouz, M. R., Kuhn, M. J., & To G. (2013). Wireless medical devices: A review of current research and commercial systems. In *2013 IEEE Topical Conference on Biomedical Wireless Technologies, Networks, and Sensing Systems*, Austin, TX, USA (pp. 16–18).
2. Gullo, C. (2016). By 2016: 100M wearable wireless sensors. *MobiHealthNews*. Retrieved August 18, 2011, from http://mobihealthnews.com/12658/by-2016-100m-wearable-wireless-sensors/
3. Saha, J., Saha, A. K., Chatterjee, A., Agrawal, S., Saha, A., Kar, A., et al. (2018). Advanced IOT based combined remote health monitoring, home automation and alarm system. In *2018 IEEE 8th Annual Computing and Communication Workshop and Conference (CCWC)*, Las Vegas, NV, USA (pp. 602–606).
4. Pardeshi, V., Sagar, S., Murmurwar, S., & Hage, P. (2017). Health monitoring systems using IoT and Raspberry Pi—a review. In *2017 International Conference on Innovative Mechanisms for Industry Applications (ICIMIA), 21–23 Feb. 2017*, Bangalore, India (pp. 134–137).
5. Surya Deekshith Gupta, M., Patchava, V., & Menezes, V. (2015). Healthcare based on IoT using Raspberry Pi. In *2015 International Conference on Green Computing and Internet of Things (ICGCIoT), 8–10 Oct. 2015*, Noida, India (pp. 796–799).

6. Udin Harun Al Rasyid, M., Pranata, A. A., Lee, B.-H., Saputra, F.A., & Sudarsono, A. (2016). Portable electrocardiogram sensor monitoring system based on Body Area Network. In *2016 IEEE International Conference on Consumer Electronics-Taiwan (ICCE-TW)*, 27–29 May 2016, Nantou, Taiwan (pp. 1–2).
7. Rajkumar, S., Srikanth, M., & Ramasubramanian, N. (2017). Health monitoring system using Raspberry PI. In *2017 International Conference on Big Data, IoT and Data Science (BID)*, 20–22 Dec. 2017. Pune, India (pp. 116–119).
8. Movassaghi, S., Abolhasan, M., Lipman, J., Smith, D., & Jamalipour, A. (2014). Wireless body area networks: a survey. *IEEE Communications Surveys and Tutorials, 16*(3), 1658–1686.
9. *Sparkfun Raspberry Pi*. Retrieved April 29, 2019, from https://cdn.sparkfun.com/assets/8/3/9/b/2/Raspberry-Pi-Model-B_-Product-Brief.pdf
10. Protocentral Homepage. Retrieved April 29, 2019, from http://healthypi.protocentral.com/
11. Texas Instruments ECG ADS1292R. Retrieved April 29, 2019, from http://www.ti.com/lit/ds/symlink/ads1292r.pdf
12. Texas Instruments Pulse Oximeter AFE4400. Retrieved April 29, 2019, from http://www.ti.com/lit/ds/symlink/ afe4400. pdf
13. Maxim Integrated Human Body Temperature MAX30205. Retrieved April 29, 2019, from https://datasheets.maximintegrated.com/en/ds/MAX30205.pdf
14. Al-Fuqaha, A., Guizani, M., Mohammadi, M., Aledhari, M., & Ayyash, M. (2015). Internet of things: A survey on enabling technologies, protocols, and applications. *IEEE Communications Surveys and Tutorials, 17*(4), 2347–2376.
15. Hunkeler, U., Truong, H. L., & Stanford-Clark, A. (2008). MQTT-S—A publish/subscribe protocol for Wireless Sensor Networks. In *2008 3rd International Conference on Communication Systems Software and Middleware and Workshops (COMSWARE '08)*, 6–10 January 2008, Bangalore, India (pp. 791–798).
16. IoTool Homepage. Retrieved April 29, 2019, from https://iotool.io/
17. Feng, G. (09 October 2006). A survey on analysis and design of model-based fuzzy control systems. *IEEE Transactions on Fuzzy Systems, 14*(5), 676–697.
18. Liu, B., & Liu, Y.-K. (2002). Expected value of fuzzy variable and fuzzy expected value models. *IEEE Transactions on Fuzzy Systems, 10*(4), 445–450.
19. Elhoushi, M. (2015). *Advanced motion mode recognition for portable navigation*. Queen's University.
20. Sakti, I. (2014). Methodology of fuzzy logic with mamdani fuzzy models applied to the microcontroller. In *2014 The 1st International Conference on Information Technology, Computer, and Electrical Engineering*, Semarang, Indonesia (pp. 93–98).
21. Ingole, A., Ambatkar, S., Kakde, & S. (2015). Implementation of health-care monitoring system using Raspberry Pi. In *2015 International Conference on Communications and Signal Processing (ICCSP)*, 2–4 April 2015, Melmaruvathur, India (pp. 1083–1086).
22. Brown, K., Secco, E. L., & Nagar, A. K. (2019). A low-cost portable health platform for monitoring of human physiological signals. In EAI International Conference on Technology, Innovation, Entrepreneurship and Education, TIE 2017 (Vol. 532, pp. 211–224). Cham: Springer.
23. Abtahi, F., Aslamy, B., Boujabir, I., Seoane, F., & Lindecrantz, K. (2014). An affordable ECG and respiration monitoring system based on Raspberry PI and ADAS1000: First step towards homecare applications. In H. Mindedal, M. Persson (Eds) 16th Nordic-Baltic Conference on Biomedical Engineering. IFMBE Proceedings, Gothenburg, Sweden (Vol. 48). Cham: Springer.

Index

© Springer Nature Switzerland AG 2020

391

M. H. Farouk, M. A. Hassanein (eds.), *Recent Advances in Engineering Mathematics and Physics*, https://doi.org/10.1007/978-3-030-39847-7

Made in the USA
Monee, IL
07 July 2026

56547418R00234